Lecture Notes in Physics

New Series m: Monographs

Springer-Verlag Berlin Heidelberg GmbH

The Editorial Policy for Monographs

The series Lecture Notes in Physics reports new developments in physical research and teaching - quickly, informally, and at a high level. The type of material considered for publication in the New Series m includes monographs presenting original research or new angles in a classical field. The timeliness of a manuscript is more important than its form, which may be preliminary or tentative. Manuscripts should be reasonably self-contained. They will often present not only results of the author(s) but also related work by other people and will provide sufficient motivation, examples, and applications.
The manuscripts or a detailed description thereof should be submitted either to one of the series editors or to the managing editor. The proposal is then carefully refereed. A final decision concerning publication can often only be made on the basis of the complete manuscript, but otherwise the editors will try to make a preliminary decision as definite as they can on the basis of the available information.
Manuscripts should be no less than 100 and preferably no more than 400 pages in length. Final manuscripts should preferably be in English, or possibly in French or German. They should include a table of contents and an informative introduction accessible also to readers not particularly familiar with the topic treated. Authors are free to use the material in other publications. However, if extensive use is made elsewhere, the publisher should be informed. Authors receive jointly 50 complimentary copies of their book. They are entitled to purchase further copies of their book at a reduced rate. As a rule no reprints of individual contributions can be supplied. No royalty is paid on Lecture Notes in Physics volumes. Commitment to publish is made by letter of interest rather than by signing a formal contract. Springer-Verlag secures the copyright for each volume.

The Production Process

The books are hardbound, and quality paper appropriate to the needs of the author(s) is used. Publication time is about ten weeks. More than twenty years of experience guarantee authors the best possible service. To reach the goal of rapid publication at a low price the technique of photographic reproduction from a camera-ready manuscript was chosen. This process shifts the main responsibility for the technical quality considerably from the publisher to the author. We therefore urge all authors to observe very carefully our guidelines for the preparation of camera-ready manuscripts, which we will supply on request. This applies especially to the quality of figures and halftones submitted for publication. Figures should be submitted as originals or glossy prints, as very often Xerox copies are not suitable for reproduction. For the same reason, any writing within figures should not be smaller than 2.5 mm. It might be useful to look at some of the volumes already published or, especially if some atypical text is planned, to write to the Physics Editorial Department of Springer-Verlag direct. This avoids mistakes and time-consuming correspondence during the production period.
As a special service, we offer free of charge LaTeX and TeX macro packages to format the text according to Springer-Verlag's quality requirements. We strongly recommend authors to make use of this offer, as the result will be a book of considerably improved technical quality.
Manuscripts not meeting the technical standard of the series will have to be returned for improvement.
For further information please contact Springer-Verlag, Physics Editorial Department II, Tiergartenstrasse 17, D-69121 Heidelberg, Germany.

Peter Kopietz

Bosonization of Interacting Fermions in Arbitrary Dimensions

Springer

Author

Peter Kopietz
Institut für Theoretische Physik
Universität Göttingen
Bunsenstrasse 9
D-37073 Göttingen, Germany

Cataloging-in-Publication Data applied for.

Die Deutsche Bibliothek - CIP-Einheitsaufnahme

Kopietz, Peter:
Bosonization of interacting fermions in arbitrary dimensions / Peter Kopietz. - Berlin ; Heidelberg ; New York ; Barcelona ; Budapest ; Hong Kong ; London ; Milan ; Paris ; Santa Clara ; Singapore ; Tokyo : Springer, 1996
(Lecture notes in physics : N.s. M, Monographs ; 48)

NE: Lecture notes in physics / M

ISBN 978-3-662-14169-4 ISBN 978-3-540-68495-4 (eBook)
DOI 10.1007/978-3-540-68495-4

Originally published by Springer-Verlag Berlin Heidelberg New York in 1997
Softcover reprint of the hardcover 1st edition 1997

Typesetting: Camera-ready by author
Cover design: *design & production* GmbH, Heidelberg
SPIN: 10550879 55/3144-543210 - Printed on acid-free paper

The creative act depends on private visions and on solitary constructions and always draws on the legacy and the resources of the community – be it in the arts, literature, technology, or the sciences.[1]

TO CORNELIA

[1] S. S. Schweber, in *QED and the men who made it: Dyson, Feynman, Schwinger, and Tomonaga*, (Princeton University Press, Princeton, 1994), p.473.

Preface

In this book I shall describe a new non-perturbative approach to the fermionic many-body problem, which can be considered as a higher-dimensional generalization of the well-known bosonization technique for fermions in one spatial dimension. This approach is based on the direct calculation of correlation functions of interacting Fermi systems with dominant forward scattering via functional integration and Hubbard–Stratonovich transformations. I shall not attempt to generalize one-dimensional operator identities between the fermionic and bosonic field operators to higher dimensions. The fundamental ideas of higher-dimensional bosonization were first formulated by A. Luther (1979) and by F. D. M. Haldane (1992). In this book I shall go beyond these ideas and develop a powerful and systematic method for including the non-linear terms in the expansion of the energy dispersion close to the Fermi surface into the bosonization procedure. This is of particular importance in dimensions larger than one, because then the *curvature* of the Fermi surface can give rise to qualitatively new effects that are not correctly described if the energy dispersion is linearized. I shall also apply this method to a number of problems of physical interest that are very difficult – and in some cases impossible – to solve by means of conventional perturbation theory.

The restriction to dominant forward scattering means that in real space the effective interaction between the fermions must be sufficiently long-range. Physical examples are the Coulomb interaction at high densities, or the effective current–current interaction mediated by transverse gauge fields. On the other hand, short-range interactions such as the local Hubbard interaction cannot be directly treated within higher-dimensional bosonization, although the perturbative inclusion of scattering processes involving large momentum transfers seems to be possible. For simplicity, I shall restrict myself to *normal* Fermi systems. However, within the functional integral approach to higher-dimensional bosonization described here it should be straightforward to take also spontaneous symmetry breaking into account. To encourage all readers to contribute to the further development of this novel non-perturbative technique, I shall mention at the end of each chapter some open research problems, which might be solvable by means of modifications or extensions of the methods developed in this book.

I would like to thank at this point everyone who – directly or indirectly – has helped me to complete this book. First of all, I am grateful to Kurt Schönhammer for numerous collaborations and discussions, for getting me interested in bosonization shortly after I moved to Göttingen, and for giving me the freedom I needed to pursue my own ideas. The formal development of the functional bosonization approach was partially carried out in collaboration with Kurt, who often saw subtleties that I overlooked at first. More recently I have been collaborating with my friend Guillermo Castilla, on whom I could always count whenever I needed encouragement, advise, or help. Although we communicate mainly via e-mail, my information exchange with Guillermo has been almost as intense as during our common time as graduate students at UCLA.

I am also grateful to Sudip Chakravarty and Konstantin Efetov for being my teachers. Under Sudip's guidance I have learnt to do independent research. He has taught me to distinguish interesting physics from empty mathematics, and his very intuitive way of thinking about physical problems has strongly influenced my personal style of choosing and solving my own research projects. I have enjoyed very much being a postdoc in Konstantin Efetov's international and very active group at the *Max-Planck-Institut für Festkörperforschung* at Stuttgart. During this time I could broaden my horizons and become familiar with the physics of disordered Fermi systems. I have greatly profited from Konstantin's profound knowledge in this field.

I would like to thank Peter Wölfle for comments on the manuscript, and for pointing out some references related to gauge fields. In one way or the other, I have also profited from discussions and collaborations with Lorenz Bartosch, Jim "Claude" Cochran, Fabian "Fabman" Essler, Jens Fricke, Lev Gehlhoff, Ralf Hannappel, Joachim Hermisson, Jens Kappey, Stefan Kettemann, Volker Meden, Walter Metzner, Jacob Morris, Ben Sauer, Peter Scharf, Axel Völker, and Roland Zeyher.

Although I sometimes tend to ignore it, I know very well that there are more important things in life other than physics. This book is dedicated to my girlfriend Cornelia Buhrke for helping me to keep in touch with the real world during the nearly two years of writing, and for much more.

Göttingen, December 1996 *Peter Kopietz*

Contents

Part I

Development of the formalism

1. Introduction

...in which we try to explain why we have written this book.

1.1 Perturbation theory and quasi-particles

Perturbation theory for the single-particle Green's function of an interacting Fermi system usually works as long as the quasi-particle picture is valid.

The long-wavelength and low-energy behavior of the single-particle Green's function $G(\boldsymbol{k},\omega)$ of an interacting many-body system is directly related to the nature of its ground state and low lying excited states [1.1–1.6]. Because the qualitative features of the low-energy spectrum of a many-body Hamiltonian are usually determined by certain universal parameters such as dimensionality, symmetries, and conservation laws [1.7], the infrared behavior of the single-particle Green's function can be used to classify interacting many-body systems. Moreover, if $G(\boldsymbol{k},\omega)$ is known for all wave-vectors $\boldsymbol{k}$ and frequencies ω, one can in principle calculate all thermodynamic properties of the system [1.6]. Unfortunately, in almost all physically interesting cases it is impossible to calculate the Green's function exactly, so that one has to resort to approximate methods. The most naive approach would be the direct expansion of $G(\boldsymbol{k},\omega)$ in powers of the interaction. It is well known, however, that even for small interactions such an expansion is not valid for all wave-vectors and frequencies, because $G(\boldsymbol{k},\omega)$ usually has poles or other singularities, in the vicinity of which a power series expansion of $G(\boldsymbol{k},\omega)$ is not possible. In many cases this problem can be avoided if one introduces the irreducible self-energy $\Sigma(\boldsymbol{k},\omega)$ via the Dyson equation,

$$[G(\boldsymbol{k},\omega)]^{-1} = [G_0(\boldsymbol{k},\omega)]^{-1} - \Sigma(\boldsymbol{k},\omega) \quad , \tag{1.1}$$

and calculates $\Sigma(\boldsymbol{k},\omega)$ instead of $G(\boldsymbol{k},\omega)$ in powers of the interaction. Here $G_0(\boldsymbol{k},\omega)$ is the Green's function of a suitably defined non-interacting system, which can be calculated exactly. It is important to stress that the Dyson equation does not simply express one unknown quantity $G(\boldsymbol{k},\omega)$ in terms of another unknown $\Sigma(\boldsymbol{k},\omega)$, but tells us that the *inverse* Green's function should be expanded in powers of the interaction.

In so-called *Landau Fermi liquids* the above perturbative approach can indeed be used to calculate the Green's function. Of course, for strong interactions infinite orders in perturbation theory have to be summed, but the integrals generated in the perturbative expansion are free of divergencies and lead to a finite expression for the self-energy. The theory of Fermi liquids was advanced by Landau [1.8] in 1956 as a phenomenological theory to describe the static and dynamic properties of a large class of interacting fermions [1.9]. The most important physical realization of a Fermi liquid are electrons in clean three-dimensional metals, but also liquid ^{3}He is a Fermi liquid [1.10]. Simultaneously with Landau's pioneering ideas the powerful machinery of quantum field theory was developed and applied to condensed matter systems [1.1–1.5], and a few years later his phenomenological theory was put on a solid theoretical basis [1.9]. The retarded single-particle Green's function[1] of a Fermi liquid is for wave-vectors $\boldsymbol{k}$ in the vicinity of the Fermi surface and small frequencies ω to a good approximation given by

$$G(\boldsymbol{k}, \omega + \mathrm{i}0^+) \approx \frac{Z_{\boldsymbol{k}}}{\omega - \tilde{\xi}_{\boldsymbol{k}} + \mathrm{i}\gamma_{\boldsymbol{k}}} \quad , \tag{1.2}$$

where the number $Z_{\boldsymbol{k}}$ is the so-called *quasi-particle residue*, and the energy $\tilde{\xi}_{\boldsymbol{k}}$ is the single-particle excitation energy. Because by definition Landau Fermi liquids are metals, the excitation energy $\tilde{\xi}_{\boldsymbol{k}}$ must be gapless. This means that there exists a surface in $\boldsymbol{k}$-space where $\tilde{\xi}_{\boldsymbol{k}} = 0$. In a Fermi liquid this equation can be used to *define* the Fermi surface. The positive energy $\gamma_{\boldsymbol{k}}$ in Eq.(1.2) can be identified with the quasi-particle damping, and is assumed to vanish faster than $\tilde{\xi}_{\boldsymbol{k}}$ when the wave-vector $\mathbf{k}$ approaches the Fermi surface. Note that in the complex ω-plane $G(\boldsymbol{k}, \omega + \mathrm{i}0^+)$ has a simple pole at $\omega = \tilde{\xi}_{\boldsymbol{k}} - \mathrm{i}\gamma_{\boldsymbol{k}}$ with residue $Z_{\boldsymbol{k}}$. Obviously, the Green's function of *non-interacting* fermions can be obtained as a special case of Eq.(1.2), namely by setting $Z_{\boldsymbol{k}} = 1$, $\gamma_{\boldsymbol{k}} = 0^+$, and identifying $\tilde{\xi}_{\boldsymbol{k}}$ with the non-interacting energy dispersion measured relative to the chemical potential. Then the pole at $\omega = \tilde{\xi}_{\boldsymbol{k}} - \mathrm{i}0^+$ with unit residue is a consequence of the undamped propagation of a particle with energy dispersion $\tilde{\xi}_{\boldsymbol{k}}$ through the system. The corresponding pole in the Green's function of an interacting Fermi liquid is associated with a so-called *quasi-particle*. The important point is that in the vicinity of the quasi-particle pole the Green's function of a Fermi liquid has *qualitatively* the same structure as the Green's function of free fermions. In renormalization group language, the interacting Fermi liquid and the free Fermi gas correspond to the same fixed point in the infinite-dimensional parameter space spanned by all possible scattering processes [1.11, 1.12]. As explained in detail in Chap. 2, in a Landau Fermi liquid the quantities $Z_{\boldsymbol{k}}$, $\tilde{\xi}_{\boldsymbol{k}}$ and $\gamma_{\boldsymbol{k}}$ can be calculated from the derivatives of the self-energy $\Sigma(\boldsymbol{k}, \omega)$.

[1] We denote the Fourier transform of the *time-ordered* Green's function at wave-vector $\boldsymbol{k}$ and frequency ω by $G(\boldsymbol{k}, \omega)$. The corresponding retarded Green's function will be denoted by $G(\boldsymbol{k}, \omega + \mathrm{i}0^+)$, and the advanced one by $G(\boldsymbol{k}, \omega - \mathrm{i}0^+)$.

In some cases, however, the application of the standard machinery of many-body theory leads to divergent integrals in the perturbative expansion of $\Sigma(\boldsymbol{k}, \omega)$. The breakdown of perturbation theory is a manifestation of the fact that the interacting Green's function is not any more related in a simple way to the non-interacting one. In this case the system cannot be a Fermi liquid. A well known example are electrons in one spatial dimension with regular interactions, which under quite general conditions show *Luttinger liquid* behavior [1.13–1.15]. In contrast to a Fermi liquid, the Green's function of a Luttinger liquid does not have simple poles in the complex frequency plane, but exhibits only branch cut singularities involving non-universal power laws[2]. As a consequence, in a Luttinger liquid $[G(\boldsymbol{k}, \omega)]^{-1}$ cannot be calculated by simple perturbation theory around $[G_0(\boldsymbol{k}, \omega)]^{-1}$. Hence, non-perturbative methods are necessary to calculate the Green's function of interacting fermions in $d = 1$ dimension. Besides the Bethe ansatz [1.16] and renormalization group methods [1.13], the bosonization approach has been applied to one-dimensional Fermi systems with great success [1.13–1.15]. Over the past 30 years numerous interesting results have been obtained with this non-perturbative method. The so-called Tomonaga-Luttinger model is a paradigm for an exactly solvable non-trivial many-body system which exhibits all the characteristic Luttinger liquid properties, such as the absence of a quasi-particle peak in the single-particle Green's function, anomalous scaling, and spin-charge separation [1.17–1.19]. Even now interesting new results on the Tomonaga-Luttinger model are reported in the literature [1.20, 1.21]. For an up-to-date overview and extensive references on bosonization in $d = 1$ we would like to refer the reader to the recent reprint volume by M. Stone [1.22]. The central topic of this book is the generalization of the bosonization approach to arbitrary dimensions.

1.2 A brief history of bosonization in $d > 1$

We apologize in advance if we should have forgotten someone. Maybe some Russians have bosonized higher-dimensional Fermi systems long time ago, and we just don't know about their work ...

The discovery of the high-temperature superconductors and Anderson and co-workers suggestion [1.23, 1.24] that the normal-state properties of these materials are a manifestation of non-Fermi liquid behavior in dimensions $d > 1$ has revived the interest to develop non-perturbative methods for analyzing interacting fermions in $d > 1$. Note, however, that for regular interactions in $d > 1$ perturbation theory is consistent in the sense that within the framework

[2] In Chap. 6.3 we shall discuss the behavior of the Green's function of Luttinger liquids in some detail.

of perturbation theory itself there is no signal for its breakdown [1.25, 1.26]. Nevertheless, consistency of perturbation theory does not imply that the perturbative result must be correct. It is therefore highly desirable to analyze interacting Fermi systems by means of a non-perturbative approach which does not assume *a priori* that the system is a Fermi liquid. The recently developed higher-dimensional generalization of bosonization seems to be the most promising analytical method which satisfies this criterion in $d > 1$.

In one dimension bosonization is based on the observation that, after proper rescaling, the operators describing density fluctuations obey canonical bosonic commutation relations [1.13–1.15]. But also in $d = 3$ density fluctuations in an interacting Fermi system behave in many respects like bosonic degrees of freedom [1.27, 1.28]. The first serious attempt to formalize this observation and exploit it to develop a generalization of the one-dimensional bosonization approach to arbitrary dimensions was due to Luther [1.29]. However, Luther's pioneering work has not received much attention until Haldane [1.30] added the grain of salt that was necessary to turn higher-dimensional bosonization into a practically useful non-perturbative approach to the fermionic many-body problem. Haldane's crucial insight was that the degrees of freedom in the vicinity of the Fermi surface should be subdivided into boxes of *finite cross section*, such that *the motion of particle-hole pairs can be described without taking momentum-transfer between different boxes into account.* In Luther's formulation only the motion normal to the Fermi surface can be described in such a simple way. The first applications of Haldane's bosonization ideas to problems of physical interest were given by Houghton, Marston and Kwon [1.31], and independently by Castro Neto and Fradkin [1.32]. These approaches follow closely the usual bosonization procedure in one-dimensional systems, and are based on higher-dimensional generalizations of the Kac-Moody algebra that is *approximately* satisfied by charge and spin current operators. Just like in $d = 1$, it is possible to map with this method the fermionic many-body Hamiltonian onto an effective non-interacting bosonic Hamiltonian. The potential of these operator bosonization approaches is certainly not yet exhausted [1.33, 1.34]. However, unlike recent claims in the literature [1.34], bosonization in $d > 1$ is *not exact.* For example, scattering processes that transfer momentum between different boxes on the Fermi surface and non-linear terms in the energy dispersion definitely give rise to corrections to the free-boson approximation for the Hamiltonian. The problem of calculating these corrections within the conventional operator approach seems to be very difficult and so far has not been solved.

In the present book we shall develop an alternative generalization of the bosonization approach to arbitrary dimensions, which is based on functional integration and Hubbard-Stratonovich transformations. In this way we avoid the algebraic considerations of commutation relations which form the basis of the operator bosonization approaches [1.31, 1.32]. The functional integral formulation of higher-dimensional bosonization has been developed by the author in collaboration with Kurt Schönhammer [1.35] during spring 1994. Since then we have considerably refined this method [1.36–1.38] and applied it to various problems of physical interest. A coherent and detailed presentation of these results will be given in this book. A similar functional bosonization method, which emphasizes more the mathematical aspects of bosonization, has been developed independently by Fröhlich and collaborators [1.39, 1.40]. In the context of the one-dimensional Tomonaga-Luttinger model the functional bosonization technique has first been discussed by Fogedby [1.41], and later by Lee and Chen [1.42].

Compared with the more conventional operator bosonization [1.31–1.34], the functional bosonization approach has several advantages. The most important advantage is that within our functional integral approach it is possible *to handle the non-linear terms in the energy dispersion* (and hence in $d > 1$ the *curvature* of the Fermi surface). Note that the linearization of the energy dispersion close to the Fermi surface is one of the crucial (and a priori uncontrolled) approximations of conventional bosonization; even in $d = 1$ it is very difficult to calculate systematically the corrections due to the non-linear terms in the expansion of the dispersion relation close to the Fermi surface [1.15, 1.43]. A practically useful method for doing this will be developed in this book. In Chap. 4 we shall explicitly calculate the leading correction to the free bosonized Hamiltonian and the density-density correlation function. Moreover, in Chap. 5.2 we shall show how the bosonization result for the single-particle Green's function for fermions with linearized energy dispersion is modified by the quadratic term in the expansion of the energy dispersion close to the Fermi surface. In this way the approximations inherent in higher-dimensional bosonization become very transparent.

Another advantage of the functional integral formulation of higher-dimensional bosonization is that it can be applied in a straightforward way to physical problems where non-locality and retardation are essential. It is well-known [1.44] that these important many-body effects can be described in the most simple and general way via functional integrals and effective actions. In fact, the complicated effective dynamics of a quantum mechanical system that is coupled to another subsystem can sometimes only be described by means of a non-local effective action, and not by a Hamiltonian [1.45]. For example, the effective retarded interaction between electrons that is mediated via phonons or photons cannot be represented in terms of a conventional Hamiltonian. It is therefore advantageous to use functional integrals and the concept of an effective action as a basis to generalize the bosonization approach to dimensions larger than one.

Alternative formulations of higher-dimensional bosonization have also been proposed by Schmelzer and Bishop [1.46], by Khveshchenko and collaborators [1.47, 1.48], and by Li [1.49]. In particular, Khveshchenko [1.48] has also developed a formal method to include the curvature of the Fermi surface into higher-dimensional bosonization. However, so far his method has not been proven to be useful in practice. We shall not further discuss the above works in this book, because we believe that our functional bosonization technique leads to a more transparent and practically more useful approach to the bosonization problem in arbitrary dimensions. Finally, it should be mentioned that recently Castellani, Di Castro and Metzner [1.50–1.52] have proposed another non-perturbative approach to the fermionic many-body problem in $d > 1$. Their method is based on Ward identities and sums exactly the same infinite number of Feynman diagrams in the perturbation series as higher-dimensional bosonization with linearized energy dispersion. We shall derive the precise relation between the Ward identity approach and bosonization in Chap. 5.1.4.

1.3 The scope of this book

We have subdivided this book into two parts. Part I comprises the first five chapters and is devoted to the formal development of the functional bosonization approach. We begin by reminding the reader in Chap. 2 of some basic facts about interacting fermions. We also describe in some detail various ways of subdividing the momentum space in the vicinity of the Fermi surface into sectors. These geometric constructions are the key to the generalization of the bosonization approach to arbitrary dimensions. In Chap. 3 we introduce two Hubbard-Stratonovich transformations which directly lead to the bosonization result for the single-particle Green's function and the boson representation of the Hamiltonian. The explicit calculation of the bosonic Hamiltonian is presented in Chap. 4, where we also show that the problem of bosonizing the Hamiltonian is essentially equivalent with the problem of calculating the density-density correlation function. We also show that the non-Gaussian terms in the bosonic Hamiltonian are closely related to the local field corrections to the random-phase approximation. Chapter 5 is devoted to the calculation of the single-particle Green's function. This is the most important chapter of this book, because here we describe in detail our non-perturbative method for including the non-linear terms in the expansion of the energy dispersion for wave-vectors close to the Fermi surface into the bosonization procedure. Note that in $d > 1$ the local *curvature* of the Fermi surface can only be described if the quadratic term in the energy dispersion is retained. Our method is based on a generalization of the Schwinger ansatz for the Green's function in a given external field, an imaginary-time eikonal expansion, and diagrammatic techniques borrowed from the theory of disordered systems.

In Part II we shall use our formalism to calculate and classify the long-wavelength and low-energy behavior of a number of normal fermionic quantum liquids. In most cases we shall concentrate on parameter regimes where conventional perturbation theory is not applicable. In particular, we discuss fermions with singular density-density interactions (Chap. 6), quasi-one-dimensional metals (Chap. 7), electron-phonon interactions (Chap. 8), electrons in a dynamic random medium (Chap. 9), and fermions that are coupled to transverse gauge fields (Chap. 10.). Finally, in the Appendix we summarize some useful results on screening and collective modes in arbitrary dimensions.

Because the method described in this book is rather new, much remains to be done to establish higher-dimensional bosonization as a generally accepted, practically useful non-perturbative tool for studying strongly correlated Fermi systems. We would like to encourage all readers to actively participate in the process of further developing this method. For this purpose we have given at the end of each chapter a brief summary of the main results, together with a list of open problems and possible directions for further research.

1.4 Notations and assumptions

Let us briefly summarize the conventions that will be used throughout this work. We shall measure temperature T and frequencies ω in units of energy, which amounts to formally setting the Boltzmann constant k_{B} and Planck's constant $\hbar$ equal to unity. Note that in these units it is not necessary to distinguish between wave-vectors and momenta. The charge of the electron will be denoted by $-e$, and the fine structure constant is $\alpha = \frac{e^2}{c} \approx \frac{1}{137}$. The velocity of light c will not be set equal to unity, because in our discussion of transverse gauge fields in Chap. 10 it is useful to explicitly see the ratio v_{F}/c, where v_{F} is the Fermi velocity. The inverse temperature will be denoted by $\beta = 1/T$, and the volume of the system by V. Although at intermediate steps the volume of space-time $V\beta$ will be held finite, we are eventually interested in the limits of infinite volume ($V \to \infty$) and zero temperature ($\beta \to \infty$). As pointed out by Kohn, Luttinger, and Ward [1.53], in case of ambiguities the limit $V \to \infty$ should be taken before the limit $\beta \to \infty$. However, we shall ignore the subtleties associated with the infinite volume limit that have recently been discussed by Metzner and Castellani [1.54]. Although we are interested in the zero-temperature limit, we shall use the Matsubara formalism and work at intermediate steps at finite temperatures. In this way we also eliminate possible unphysical "anomalous" terms [1.53] which sometimes appear in a zero-temperature formalism, but are avoided if the Matsubara sums are performed at finite temperature and the $T \to 0$ limit is carefully taken afterwards.

We shall denote bosonic Matsubara frequencies by $\omega_m = 2\pi m T$, $m = 0, \pm 1, \pm 2, \ldots$, and put an extra tilde over fermionic ones, $\tilde{\omega}_n = 2\pi[n + \frac{1}{2}]T$, $n = 0, \pm 1, \pm 2, \ldots$. To simplify the notation, we introduce composite labels for wave-vectors and Matsubara frequencies: $k \equiv [\boldsymbol{k}, \mathrm{i}\tilde{\omega}_n]$, $q \equiv [\boldsymbol{q}, \mathrm{i}\omega_m]$, and $\tilde{q} \equiv [\boldsymbol{q}, \mathrm{i}\tilde{\omega}_n]$. Note that the label q is associated with bosonic frequencies, whereas k and $\tilde{q}$ involve fermionic frequencies.

2. Fermions and the Fermi surface

We summarize some basic facts about interacting fermions and introduce notations that will be used throughout this book. We also describe Haldane's way of partitioning the Fermi surface into patches and generalize it such that the curvature of the Fermi surface can be taken into account.

2.1 The generic many-body Hamiltonian

We first introduce the many-body Hamiltonian for interacting fermions and point out some subtleties associated with ultraviolet cutoffs.

The starting point of conventional many-body theory is a second-quantized Hamiltonian of the form

$$\hat{H}_{\text{mat}} = \hat{H}_0 + \hat{H}_{\text{int}} \quad , \tag{2.1}$$

$$\hat{H}_0 = \sum_{\boldsymbol{k}} \sum_{\sigma} \epsilon_{\boldsymbol{k}} \hat{\psi}^{\dagger}_{\boldsymbol{k}\sigma} \hat{\psi}_{\boldsymbol{k}\sigma} \quad , \tag{2.2}$$

$$\hat{H}_{\text{int}} = \frac{1}{2V} \sum_{\boldsymbol{q}\boldsymbol{k}\boldsymbol{k}'} \sum_{\sigma\sigma'} f_{\boldsymbol{q}}^{\boldsymbol{k}\sigma\boldsymbol{k}'\sigma'} \hat{\psi}^{\dagger}_{\boldsymbol{k}+\boldsymbol{q}\sigma} \hat{\psi}^{\dagger}_{\boldsymbol{k}'-\boldsymbol{q}\sigma'} \hat{\psi}_{\boldsymbol{k}'\sigma'} \hat{\psi}_{\boldsymbol{k}\sigma} \quad , \tag{2.3}$$

where $\hat{\psi}_{\boldsymbol{k}\sigma}$ and $\hat{\psi}^{\dagger}_{\boldsymbol{k}\sigma}$ are canonical annihilation and creation operators for fermions with wave-vector $\boldsymbol{k}$ and spin σ, which satisfy the anti-commutation relations

$$[\hat{\psi}_{\boldsymbol{k}\sigma}, \hat{\psi}^{\dagger}_{\boldsymbol{k}'\sigma'}]_{+} = \hat{\psi}_{\boldsymbol{k}\sigma} \hat{\psi}^{\dagger}_{\boldsymbol{k}'\sigma'} + \hat{\psi}^{\dagger}_{\boldsymbol{k}'\sigma'} \hat{\psi}_{\boldsymbol{k}\sigma} = \delta_{\boldsymbol{k}\boldsymbol{k}'} \delta_{\sigma\sigma'} \quad . \tag{2.4}$$

The quantities $f_{\boldsymbol{q}}^{\boldsymbol{k}\sigma\boldsymbol{k}'\sigma'}$ are the so-called *Landau interaction parameters*, describing the scattering of two particles from initial states with quantum numbers $(\boldsymbol{k}, \sigma)$ and $(\boldsymbol{k}', \sigma')$ into final states with quantum numbers $(\boldsymbol{k} + \boldsymbol{q}, \sigma)$ and $(\boldsymbol{k}' - \boldsymbol{q}, \sigma')$. This process can be represented graphically by the Feynman diagram shown in Fig. 2.1. Quantum many-body theory is usually formulated in the grand canonical ensemble, where the relevant combination is $\hat{H}_{\text{mat}} - \mu \hat{N}$. Here $\hat{N} = \sum_{\boldsymbol{k}} \sum_{\sigma} \hat{\psi}^{\dagger}_{\boldsymbol{k}\sigma} \hat{\psi}_{\boldsymbol{k}\sigma}$ is the particle number operator, and μ is the chemical potential. Thus, the energy dispersion $\epsilon_{\boldsymbol{k}}$ appears exclusively in the combination

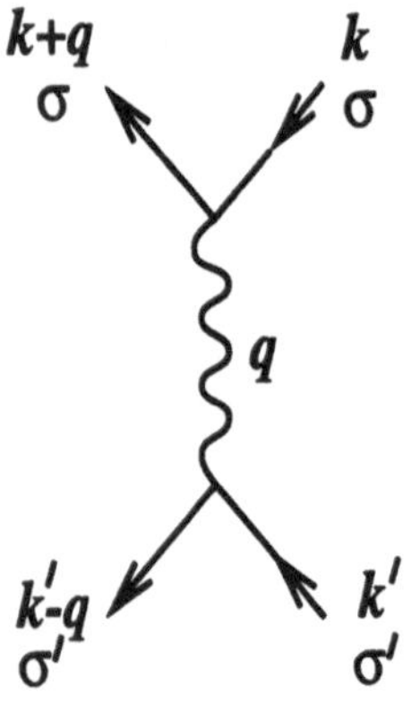

Fig. 2.1. Feynman diagram representing the interaction $f_{q}^{k\sigma k'\sigma'}$ in Eq.(2.3).

$$\xi_{\boldsymbol{k}} \equiv \epsilon_{\boldsymbol{k}} - \mu \quad . \tag{2.5}$$

The value of μ at zero temperature is also called the Fermi energy E_{F}. Although in most physical applications we are interested in three dimensions, it is very useful and instructive to formulate the theory in arbitrary dimension d. Then the equation

$$\xi_{\boldsymbol{k}} = 0 \tag{2.6}$$

defines a $d-1$-dimensional surface in momentum space, the ***non-interacting*** Fermi surface. The precise definition of the *interacting* Fermi surface will be given in Sect. 2.2. Note that in $d = 1$ the non-interacting Fermi surface consists of two distinct points $\pm k_{\mathrm{F}}$, where k_{F} is the Fermi wave-vector. In higher dimensions the Fermi surface is a $d-1$-dimensional manifold, the topology of which depends on the form of $\xi_{\boldsymbol{k}}$. There is actually a subtle point hidden in the above definition: although the energy $\epsilon_{\boldsymbol{k}}$ is a parameter of the non-interacting Hamiltonian $\hat{H}_0$, the chemical potential μ is by definition *the exact chemical potential of the interacting many-body system.* Of course, the precise value of μ remains unknown unless we can solve the many-body problem, but fortunately it is not necessary to know μ in order to calculate physical correlation functions. By defining μ to be the chemical potential of the interacting many-body system, one implicitly adds a suitable counter-term to the bare chemical potential which eliminates, order by order in perturbation theory, all terms which would otherwise contribute to $\Sigma(\boldsymbol{k}, 0)$ for wave-vectors $\boldsymbol{k}$ on the Fermi surface. In particular, all Feynman diagrams of the Hartree type are cancelled by the counter-term. Such a procedure is familiar from perturbative quantum field theory [2.1]. The consistency for such a construction is by no means obvious, and has recently been questioned by Anderson [2.2]. For a thorough discussion and partial solution of this problem see [2.3].

It should be emphasized that Eqs.(2.1)–(2.3) can be interpreted in three distinct ways, which can be classified according to the effective ultraviolet cutoff for the wave-vector sums.

(a) *Homogeneous electron gas.* First of all, we may define $\hat{H}_{\mathrm{mat}}$ to be the Hamiltonian of the homogeneous electron gas in d dimensions. For example the Coulomb-interaction in $d = 3$ dimensions corresponds to $\epsilon_{\boldsymbol{k}} = \boldsymbol{k}^2/(2m)$ and $f_{\boldsymbol{q}}^{\boldsymbol{k}\sigma\boldsymbol{k}'\sigma'} = 4\pi e^2/\boldsymbol{q}^2$, where m is the mass of the electrons. In this case there is no intrinsic short-distance cutoff for the wave-vector sums.

(b) *Relevant band of a lattice model.* Because in realistic materials the electrons feel the periodic potential due to the ions, the allowed energies in the absence of interactions are subdivided into energy bands, and the interaction has interband matrix elements. But if there exists only a single band in the vicinity of the Fermi surface, then it is allowed to ignore all other bands as long as one is interested in energy scales small compared with the interband gap. In this case the Hamiltonian defined in Eqs.(2.1)–(2.3) should be considered as the *effective Hamiltonian* for the band in the vicinity of the Fermi energy. In this model the wave-vector sums have a cutoff of the order of $2\pi/a$, where a is the distance between the ions. The energy dispersion $\epsilon_{\boldsymbol{k}}$ in Eq.(2.2) incorporates then by definition the effects of the underlying lattice, which in general leads also to a renormalization of the effective mass of the electrons.

(c) *Effective Hamiltonian for degrees of freedom close to the Fermi surface.* Finally, we may define $\hat{H}_{\mathrm{mat}}$ to be the effective Hamiltonian for the low-energy degrees of freedom in the vicinity of the Fermi surface, assuming that all degrees of freedom outside a thin shell with radial thickness $\lambda \ll k_{\mathrm{F}}$ have been integrated out via functional integration and renormalization group methods [1.12]. Of course, the operation of integrating out the high-energy degrees of freedom will also generate three-body and higher order interactions, which are ignored in Eqs.(2.1)–(2.3). The quantities $\epsilon_{\boldsymbol{k}}$ and $f_{\boldsymbol{q}}^{\boldsymbol{k}\sigma\boldsymbol{k}'\sigma'}$ should then be considered as effective parameters, which take the finite renormalizations due to the high-energy degrees of freedom into account. In this picture the $\boldsymbol{k}$- and $\boldsymbol{k}'$-sums in Eqs.(2.2) and (2.3) are confined to a thin shell of thickness λ around the Fermi surface, while the $\boldsymbol{q}$-sum in Eq.(2.3) is restricted to the regime $|\boldsymbol{q}| \leq \lambda$.

All three interpretations of the many-body Hamiltonian (2.1)–(2.3) are useful. First of all, the model (a) has the advantage that it contains no free parameters, so that it can be the starting point of a first principles microscopic calculation. The model (b) is more realistic, although the effects of the underlying lattice are only included on a phenomenological level. Finally, the model (c) has the advantage that it contains explicitly only the low-energy degrees of freedom close to the Fermi surface, so that, to a first approximation, we may locally linearize the energy dispersion at the Fermi surface. Evidently the model (c) cannot be used for the calculation of the precise numerical value of physically measurable quantities that depend on fluctuations on all length scales. Furthermore, the integration over the degrees of freedom far away from the Fermi surface usually cannot be explicitly carried out.

2.2 The single-particle Green's function

We define the single-particle Green's function and the Fermi surface of an interacting Fermi system. We then discuss in some detail the low-energy behavior of the Green's function in a Landau Fermi liquid.

Because in the rest of this book the spin degree of freedom will not play any role, we shall from now on simply ignore the spin index. Formally, the spin is easily taken into account by defining $\boldsymbol{k}$ and $\boldsymbol{k}'$ to be collective labels for wave-vector and spin. For practical calculations we prefer to work with the Matsubara formalism, because in this way we avoid the problem of regularizing formally divergent integrals by means of pole prescriptions, which arises in the real time zero-temperature formulation of quantum many-body theory. Furthermore, the Matsubara Green's function can be represented as an imaginary time functional integral [2.4–2.7], so that the entire many-body problem can be reformulated in the language of path integrals. In this work we shall make extensive use of this modern approach to the many-body problem.

2.2.1 Definition of the Green's function

The single particle Matsubara Green's function $G(k)$ of an interacting Fermi system is defined by

$$G(k) \equiv G(\boldsymbol{k}, \mathrm{i}\tilde{\omega}_n) = -\frac{1}{\beta}\int_0^\beta \mathrm{d}\tau \int_0^\beta \mathrm{d}\tau' \mathrm{e}^{-\mathrm{i}\tilde{\omega}_n(\tau-\tau')} < \mathcal{T}\left[\hat{\psi}_{\boldsymbol{k}}(\tau)\hat{\psi}^\dagger_{\boldsymbol{k}}(\tau')\right] > , \tag{2.7}$$

where for fermions the time-ordering operator $\mathcal{T}$ in imaginary time is defined by

$$\begin{aligned}\mathcal{T}\left[\hat{\psi}_{\boldsymbol{k}}(\tau)\hat{\psi}^\dagger_{\boldsymbol{k}}(\tau')\right] &= \Theta(\tau - \tau' - 0^+)\hat{\psi}_{\boldsymbol{k}}(\tau)\hat{\psi}^\dagger_{\boldsymbol{k}}(\tau') \\ &- \Theta(\tau' - \tau + 0^+)\hat{\psi}^\dagger_{\boldsymbol{k}}(\tau')\hat{\psi}_{\boldsymbol{k}}(\tau) \ ,\end{aligned} \tag{2.8}$$

and the average in Eq.(2.7) denotes grand canonical thermal average with respect to all degrees of freedom in the system. For any operator $\hat{O}$ the time evolution in imaginary time is defined by

$$\hat{O}(\tau) = \mathrm{e}^{\tau(\hat{H}_{\mathrm{mat}}-\mu\hat{N})}\hat{O}\mathrm{e}^{-\tau(\hat{H}_{\mathrm{mat}}-\mu\hat{N})} \ , \tag{2.9}$$

where $\hat{H}_{\mathrm{mat}}$ is given in Eqs.(2.1)–(2.3). The Matsubara Green's function of a system of non-interacting fermions with Hamiltonian $\hat{H}_0$ (see Eq.(2.2)) is given by

$$G_0(k) = \frac{1}{\mathrm{i}\tilde{\omega}_n - \xi_{\boldsymbol{k}}} \ , \tag{2.10}$$

where the subscript $_0$ indicates the absence of interactions. Once the imaginary-frequency Green's function is known, we can obtain the corresponding retarded zero-temperature Green's function by analytic continuation in the complex frequency plane just above the real axis, $\mathrm{i}\tilde{\omega}_n \to \omega + \mathrm{i}0^+$. For the non-interacting retarded Green's function we obtain

$$G_0(\boldsymbol{k}, \omega + \mathrm{i}0^+) = \frac{1}{\omega - \xi_{\boldsymbol{k}} + \mathrm{i}0^+} \quad . \tag{2.11}$$

This function has a pole at $\omega = \xi_{\boldsymbol{k}} - \mathrm{i}0^+$ with residue $Z_{\boldsymbol{k}} = 1$. The infinitesimal imaginary part shifts the pole below the real axis, so that the retarded Green's function is analytic in the upper half of the complex frequency plane [1.5,1.6]. The corresponding advanced Green's function $G_0(\boldsymbol{k}, \omega - \mathrm{i}0^+)$ is analytic in the lower half of the frequency plane, while the time-ordered Green's function,

$$G_0(\boldsymbol{k}, \omega) = \frac{1}{\omega - \xi_{\boldsymbol{k}} + \mathrm{i}0^+ \mathrm{sgn}(\omega)} \quad , \tag{2.12}$$

agrees for $\omega > 0$ with the retarded Green's function, and for $\omega < 0$ with the advanced one. The analytic structure of the time-ordered Green's function $G(\boldsymbol{k}, \omega)$ of the interacting many-body system is similar [1.5, 1.6]: It has cuts above the real negative axis and below the real positive axis, a branch point at $\omega = 0$, and poles in the neighboring Riemann sheets. The simple pole structure of the non-interacting Matsubara Green's function $G_0(k)$ makes the analytic continuation trivial. In general it can be quite difficult to perform the analytic continuation of the interacting Matsubara Green's function to obtain the corresponding real frequency function. Nevertheless, we prefer to work with the Matsubara formalism, because Euclidean time-ordering leads to the very simple result (2.10) for the non-interacting Green's function. Note that the denominator in Eq.(2.10) can never vanish, so that we avoid in this way the singular integrands with poles on the real frequency axis that appear in a zero-temperature formalism.

2.2.2 Definition of the interacting Fermi surface

We define the Fermi surface of an interacting Fermi system as the set of points in momentum space where, in the limit of zero temperature, the momentum distribution $n_{\boldsymbol{k}}$ has some kind of non-analyticity. The momentum distribution can be expressed in terms of the exact Matsubara Green's function as

$$n_{\boldsymbol{k}} = \frac{1}{\beta} \sum_n G(\boldsymbol{k}, \mathrm{i}\tilde{\omega}_n) \quad . \tag{2.13}$$

In the absence of interactions we have $n_{\boldsymbol{k}} = f(\xi_{\boldsymbol{k}})$, where

$$f(E) = \frac{1}{\beta} \sum_n \frac{1}{\mathrm{i}\tilde{\omega}_n - E} = \frac{1}{\mathrm{e}^{\beta E} + 1} \tag{2.14}$$

is the Fermi function[1]. In the zero-temperature limit $f(E) \to \Theta(-E)$, so that the momentum distribution reduces to a step function, $n_{\boldsymbol{k}} = \Theta(-\xi_{\boldsymbol{k}})$. Because $\Theta(x)$ is not analytic at $x = 0$, we recover in the absence of interactions the definition of the non-interacting Fermi surface given in Eq.(2.6). We would like to emphasize that it is by no means clear that the momentum distribution of an interacting Fermi system has always non-analyticities. In fact, in Chap. 6.2.5 we shall give an example for a quantum liquid where $n_{\boldsymbol{k}}$ is analytic. In this case the interacting system simply does not have a sharp Fermi surface. To avoid misunderstandings, we shall from now on reserve the word *Fermi surface* for the non-interacting Fermi surface, as defined in Eq.(2.6).

2.2.3 Landau Fermi liquids

As already mentioned in Chap. 1, for wave-vectors $\boldsymbol{k}$ sufficiently close to the Fermi surface and sufficiently small energies, the Green's function of a *Landau Fermi liquid* has qualitatively the same pole structure as the non-interacting Green's function. The pole represents an elementary excitation of the system which approximately behaves like a free particle. This is the ***quasi-particle.*** To formulate the quasi-particle concept in precise mathematical language, consider a point $\boldsymbol{k}^{\alpha}$ on the Fermi surface (i.e. $\xi_{\boldsymbol{k}^{\alpha}} = 0$), and let us measure wave-vectors locally with respect to this point. The geometry is shown in Fig. 2.2. The energy dispersion is then given by

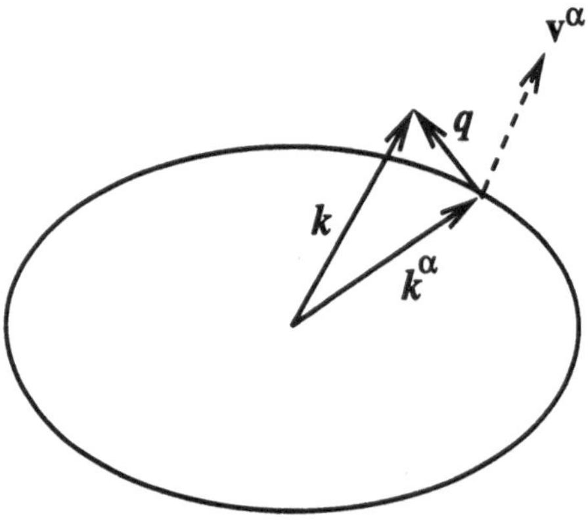

Fig. 2.2. Local coordinate system centered at point $\boldsymbol{k}^{\alpha}$ on an elliptic Fermi surface. Note that in general the local Fermi velocity $\boldsymbol{v}^{\alpha}$ is not parallel to $\boldsymbol{k}^{\alpha}$.

$$\xi_{\boldsymbol{q}}^{\alpha} \equiv \xi_{\boldsymbol{k}^{\alpha}+\boldsymbol{q}} \quad . \tag{2.15}$$

We now expand for small $\boldsymbol{q}$,

$$\xi_{\boldsymbol{q}}^{\alpha} = \boldsymbol{v}^{\alpha} \cdot \boldsymbol{q} + \frac{1}{2} \sum_{ij=1}^{d} q_i c_{ij}^{\alpha} q_j + O(|\boldsymbol{q}|^3) \quad , \tag{2.16}$$

[1] The Matsubara sum in Eq.(2.14) is formally divergent, and should be regularized by inserting a convergence factor $e^{i\tilde{\omega}_n 0^+}$, see [1.6].

where

$$\boldsymbol{v}^{\alpha} = \nabla_{\boldsymbol{k}} \epsilon_{\boldsymbol{k}} \big|_{\boldsymbol{k}=\boldsymbol{k}^{\alpha}} \quad , \quad c_{ij}^{\alpha} = \frac{\partial^2 \epsilon_{\boldsymbol{k}}}{\partial k_i \partial k_j}\bigg|_{\boldsymbol{k}=\boldsymbol{k}^{\alpha}} \quad . \tag{2.17}$$

Similarly, we expand the retarded self-energy $\Sigma(\boldsymbol{k}, \omega + \mathrm{i}0^+)$ defined in Eq.(1.1),

$$\begin{aligned} \Sigma(\boldsymbol{k}^{\alpha} + \boldsymbol{q}, \omega + \mathrm{i}0^+) = \boldsymbol{q} \cdot \nabla_{\boldsymbol{k}} \Sigma(\boldsymbol{k}, \mathrm{i}0^+)\big|_{\boldsymbol{k}=\boldsymbol{k}^{\alpha}} + \omega \left. \frac{\partial \Sigma(\boldsymbol{k}^{\alpha}, \omega + \mathrm{i}0^+)}{\partial \omega} \right|_{\omega=0} \\ + \delta \Sigma^{\alpha}(\boldsymbol{q}, \omega + \mathrm{i}0^+) \quad , \end{aligned} \tag{2.18}$$

where in a Fermi liquid $\delta \Sigma^{\alpha}(\boldsymbol{q}, \omega + \mathrm{i}0^+)$ is by assumption for small $\boldsymbol{q}$ and ω quadratic in these quantities. Note that in this expansion we have set

$$\Sigma(\boldsymbol{k}^{\alpha}, \mathrm{i}0^+) = 0 \quad , \tag{2.19}$$

assuming that the chemical potential μ is chosen such that Eq.(2.19) is satisfied for all points $\boldsymbol{k}^{\alpha}$ on the Fermi surface. As already mentioned in Sect. 2.1, this is a non-trivial assumption [2.2, 2.3].

The quasi-particle residue

Substituting Eq.(2.18) into the Dyson equation (1.1), we see that the Green's function of the interacting system can be written as

$$G(\boldsymbol{k}^{\alpha} + \boldsymbol{q}, \omega + \mathrm{i}0^+) = \frac{Z^{\alpha}}{\omega - \xi_{\boldsymbol{q}}^{\alpha} - \delta \boldsymbol{v}^{\alpha} \cdot \boldsymbol{q} - Z^{\alpha} \delta \Sigma^{\alpha}(\boldsymbol{q}, \omega + \mathrm{i}0^+)} \quad , \tag{2.20}$$

where the so called *quasi-particle residue* Z^{α} is given by

$$Z^{\alpha} = \frac{1}{1 - \left. \frac{\partial \Sigma(\boldsymbol{k}^{\alpha}, \omega + \mathrm{i}0^+)}{\partial \omega} \right|_{\omega=0}} \quad , \tag{2.21}$$

and the renormalization of the Fermi velocity at point $\boldsymbol{k}^{\alpha}$ is

$$\delta \boldsymbol{v}^{\alpha} = (Z^{\alpha} - 1) \boldsymbol{v}^{\alpha} + Z^{\alpha} \, \nabla_{\boldsymbol{k}} \Sigma(\boldsymbol{k}, \mathrm{i}0^+)\big|_{\boldsymbol{k}=\boldsymbol{k}^{\alpha}} \quad . \tag{2.22}$$

Thus, the effective Fermi velocity at $\boldsymbol{k}^{\alpha}$ is

$$\begin{aligned} \tilde{\boldsymbol{v}}^{\alpha} = \boldsymbol{v}^{\alpha} + \delta \boldsymbol{v}^{\alpha} &= Z^{\alpha} \left[\boldsymbol{v}^{\alpha} + \nabla_{\boldsymbol{k}} \Sigma(\boldsymbol{k}, \mathrm{i}0^+)\big|_{\boldsymbol{k}=\boldsymbol{k}^{\alpha}} \right] \\ &= \frac{\boldsymbol{v}^{\alpha} + \nabla_{\boldsymbol{k}} \Sigma(\boldsymbol{k}, \mathrm{i}0^+)\big|_{\boldsymbol{k}=\boldsymbol{k}^{\alpha}}}{1 - \left. \frac{\partial \Sigma(\boldsymbol{k}^{\alpha}, \omega + \mathrm{i}0^+)}{\partial \omega} \right|_{\omega=0}} \quad . \end{aligned} \tag{2.23}$$

The finite temperature generalization of Eq.(2.21) is [2.8]

$$Z^{\alpha}(T) = \frac{1}{1 - \frac{\mathrm{Im} \Sigma(\boldsymbol{k}^{\alpha}, \mathrm{i}\tilde{\omega}_0)}{\tilde{\omega}_0}} \quad , \tag{2.24}$$

where $\tilde{\omega}_0 = \pi T$ is the zeroth fermionic Matsubara frequency. The quasi-particle residue determines at $T = 0$ the discontinuity of the momentum distribution $n_{\boldsymbol{k}}$ when $\boldsymbol{k}$ crosses the Fermi surface. To calculate the change in the momentum distribution at point $\boldsymbol{k}^{\alpha}$ on the Fermi surface, consider

$$\delta n_{\boldsymbol{q}}^{\alpha} = n_{\boldsymbol{k}^{\alpha}-\boldsymbol{q}} - n_{\boldsymbol{k}^{\alpha}+\boldsymbol{q}} \quad . \tag{2.25}$$

For small enough $\boldsymbol{q}$ we may approximate $\xi_{\boldsymbol{q}}^{\alpha} \approx \boldsymbol{v}^{\alpha} \cdot \boldsymbol{q}$ and ignore the correction term $\delta\Sigma^{\alpha}$ in Eq.(2.20). At finite temperatures we obtain then

$$\delta n_{\boldsymbol{q}}^{\alpha} = Z^{\alpha}(T)\left[f(-\tilde{\boldsymbol{v}}^{\alpha} \cdot \boldsymbol{q}) - f(\tilde{\boldsymbol{v}}^{\alpha} \cdot \boldsymbol{q})\right] \quad . \tag{2.26}$$

In the zero-temperature limit $f(E) \rightarrow \Theta(-E)$ and $Z^{\alpha}(T) \rightarrow Z^{\alpha}$, so that $\delta n_{\boldsymbol{q}}^{\alpha} = Z^{\alpha}\mathrm{sgn}(\tilde{\boldsymbol{v}}^{\alpha} \cdot \boldsymbol{q})$. Note that $\delta n_{\boldsymbol{q}}^{\alpha}$ depends only on the projection of $\boldsymbol{q}$ that is normal to the Fermi surface, because this corresponds to a crossing of the Fermi surface and can thus give rise to a discontinuity.

The effective mass renormalization

If $\nabla_{\boldsymbol{k}}\Sigma(\boldsymbol{k}, \mathrm{i}0^{+})|_{\boldsymbol{k}=\boldsymbol{k}^{\alpha}}$ is parallel to $\boldsymbol{v}^{\alpha}$ (for example, for spherical Fermi surfaces and rotationally invariant interactions this is the case), we see from Eq.(2.23) that the renormalized Fermi velocity $\tilde{\boldsymbol{v}}^{\alpha}$ associated with point $\boldsymbol{k}^{\alpha}$ on the Fermi surface can be written as

$$\tilde{\boldsymbol{v}}^{\alpha} = Z_{\mathrm{m}}^{\alpha}\boldsymbol{v}^{\alpha} \quad , \tag{2.27}$$

where the *effective mass renormalization factor* Z_{m}^{α} is given by

$$\begin{aligned} Z_{\mathrm{m}}^{\alpha} &= Z^{\alpha}\left[1 + \frac{\hat{\boldsymbol{v}}^{\alpha} \cdot \nabla_{\boldsymbol{k}}\, \Sigma(\boldsymbol{k}, \mathrm{i}0^{+})|_{\boldsymbol{k}=\boldsymbol{k}^{\alpha}}}{|\boldsymbol{v}^{\alpha}|}\right] \\ &= \frac{1 + \frac{\hat{\boldsymbol{v}}^{\alpha} \cdot \nabla_{\boldsymbol{k}}\, \Sigma(\boldsymbol{k}, \mathrm{i}0^{+})|_{\boldsymbol{k}=\boldsymbol{k}^{\alpha}}}{|\boldsymbol{v}^{\alpha}|}}{1 - \left.\frac{\partial\Sigma(\boldsymbol{k}^{\alpha}, \omega + \mathrm{i}0+)}{\partial\omega}\right|_{\omega=0}} \quad , \end{aligned} \tag{2.28}$$

with $\hat{\boldsymbol{v}}^{\alpha} = \boldsymbol{v}^{\alpha}/|\boldsymbol{v}^{\alpha}|$. At finite temperatures, Eq.(2.28) should again be generalized as follows,

$$Z_{\mathrm{m}}^{\alpha}(T) = \frac{1 + \frac{\hat{\boldsymbol{v}}^{\alpha} \cdot \nabla_{\boldsymbol{k}}\, \mathrm{Re}\Sigma(\boldsymbol{k}, \mathrm{i}\tilde{\omega}_0)|_{\boldsymbol{k}=\boldsymbol{k}^{\alpha}}}{|\boldsymbol{v}^{\alpha}|}}{1 - \frac{\mathrm{Im}\Sigma(\boldsymbol{k}^{\alpha}, \mathrm{i}\tilde{\omega}_0)}{\tilde{\omega}_0}} \quad . \tag{2.29}$$

The effective mass $\tilde{m}^{\alpha}$ is defined in terms of the bare mass m via $\tilde{m}^{\alpha}\tilde{\boldsymbol{v}}^{\alpha} = m\boldsymbol{v}^{\alpha}$, so that

$$Z_{\mathrm{m}}^{\alpha} = \frac{m}{\tilde{m}^{\alpha}} = \frac{|\tilde{\boldsymbol{v}}^{\alpha}|}{|\boldsymbol{v}^{\alpha}|} \quad . \tag{2.30}$$

In other words, a *small* value of Z_{m}^{α} corresponds to a *large* effective mass. One of the fundamental properties of a Fermi liquid is that the renormalization factors Z^{α} and Z_{m}^{α} are finite[2].

---

[2] This working definition is sufficient for most physically interesting systems, although in some rather special cases it is not accurate enough. For example, if we retain only so-called g_4-processes in the one-dimensional Tomonaga-Luttinger model [1.17, 1.18] with spin and set $g_2 = 0$ (the spinless model is discussed in Chap. 6.3), then Z^{α} and Z_{m}^{α} are finite, but the Green's function exhibits spin-charge separation, which does not occur in Fermi liquids. I would like to thank Walter Metzner for pointing this out to me.

The quasi-particle damping

Eq.(2.20) is formally exact, provided Eq.(2.18) is taken as the *definition* of $\delta\Sigma^\alpha(\boldsymbol{q},\omega+\mathrm{i}0^+)$, expressing one unknown quantity in terms of another one. Of course, this parameterization is only useful if the correction $\delta\Sigma^\alpha$ becomes negligible, at least for wave-vectors in the vicinity of the Fermi surface. In Landau Fermi liquids $\delta\Sigma^\alpha(\boldsymbol{q},\omega+\mathrm{i}0^+)$ is by assumption analytic, so that for small $\boldsymbol{q}$ and for frequencies ω close to $\tilde{\boldsymbol{v}}^\alpha\cdot\boldsymbol{q}$ we may approximate

$$Z^\alpha\delta\Sigma^\alpha(\boldsymbol{q},\omega+\mathrm{i}0^+)\approx Z^\alpha\delta\Sigma^\alpha(\boldsymbol{q},\tilde{\boldsymbol{v}}^\alpha\cdot\boldsymbol{q}+\mathrm{i}0^+)\approx\frac{1}{2}\sum_{ij=1}^{d}q_i\delta c^\alpha_{ij}q_j \quad , \tag{2.31}$$

where δc^α_{ij} is a complex matrix that is determined by the various second partial derivatives of the self-energy. Defining the renormalized second-derivative matrix

$$\tilde{c}^\alpha_{ij}=c^\alpha_{ij}+\delta c^\alpha_{ij} \quad , \tag{2.32}$$

the real part of the renormalized energy dispersion for wave-vectors close to $\boldsymbol{k}^\alpha$ is

$$\tilde{\xi}^\alpha_{\boldsymbol{q}}=\tilde{\boldsymbol{v}}^\alpha\cdot\boldsymbol{q}+\frac{1}{2}\sum_{ij=1}^{d}q_i[\mathrm{Re}\tilde{c}^\alpha_{ij}]q_j+O(|\boldsymbol{q}|^3) \quad . \tag{2.33}$$

Although c^α_{ij} is real, the matrix δc^α_{ij} is in general complex, so that the renormalized energy dispersion acquires an imaginary part due to the interactions. Defining

$$\gamma^\alpha_{\boldsymbol{q}}=-\frac{1}{2}\sum_{ij=1}^{d}q_i[\mathrm{Im}\delta c^\alpha_{ij}]q_j \quad , \tag{2.34}$$

the interacting retarded Green's function of the many-body system is for sufficiently small $\boldsymbol{q}$ and ω given by

$$G(\boldsymbol{k}^\alpha+\boldsymbol{q},\omega+\mathrm{i}0^+)\approx\frac{Z^\alpha}{\omega-\tilde{\xi}^\alpha_{\boldsymbol{q}}+\mathrm{i}\gamma^\alpha_{\boldsymbol{q}}} \quad , \tag{2.35}$$

which is equivalent[3] with Eq.(1.2). This expression has a pole in the complex frequency plane at $\omega=\tilde{\xi}^\alpha_{\boldsymbol{q}}-\mathrm{i}\gamma^\alpha_{\boldsymbol{q}}$ with residue given by Z^α. By contour integration [1.6] it is easy to see that the pole contribution to the real time Fourier transform of the Green's function is

$$\begin{aligned}G(\boldsymbol{k}^\alpha+\boldsymbol{q},t)&=\int_{-\infty}^{\infty}\frac{\mathrm{d}\omega}{2\pi}\mathrm{e}^{-\mathrm{i}\omega t}G(\boldsymbol{k}^\alpha+\boldsymbol{q},\omega+\mathrm{i}0^+)\\&=-\mathrm{i}\Theta(t)Z^\alpha\mathrm{e}^{-\mathrm{i}\tilde{\xi}^\alpha_{\boldsymbol{q}}t}\mathrm{e}^{-\gamma^\alpha_{\boldsymbol{q}}t} \quad .\end{aligned} \tag{2.36}$$

[3] Recall that wave-vectors are now measured with respect to the local coordinate system centered at $\boldsymbol{k}^\alpha$ on the Fermi surface, so that in Eqs.(1.2) and (2.35) we should identify $Z_{\boldsymbol{k}^\alpha}=Z^\alpha$, $\tilde{\xi}_{\boldsymbol{k}^\alpha+\boldsymbol{q}}=\tilde{\xi}^\alpha_{\boldsymbol{q}}$, and $\gamma_{\boldsymbol{k}^\alpha+\boldsymbol{q}}=\gamma^\alpha_{\boldsymbol{q}}$.

If the damping γ_q^α is small compared with the real part $\tilde{\xi}_q^\alpha$ of the energy, then the behavior of the interacting Green's function is, up to times of order $1/\gamma_q^\alpha$, qualitatively similar to the behavior of the non-interacting Green's function. The pole is therefore said to represent a ***quasi-particle***. Actually, at times shorter than $1/\tilde{\xi}_q^\alpha$ it is not allowed to keep only the pole contribution in Eq.(2.36), so that quasi-particle behavior can only be observed in the intermediate time domain [1.6]

$$1/\tilde{\xi}_q^\alpha \ll t \lesssim 1/\gamma_q^\alpha \quad . \tag{2.37}$$

2.3 The density-density correlation function

We define the density-density correlation function $\Pi(q)$, the dynamic structure factor $S(\mathbf{q},\omega)$, and the dielectric function $\epsilon(q)$ of an interacting Fermi system. We also explain what is meant by "random-phase approximation".

Besides the single-particle Green's function, we are interested in the density-density correlation function $\Pi(q) \equiv \Pi(\mathbf{q}, \mathrm{i}\omega_m)$, which is for $\mathbf{q} \neq 0$ defined by[4]

$$\Pi(q) = \frac{1}{\beta V}\int_0^\beta \mathrm{d}\tau \int_0^\beta \mathrm{d}\tau' \mathrm{e}^{-\mathrm{i}\omega_m(\tau-\tau')} \langle \mathcal{T}\left[\hat{\rho}_{\mathbf{q}}(\tau)\hat{\rho}_{-\mathbf{q}}(\tau')\right]\rangle \quad , \tag{2.38}$$

where the operator

$$\hat{\rho}_{\mathbf{q}} = \sum_{\mathbf{k}} \hat{\psi}_{\mathbf{k}}^\dagger \hat{\psi}_{\mathbf{k}+\mathbf{q}} \tag{2.39}$$

represents the Fourier components of the total density, and $\mathcal{T}$ denotes ***bosonic*** time-ordering, i.e.

$$\begin{aligned}\mathcal{T}\left[\hat{\rho}_{\mathbf{q}}(\tau)\hat{\rho}_{-\mathbf{q}}(\tau')\right] &= \Theta(\tau-\tau'-0^+)\hat{\rho}_{\mathbf{q}}(\tau)\hat{\rho}_{-\mathbf{q}}(\tau') \\ &\quad + \Theta(\tau'-\tau+0^+)\hat{\rho}_{-\mathbf{q}}(\tau')\hat{\rho}_{\mathbf{q}}(\tau) \quad .\end{aligned} \tag{2.40}$$

Note that, in contrast to Eq.(2.8), there is no minus sign associated with a permutation, so that $\Pi(q)$ depends on bosonic Matsubara frequencies $\omega_m = 2\pi m T$. We shall also refer to $\Pi(q)$ as the ***polarization function***, or simply the ***polarization***. In the absence of interactions $\Pi(q)$ reduces to the imaginary frequency Lindhard function,

$$\Pi_0(q) = -\frac{1}{\beta V}\sum_k G_0(k)G_0(k+q) = -\frac{1}{V}\sum_{\mathbf{k}}\frac{f(\xi_{\mathbf{k}+\mathbf{q}})-f(\xi_{\mathbf{k}})}{\xi_{\mathbf{k}+\mathbf{q}}-\xi_{\mathbf{k}}-\mathrm{i}\omega_m} \quad . \tag{2.41}$$

---

[4] At $\mathbf{q}=0$ we should subtract from the time-ordered product in Eq.(2.38) the term $<\hat{\rho}_{\mathbf{q}}(\tau)><\hat{\rho}_{-\mathbf{q}}(\tau')>$, which in a translationally invariant system vanishes for any $\mathbf{q}\neq 0$. Because in the present work we are only interested in the $\mathbf{q}\neq 0$ part of the density-density correlation function, we shall omit this term.

The corresponding real frequency function can be obtained via analytic continuation. The discontinuity of $\Pi(\boldsymbol{q}, z)$ across the real axis defines the ***dynamic structure factor*** $S(\boldsymbol{q}, \omega)$ [1.7]

$$\mathrm{Im}\Pi(\boldsymbol{q}, \omega + \mathrm{i}0^+) = \pi \left[S(\boldsymbol{q}, \omega) - S(\boldsymbol{q}, -\omega) \right] \quad . \tag{2.42}$$

In terms of the exact eigenstates $|n\rangle$ and eigen-energies E_n of the operator $\hat{H}_{\mathrm{mat}} - \mu \hat{N}$ defined in Eqs.(2.1)–(2.3), $S(\boldsymbol{q}, \omega)$ has the spectral representation

$$S(\boldsymbol{q}, \omega) = \frac{1}{V} \sum_{nm} \frac{\mathrm{e}^{-\beta E_m}}{\mathcal{Z}} |\langle n | \hat{\rho}_{\boldsymbol{q}}^{\dagger} | m \rangle|^2 \delta(\omega - (E_n - E_m)) \quad , \tag{2.43}$$

where $\mathcal{Z}$ is the exact grand canonical partition function. From this expression it is obvious that $S(\boldsymbol{q}, \omega)$ is real and positive, and satisfies the detailed balance condition

$$S(\boldsymbol{q}, -\omega) = \mathrm{e}^{-\beta\omega} S(\boldsymbol{q}, \omega) \quad . \tag{2.44}$$

Using $\frac{1}{1 - \mathrm{e}^{-\beta\omega}} = 1 + \frac{1}{\mathrm{e}^{\beta\omega} - 1}$, it is easy to see that the imaginary part of $\Pi(\boldsymbol{q}, \omega + \mathrm{i}0^+)$ and the dynamic structure factor are related via

$$S(\boldsymbol{q}, \omega) = \left[1 + \frac{1}{\mathrm{e}^{\beta\omega} - 1} \right] \frac{1}{\pi} \mathrm{Im}\Pi(\boldsymbol{q}, \omega + \mathrm{i}0^+) \quad . \tag{2.45}$$

This relation is called the ***fluctuation-dissipation theorem***. For arbitrary complex frequencies z we have [1.7],

$$\begin{aligned} \Pi(\boldsymbol{q}, z) &= \frac{1}{\pi} \int_{-\infty}^{\infty} \mathrm{d}\omega \frac{\mathrm{Im}\Pi(\boldsymbol{q}, \omega + \mathrm{i}0^+)}{\omega - z} \\ &= \int_0^{\infty} \mathrm{d}\omega [1 - \mathrm{e}^{-\beta\omega}] S(\boldsymbol{q}, \omega) \left[\frac{1}{\omega - z} + \frac{1}{\omega + z} \right] \\ &= \int_0^{\infty} \mathrm{d}\omega [1 - \mathrm{e}^{-\beta\omega}] S(\boldsymbol{q}, \omega) \frac{2\omega}{\omega^2 - z^2} \quad . \end{aligned} \tag{2.46}$$

A widely used approximation for the density-density correlation function is the so-called ***random-phase approximation*** [2.9], which we shall abbreviate by RPA. If the quasi-particle interaction in Eq.(2.1) depends only on the momentum-transfer $\boldsymbol{q}$ (and not on the momenta $\boldsymbol{k}$ and $\boldsymbol{k}'$ of the incoming particles), the density-density correlation function within RPA is approximated by

$$\Pi_{\mathrm{RPA}}(q) = \frac{\Pi_0(q)}{1 + f_{\boldsymbol{q}} \Pi_0(q)} \quad , \tag{2.47}$$

or equivalently

$$[\Pi_{\mathrm{RPA}}(q)]^{-1} = [\Pi_0(q)]^{-1} + f_{\boldsymbol{q}} \quad . \tag{2.48}$$

Corrections to the RPA are usually parameterized in terms of a local field correction $g(q)$, which is defined by writing the exact $\Pi(q)$ as [2.10, 2.11]

$$[\Pi(q)]^{-1} = [\Pi_0(q)]^{-1} + f_{\boldsymbol{q}} - g(q) \quad . \tag{2.49}$$

Defining the ***proper polarization*** $\Pi_*(q)$ via

$$[\Pi_*(q)]^{-1} = [\Pi_0(q)]^{-1} - g(q) \quad , \tag{2.50}$$

we have

$$\Pi(q) = \frac{\Pi_*(q)}{1 + f_q \Pi_*(q)} = \frac{\Pi_*(q)}{\epsilon(q)} \quad , \tag{2.51}$$

where the dimensionless function

$$\epsilon(q) = 1 + f_q \Pi_*(q) \tag{2.52}$$

is called the ***dielectric function.*** Using Eqs.(2.51) and (2.46), we may also write

$$\frac{1}{\epsilon(q)} = 1 - f_q \Pi(q) = 1 - f_q \int_0^\infty \mathrm{d}\omega [1 - \mathrm{e}^{-\beta\omega}] S(\boldsymbol{q}, \omega) \frac{2\omega}{\omega^2 + \omega_m^2} \quad . \tag{2.53}$$

Note that Eq.(2.50) has the structure $G^{-1} = G_0^{-1} - \Sigma$, i.e. it resembles the Dyson equation for the single-particle Green's function of a bosonic problem, with the proper polarization and the local field factor playing the role of the exact Green's function and the irreducible self-energy. Although this analogy has been noticed many times in the literature [2.10–2.15], it has not been thoroughly exploited as a guide to develop systematic methods to calculate corrections to the RPA. In Chap. 4 we shall show that higher-dimensional bosonization gives a natural explanation for this analogy and yields a new procedure for calculating the dielectric function beyond the RPA.

2.4 Patching the Fermi surface

We now discuss Haldane's version of subdividing the degrees of freedom close to the Fermi surface into boxes. This geometric construction opens the way for generalizing the bosonization approach to arbitrary dimensions.

2.4.1 Definition of the patches and boxes

This leads to the bosonization of the potential energy in arbitrary dimensions.

Following Haldane [1.30], we partition the Fermi surface into a finite number M of disjoint patches with volume Λ^{d-1}. The precise shape of the patches and the size of Λ should be chosen such that, to a first approximation, within a given patch the curvature of the Fermi surface can be locally neglected. We introduce a label α to enumerate the patches in some convenient ordering and denote the patch with label α by P^α_Λ. For example, a possible subdivision of a two-dimensional Fermi surface into $M = 12$ patches is shown in Fig. 2.3. By definition P^α_Λ is a subset of the Fermi surface, i.e. a $d-1$-dimensional

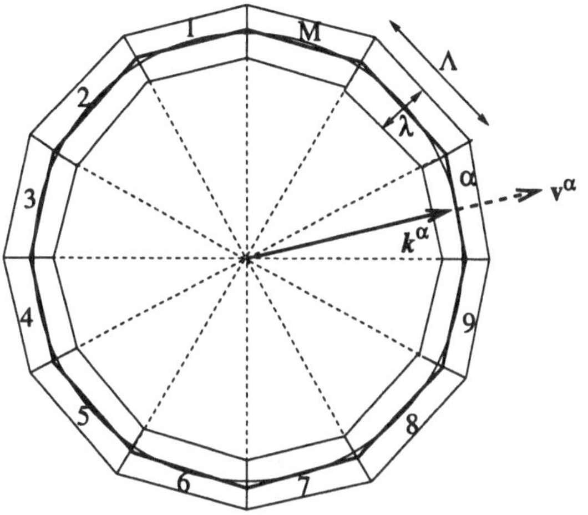

Fig. 2.3.
Subdivision of a two-dimensional spherical Fermi surface into $M = 12$ patches P_Λ^α, $\alpha = 1, \ldots, 12$, and associated boxes $K_{\Lambda,\lambda}^\alpha$. The vector $\boldsymbol{k}^\alpha$ has length k_F and points to the center of the patch P_Λ^α. The dashed arrow represents the local Fermi velocity $\boldsymbol{v}^\alpha$ associated with patch P_Λ^α.

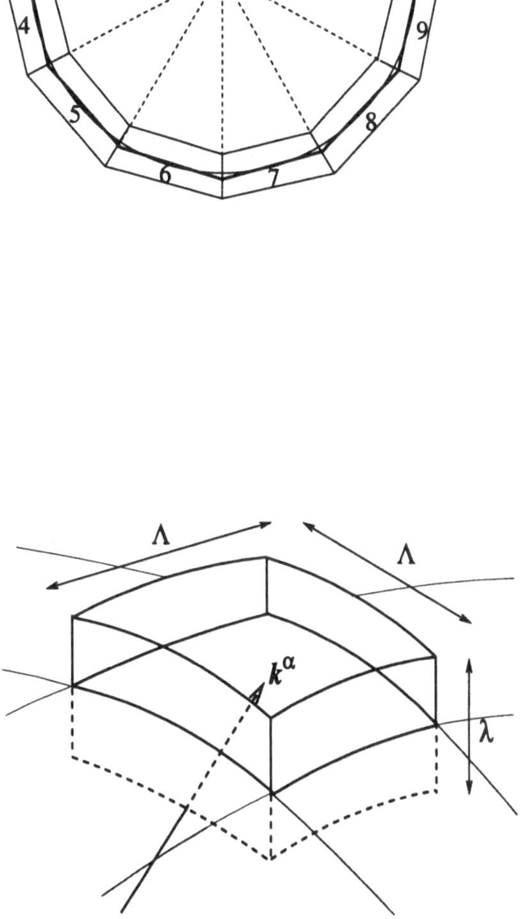

Fig. 2.4. Graph of a squat box $K_{\Lambda,\lambda}^\alpha$ with patch cutoff Λ and radial cutoff λ in three dimensions. $\boldsymbol{k}^\alpha$ points to the center of patch P_Λ^α on the Fermi surface.

manifold. To cover the entire momentum space in the vicinity of the Fermi surface, each patch P^{α}_{Λ} is extended into a d-dimensional box (or sector) $K^{\alpha}_{\Lambda,\lambda}$ such that the union $\bigcup_{\alpha=1}^{M} K^{\alpha}_{\Lambda,\lambda}$ comprises all degrees of freedom in the system.

The definition of the boxes requires the introduction of an additional radial cutoff λ. If we assume that the degrees of freedom with wave-vectors outside a thin shell of radial thickness $\lambda \ll k_{\mathrm{F}}$ around the Fermi surface have been integrated out, then in two dimensions the sectors $K^{\alpha}_{\Lambda,\lambda}$ can be chosen in form of the trapezoids shown in Fig. 2.3, while in $d = 3$ a convenient choice of the $K^{\alpha}_{\Lambda,\lambda}$ are the squat boxes shown in Fig. 2.4. The difference between Haldane's [1.30] and Luther's [1.29] way of subdividing the degrees of freedom close to the Fermi surface is that Luther takes $\Lambda = O(V^{-1/d})$, so that his sectors are actually thin tubes, with a cross section that covers only a few discrete $\boldsymbol{k}$-points. This has the obvious disadvantage that the motion *parallel to the Fermi surface* cannot be described without taking scattering between different tubes into account. Haldane's crucial idea was to choose boxes with finite cross section. In this case scattering processes that transfer momentum between different sectors[5] can be ignored *as long as the width Λ of the boxes is large compared with the typical momentum-transfer $|\boldsymbol{q}|$ of the interaction.*

To bosonize the potential energy, we decompose the Fourier components $\hat{\rho}_{\boldsymbol{q}}$ of the density operator into the contributions from the various boxes,

$$\hat{\rho}_{\boldsymbol{q}} = \sum_{\alpha=1}^{M} \hat{\rho}^{\alpha}_{\boldsymbol{q}} \quad , \tag{2.54}$$

where $\hat{\rho}^{\alpha}_{\boldsymbol{q}}$ is the contribution from wave-vectors $\boldsymbol{k}$ in sector $K^{\alpha}_{\Lambda,\lambda}$ to the total density,

$$\hat{\rho}^{\alpha}_{\boldsymbol{q}} = \sum_{\boldsymbol{k}} \Theta^{\alpha}(\boldsymbol{k}) \hat{\psi}^{\dagger}_{\boldsymbol{k}} \hat{\psi}_{\boldsymbol{k}+\boldsymbol{q}} \quad . \tag{2.55}$$

The cutoff function $\Theta^{\alpha}(\boldsymbol{k})$ is defined by

$$\Theta^{\alpha}(\boldsymbol{k}) = \begin{cases} 1 & \text{if } \boldsymbol{k} \in K^{\alpha}_{\Lambda,\lambda} \\ 0 & \text{else} \end{cases} \quad , \tag{2.56}$$

and satisfies

$$\sum_{\alpha=1}^{M} \Theta^{\alpha}(\boldsymbol{k}) = 1 \quad , \tag{2.57}$$

because by construction the union of all $K^{\alpha}_{\Lambda,\lambda}$ agrees with the total relevant $\boldsymbol{k}$-space. We shall refer to $\hat{\rho}^{\alpha}_{\boldsymbol{q}}$ as *sector density.* Note that Eq.(2.57) insures that the sum of all sector densities yields again the full density $\hat{\rho}_{\boldsymbol{q}}$, see Eq.(2.54). In terms of the sector density operators the interaction part (2.3) of the many-body Hamiltonian can be written as

[5] These so-called around-the-corner processes are difficult to handle within higher-dimensional bosonization, see Sect. 2.4.3 and Chap. 5.1.1.

$$\hat{H}_{\text{int}} = \frac{1}{2V} \sum_{q} \sum_{\alpha\alpha'} f_q^{\alpha\alpha'} : \hat{\rho}_{-q}^{\alpha} \hat{\rho}_q^{\alpha'} : \quad , \tag{2.58}$$

where : ... : denotes normal ordering, and it is assumed that the variations of $f_q^{kk'}$ are negligible if k and k' are restricted to given boxes, so that it is allowed to introduce coarse-grained interaction functions

$$f_q^{\alpha\alpha'} = \frac{\sum_{kk'} \Theta^{\alpha}(k) \Theta^{\alpha'}(k') f_q^{kk'}}{\sum_{kk'} \Theta^{\alpha}(k) \Theta^{\alpha'}(k')} \quad . \tag{2.59}$$

The motivation for introducing the operators $\hat{\rho}_q^{\alpha}$ is that these operators obey approximately (up to an overall scale factor) bosonic commutation relations among each other [1.31, 1.32]. *Thus, Eq.(2.58) is already the bosonized potential energy.*

It should be mentioned that the usefulness of the geometric construction described above is not restricted to higher-dimensional bosonization. A very similar construction has recently been used by Feldman *et al.* [2.16] to devise a $1/M$-expansion for interacting Fermi systems. Furthermore, sectorizations of this type play an important role in modern renormalization group approaches to the fermionic many-body problem [1.40].

2.4.2 Linearization of the energy dispersion

In order to bosonize the full Hamiltonian, we should also obtain a boson representation for the kinetic energy. This is only possible in a simple way if the energy dispersion is linearized at the Fermi surface.

The crucial advantage of the subdivision of the Fermi surface into patches is that it opens the way for a *linearization of the non-interacting energy dispersion.* In first-quantized notation this means that the kinetic energy operator $\hat{H}_0$ is replaced by an operator involving only first order spatial derivatives. Then it is not difficult to show that the operators $\hat{\rho}_q^{\alpha}$ defined in Eq.(2.55) have in the high-density limit simple commutation relations with the kinetic energy operator $\hat{H}_0$. Together with the bosonized potential energy in Eq.(2.58), this directly leads to the free boson representation of the Hamiltonian [1.31, 1.32]. In Chap. 4 we shall discuss the bosonization of the Hamiltonian and the underlying approximations within the framework of our functional integral approach.

Due to the non-trivial topology of the Fermi surface, it is impossible to linearize the energy dispersion globally in a fixed coordinate system. However, if the size of the patches is chosen sufficiently small, we may *locally* linearize the energy dispersion within each sector separately. To do this, let us denote by k^{α}, $\alpha = 1, \ldots, M$, the set of vectors on the Fermi surface ($\xi_{k^{\alpha}} = 0$) that point to the centers of the patches P_{Λ}^{α} (see Figs. 2.3 and 2.4). Let us then identify the vectors k^{α} with the origins of local coordinate systems on

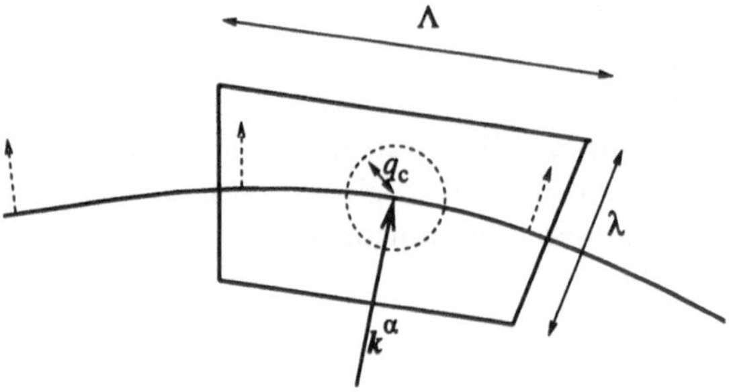

Fig. 2.5. Proper choice of the cutoffs. The patch cutoff Λ should be chosen small enough so that within a given box the variations in the direction of the local normal vectors to the Fermi surface (dashed arrows) can be neglected. On the other hand, both cutoffs Λ and λ should be chosen large compared with the range q_c of the interaction in momentum space (dashed circle). Only in this case physical correlation functions at distances $|\boldsymbol{r}| \gtrsim q_c^{-1}$ do not explicitly depend on the unphysical cutoffs Λ and λ.

the Fermi surface[6], and measure any given wave-vector $\boldsymbol{k}$ with respect to that reference point $\boldsymbol{k}^\alpha$ for which $|\boldsymbol{k} - \boldsymbol{k}^\alpha|$ assumes a minimum. The corresponding geometry has already been discussed in Sect. 2.2.3, see Fig. 2.2. Formally, we use Eq.(2.57) and write

$$\epsilon_{\boldsymbol{k}} - \mu \equiv \xi_{\boldsymbol{k}} = \sum_\alpha \Theta^\alpha(\boldsymbol{k}) \xi_{\boldsymbol{k}} = \sum_\alpha \Theta^\alpha(\boldsymbol{k}) \xi^\alpha_{\boldsymbol{k}-\boldsymbol{k}^\alpha} \quad , \tag{2.60}$$

where the functions $\xi^\alpha_{\boldsymbol{q}}$ are simply defined by $\xi^\alpha_{\boldsymbol{q}} = \xi_{\boldsymbol{k}^\alpha + \boldsymbol{q}}$, see Eqs.(2.5) and (2.15). Suppose now that the cutoff Λ is chosen sufficiently small so that within a given patch the curvature of the Fermi surface can be neglected. As shown in Fig. 2.5, this means that the variations in the direction of the local normal vector to the Fermi surface must be small within a given patch. In general Λ should be chosen small compared with the typical momentum scale on which the Fermi surface changes its shape. For spherical Fermi surfaces this means that

$$\Lambda \ll k_{\mathrm{F}} \quad . \tag{2.61}$$

On the other hand, for intrinsically flat Fermi surfaces the size of Λ can be chosen comparable to k_{F}. We shall discuss Fermi surfaces of this type in some detail in Chap. 7. In the opposite limit, when the Fermi surface has certain critical areas where its shape changes on some other characteristic scale $k_0 \ll k_{\mathrm{F}}$, we should choose $\Lambda \ll k_0$. Note that in the case of Van Hove singularities $k_0 \to 0$, so that we have to exclude this possibility if we insist on the linearization of the energy dispersion. For sufficiently small $|\boldsymbol{q}| = |\boldsymbol{k} - \boldsymbol{k}^\alpha|$ we may then ignore the quadratic and higher order corrections in Eq.(2.16), and approximate

[6] Such a collection of coordinate systems is also called an *atlas* [2.17].

$$\xi_q^\alpha \approx v^\alpha \cdot q \quad . \tag{2.62}$$

Note that for energy dispersions that are intrinsically almost linear[7] the quadratic corrections to Eq.(2.62) are small even for $|q| = O(k_F)$. In most cases, however, Eq.(2.62) will only be a good approximation for the calculation of quantities that are determined by the degrees of freedom in the vicinity of the Fermi surface.

2.4.3 Around-the-corner processes and the proper choice of the cutoffs

The sector cutoffs Λ and λ should not be chosen too small, but also not too large. The proper choice depends on the shape of the Fermi surface and on the nature of the interaction.

Although the variations in the direction of the local normal vector can always be reduced by choosing a sufficiently small patch cutoff Λ, this cutoff cannot be made arbitrarily small. The reason is that for practical calculations the sectorization turns out to be only useful if scattering processes that transfer momentum between different boxes (so called *around-the-corner processes*) can be neglected. This will only be the case if the Fourier transform of the interaction is dominated by momentum-transfers $|q| \lesssim q_c$, where q_c is some physical interaction cutoff satisfying

$$q_c \ll \min\{\Lambda, \lambda\} \quad . \tag{2.63}$$

In other words, the interaction must be dominated by forward scattering. As illustrated in Fig. 2.5, the volume in momentum space swept out by the interaction is then small compared with the volume $\Lambda^{d-1}\lambda$ of the boxes, so that boundary effects can be neglected. For example, in case of the long-range part of the Coulomb potential the cutoff q_c can be identified with the usual Thomas-Fermi screening wave-vector κ. In this case the condition $\kappa \ll k_F$ is satisfied at high densities (see Chap. 6.2.3 and Appendix A.3.1). Of course, the Coulomb potential has also a non-vanishing short-range part, which cannot be treated explicitly within our bosonization approach. Fortunately, there exist physically interesting quantities (for example the quasi-particle residue or the leading behavior of the momentum distribution in the vicinity of the Fermi surface, see Chap. 6.2) which are completely determined by long-wavelength fluctuations with wave-vectors $|q| \lesssim \kappa$. In this case our bosonization approach leads to cutoff-independent results that involve only physical quantities, because the condition (2.63) insures that the numerical value of momentum integrals is independent of the unphysical cutoffs Λ and λ.

---

[7] For example, for some peculiar form of the band structure the coefficients c_{ij}^α in Eq.(2.17) might be small.

Finally, let us consider the radial cutoff λ. If we would like to linearize the energy dispersion, then we should choose λ small enough such that it does not matter whether the energy dispersion is linearized precisely at the Fermi surface, or at the top (or bottom) of the boxes $K^{\alpha}_{\Lambda,\lambda}$. For a spherical Fermi surface this condition is satisfied if

$$\lambda \ll k_{\mathrm{F}} \quad . \tag{2.64}$$

However, by introducing such a radial cutoff we are assuming that the high-energy degrees of freedom have already been integrated out. As discussed in Sect. 2.1, the parameters which define our model (such as the local Fermi velocities $\boldsymbol{v}^{\alpha}$ or the physical cutoff q_{c}) must then incorporate the finite renormalizations due to the high-energy degrees of freedom. Therefore these parameters ***depend implicitly on the cutoff*** λ. Although the precise form of this cutoff dependence remains unknown unless we can explicitly perform the integration over the high-energy degrees of freedom, these physical parameters can in principle be determined from experiments, for example by measuring the density of states at the Fermi energy or the screening length. Such a procedure is familiar from renormalizable quantum field theories, where all cutoff dependence can be lumped onto a finite number of experimentally measurable parameters [2.1,2.18]. But also in field theory approaches to condensed matter systems this strategy has been adopted with great success [2.19].

2.5 Curved patches and reduction of the patch number

If we do not require that the energy dispersion should be linearized, we are free to subdivide the Fermi surface into a small number of curved patches. In some special cases we may complete abandon the patching construction, and formally identify the entire momentum space with a single sector. Then the around-the-corner processes simply do not exist.

Because in this book we shall develop a systematic method for including the non-linear terms of the energy dispersion into higher-dimensional bosonization, we shall ultimately drop the requirement that the variation of the local normal vector within a given patch must be negligible. We then have the freedom of choosing much larger patches P^{α}_{Λ} and sectors $K^{\alpha}_{\Lambda,\lambda}$ than for linearized energy dispersion. For example, in Fig. 2.6 we show a sector $K^{\alpha}_{\Lambda,\lambda}$ that is constructed from five smaller boxes. Clearly, by choosing larger boxes with finite curvature we automatically take into account all around-the-corner processes between the smaller sub-boxes used in the linearized theory! Note that the ***curvature*** of the Fermi surface is described by the non-linear terms in the expansion of the non-interacting energy dispersion close to the Fermi surface, see Eq.(2.16). For our purpose, it will be sufficient to assume that the expansion of $\epsilon_{\boldsymbol{k}^{\alpha}+\boldsymbol{q}}$ for small $\boldsymbol{q}$ truncates at the quadratic order. By a proper

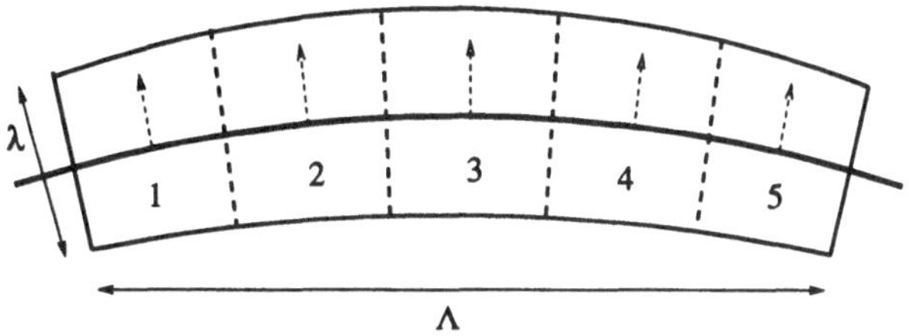

Fig. 2.6. Sector $K^{\alpha}_{\Lambda,\lambda}$ on the Fermi surface (thick solid line) with non-negligible curvature. The dashed arrows are the local normal vectors at the Fermi surface. The dashed lines separate the smaller boxes which are more appropriate if the energy dispersion is linearized. Note that around-the-corner processes corresponding to momentum transfer between the smaller boxes $1 \leftrightarrow 2$, $2 \leftrightarrow 3$, $3 \leftrightarrow 4$ and $4 \leftrightarrow 5$ are automatically taken into account in $K^{\alpha}_{\Lambda,\lambda}$.

orientation of the axes of the local coordinate system centered at $\boldsymbol{k}^{\alpha}$, we can always diagonalize the second-derivative tensor c^{α}_{ij} in Eq.(2.16), so that the energy dispersion relative to the chemical potential for wave-vectors close to $\boldsymbol{k}^{\alpha}$ becomes

$$\epsilon_{\boldsymbol{k}^{\alpha}+\boldsymbol{q}} - \mu = \epsilon_{\boldsymbol{k}^{\alpha}} - \mu + \xi^{\alpha}_{\boldsymbol{q}} \quad , \tag{2.65}$$

where

$$\xi^{\alpha}_{\boldsymbol{q}} = \boldsymbol{v}^{\alpha} \cdot \boldsymbol{q} + \sum_{i=1}^{d} \frac{q_i^2}{2m_i^{\alpha}} \tag{2.66}$$

is the excitation energy relative to $\epsilon_{\boldsymbol{k}^{\alpha}}$, and the inverse effective masses $1/m_i^{\alpha}$ are the eigenvalues of the second derivative tensor c^{α}_{ij} defined in Eq.(2.17). So far we have always chosen $\boldsymbol{k}^{\alpha}$ such that $\epsilon_{\boldsymbol{k}^{\alpha}} - \mu = 0$, in which case the first two terms on the right-hand side of Eq.(2.65) cancel and $\epsilon_{\boldsymbol{k}^{\alpha}+\boldsymbol{q}} - \mu = \xi^{\alpha}_{\boldsymbol{q}}$. More generally, we may subdivide the entire momentum space into sectors centered at points $\boldsymbol{k}^{\alpha}$ which are not necessarily located on the Fermi surface. Of course, in this case $\epsilon_{\boldsymbol{k}^{\alpha}} - \mu$ does in general not vanish, so that we should distinguish between the quantities $\epsilon_{\boldsymbol{k}^{\alpha}+\boldsymbol{q}} - \mu$ and the *excitation energy* $\xi^{\alpha}_{\boldsymbol{q}} = \epsilon_{\boldsymbol{k}^{\alpha}+\boldsymbol{q}} - \epsilon_{\boldsymbol{k}^{\alpha}}$ given in Eq.(2.66). However, as long as we keep track of this difference, we may partition all degrees of freedom into sectors as shown in Fig. 2.7. Note that in general it will also be convenient to allow for sector-dependent cutoffs Λ^{α} and λ^{α} in order to match the special geometry of the Fermi surface. As discussed in Sect. 2.4, the sectors cutoffs should be chosen large compared with the range $q_{\rm c}$ of the interaction in momentum space, so that the final result for the Green's function at distances large compared with $q_{\rm c}^{-1}$ is independent of the unphysical sector cutoffs. In fact, it is advantageous to choose the sectors as large as possible in order to avoid corrections due to the around-the-corner processes. Hence, as soon as we include the non-linear terms in the energy dispersion, the only condition which puts an upper limit to the sector size is the requirement that within a given sector $K^{\alpha}_{\Lambda,\lambda}$ the effective masses m_i^{α} and

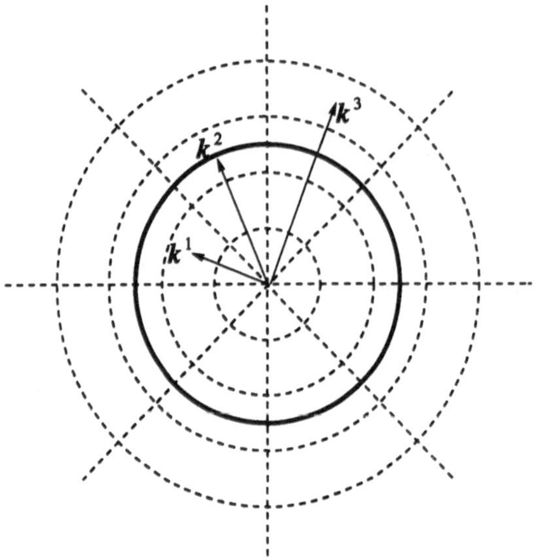

Fig. 2.7. Subdivision of the entire momentum space of a two-dimensional system with a spherical Fermi surface (thick solid circle) into sectors. The solid arrows point to the origins $\boldsymbol{k}^{\alpha}$ of local coordinate systems associated with the sectors. Note that only for sectors at the Fermi surface we may choose $\boldsymbol{k}^{\alpha}$ such that $\epsilon_{\boldsymbol{k}^{\alpha}} = \mu$. For example $\epsilon_{\boldsymbol{k}^2} = \mu$, but $\epsilon_{\boldsymbol{k}^1}$ and $\epsilon_{\boldsymbol{k}^3}$ are different from μ.

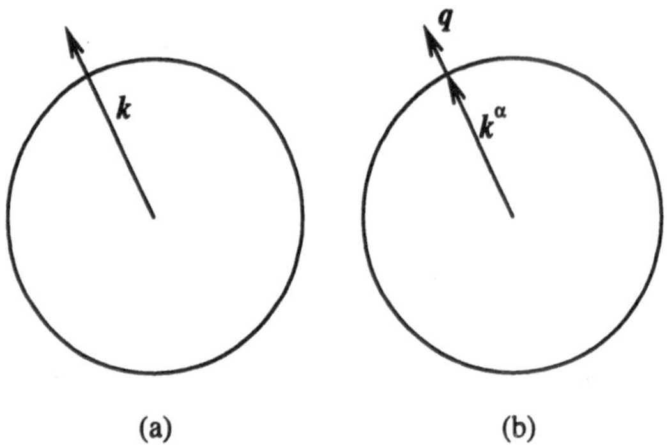

Fig. 2.8. (a) Spherical Fermi surface and wave-vector $\boldsymbol{k}$ close to the Fermi surface. (b) If we are interested in $G(\boldsymbol{k}, \mathrm{i}\tilde{\omega}_n)$, we choose the coordinate origin $\boldsymbol{k}^{\alpha}$ such that $\boldsymbol{k} = \boldsymbol{k}^{\alpha} + \boldsymbol{q}$, with $\boldsymbol{q}$ parallel to $\boldsymbol{k}^{\alpha}$. Note that in this case $|\boldsymbol{k} - \boldsymbol{k}^{\alpha}|$ assumes the smallest possible value.

the coarse-grained Landau parameters $f_{\boldsymbol{q}}^{\alpha\alpha'}$ (see Eq.(2.59)) should be well-defined. If these conditions are satisfied, we may use our formalism to calculate the single-particle Green's function $G(\boldsymbol{k}^{\alpha} + \boldsymbol{q}, \mathrm{i}\tilde{\omega}_n)$ for all wave-vectors $\boldsymbol{q}$ that are small compared with the sector cutoffs. Obviously, in the extreme case of Fermi surfaces that have constant curvature (at least within the range q_{c} of the interaction), and for Landau parameters that are independent of the momenta of the incoming particles (i.e. $f_{\boldsymbol{q}}^{\alpha\alpha'} = f_{\boldsymbol{q}}$, such as the long-range tail of the Coulomb interaction), we may identify the entire momentum space with a single sector. In other words, there is no need any more for subdividing the degrees of freedom into several sectors. In this case the problem of around-the-corner processes is solved trivially. However, given the fact that our main interest is the calculation of the single particle Green's function in the vicinity of the Fermi surface, it is still advantageous to work with a coordinate system centered on the Fermi surface, as shown in Fig. 2.8. Once we know the function

$G^{\alpha}(\boldsymbol{q}, \mathrm{i}\tilde{\omega}_n) \equiv G(\boldsymbol{k}^{\alpha} + \boldsymbol{q}, \mathrm{i}\tilde{\omega}_n)$ for wave-vectors of the form $\boldsymbol{q} = q_{\parallel}^{\alpha} \hat{\boldsymbol{k}}^{\alpha}$ (where $\hat{\boldsymbol{k}}^{\alpha}$ is a unit vector in the direction of $\boldsymbol{k}^{\alpha}$), we may use the symmetry of the Fermi surface to reconstruct $G(\mathbf{k}, \mathrm{i}\tilde{\omega}_n)$. For spherically symmetric systems we simply have to substitute $q_{\parallel}^{\alpha} \to |\boldsymbol{k}| - k_{\mathrm{F}}$ in the result for $G^{\alpha}(q_{\parallel}^{\alpha} \hat{\boldsymbol{k}}^{\alpha}, \mathrm{i}\tilde{\omega}_n)$.

2.6 Summary and outlook

In the first three sections of this chapter we have summarized some basic facts about the fermionic many-body problem, mainly to introduce the notation and to set the stage for the calculations that follow. In Sect.2.4 we have given a detailed description of the geometric patching construction in momentum space, which forms the basis of higher-dimensional bosonization with linearized energy dispersion. This construction has first been suggested by Haldane [1.30], and has been discussed in some detail in the work of Houghton, Kwon, and Marston [1.31, 1.33], and later in [1.36]. Note that Haldane's way of subdividing the degrees of freedom close to the Fermi surface into sectors is a generalization of an earlier suggestion due to Luther [1.29], who used thin tubes.

We have also pointed out that scattering processes that transfer momentum between different sectors (the around-the-corner processes) are difficult to handle within higher-dimensional bosonization. It is therefore desirable to choose the size of the sectors as large as possible. In Sect.2.5 we have further generalized the patching construction by defining larger patches with finite curvature, anticipating that in this book we shall present a systematic method for including curvature effects into bosonization. In other words, Haldane's boxes, which can be considered as the union of a large number of Luther's narrow tubes, have merged into a small number of sectors, within which the curvature of the Fermi surface cannot be neglected. In the case of a spherical Fermi surface and rotationally invariant interactions we shall formally identify the entire momentum space with a single sector, thus completely abandoning the patching construction.

Finally, we would like to draw the attention of the reader to the problem of **Van Hove singularities**, which will not be further discussed in this book, although non-perturbative methods for analyzing this problem will be developed in the following chapters[8]. As discussed at the end of Sect. 2.4.2, at a Van Hove singularity the local Fermi velocity $\boldsymbol{v}^{\alpha}$ vanishes, so that the leading term in the expansion of the energy dispersion close to the Fermi surface is quadratic. Obviously, the effect of Van Hove singularities on the low-energy

[8] The fact that so far I have not studied this problem by myself with the help of the technique described in this book does not necessarily mean that this problem is very difficult or requires conceptually new ideas. I simply have not found the time to work on this problem. This is also true for the other research problems mentioned in the concluding sections of the following chapters.

behavior of the Green's function cannot be studied within an approximation that relies on the linearization of the energy dispersion close to the Fermi surface. However, our more refined functional bosonization approach developed in Chaps. 4.3 and 5.2 retains the quadratic term in the energy dispersion, so that our method might shed some new light on the problem of Van Hove singularities in strongly correlated Fermi systems.

3. Hubbard-Stratonovich transformations

Our functional bosonization approach is based on two Hubbard-Stratonovich transformations, which are described in detail in this chapter.

We start with the imaginary time functional integral formulation of quantum statistical mechanics. This modern approach to the many-body problem has recently been described in excellent textbooks [2.4–2.7], so that we can be rather brief here and simply summarize the relevant representations of fermionic correlation functions as Grassmannian functional integrals. We then eliminate the Grassmann fields in favour of collective bosonic fields by means of suitable Hubbard-Stratonovich transformations [3.1]. These can be viewed as a clever change of variables to collective coordinates in functional integrals, which exhibit the physically most relevant degrees of freedom. The associated Jacobians define the effective actions for the Hubbard-Stratonovich fields. It turns out that the non-perturbative bosonization result for the single-particle Green's function can be obtained with the help of a conventional Hubbard-Stratonovich transformation that involves a space- and time-dependent auxiliary field ϕ^{α}. This transformation will be discussed in Sect. 3.2. On the other hand, for the calculation of the boson representation of the Hamiltonian or the density-density correlation function we need a generalized two-field Hubbard-Stratonovich transformation, which involves besides the ϕ^{α}-field another bosonic field $\tilde{\rho}^{\alpha}$. Section 3.3 is devoted to a detailed description of this transformation.

3.1 Grassmannian functional integrals

Fermionic correlation functions can be represented as Grassmannian functional integrals. These representations are particularly convenient for our purpose, because they can be directly manipulated via Hubbard-Stratonovich transformations.

The grand canonical partition function $\mathcal{Z}$ of our many-body Hamiltonian defined in Eqs.(2.1), (2.2) and (2.58) can be written as an imaginary time (i.e. Euclidean) functional integral over a Grassmann field ψ [2.4–2.7],

$$\frac{\mathcal{Z}}{\mathcal{Z}_0} = \frac{\int \mathcal{D}\{\psi\} e^{-S_{\text{mat}}\{\psi\}}}{\int \mathcal{D}\{\psi\} e^{-S_0\{\psi\}}} \quad , \tag{3.1}$$

where $\mathcal{Z}_0$ is the grand canonical partition function in the absence of interactions, and the Euclidean action $S_{\text{mat}}\{\psi\}$ is given by

$$S_{\text{mat}}\{\psi\} = S_0\{\psi\} + S_{\text{int}}\{\psi\} \quad , \tag{3.2}$$

$$S_0\{\psi\} = \beta \sum_k [-i\tilde{\omega}_n + \xi_k] \psi_k^\dagger \psi_k \quad , \tag{3.3}$$

$$S_{\text{int}}\{\psi\} = \frac{\beta}{2V} \sum_q \sum_{\alpha\alpha'} f_q^{\alpha\alpha'} \rho_{-q}^{\alpha} \rho_q^{\alpha'} \quad . \tag{3.4}$$

Here

$$\rho_q^\alpha = \sum_k \Theta^\alpha(\boldsymbol{k}) \psi_k^\dagger \psi_{k+q} \tag{3.5}$$

is the Grassmann representation of the sector density operator $\hat{\rho}_{\boldsymbol{q}}^\alpha$ defined in Eq.(2.55). Note that the k- and q-sums in these expressions are over wave-vectors *and* Matsubara frequencies. Although the Landau parameters $f_{\boldsymbol{q}}^{\alpha\alpha'}$ that appear in the Hamiltonian $\hat{H}_{\text{int}}$ in Eq.(2.58) depend only on the wave-vector $\boldsymbol{q}$, we have replaced them in Eq.(3.4) by more general frequency-dependent parameters $f_q^{\alpha\alpha'} \equiv f_{\boldsymbol{q},i\omega_m}^{\alpha\alpha'}$. In our functional integral approach the frequency-dependence does not introduce any additional complications. In physical applications the frequency-dependence is due to the fact that the underlying microscopic mechanism responsible for the effective interaction between the electrons is the exchange of some particle with a finite velocity, such as phonons[1]. Moreover, even in the case of electromagnetism the effective interaction becomes frequency-dependent if the corrections of higher order in v_F/c are retained. The static Coulomb potential is just the $v_F/c = 0$ limit. The leading correction is a retarded current-current interaction mediated by the transverse radiation field, which will be discussed in Chap. 10.

The time-ordered Matsubara Green's function defined in Eq.(2.7) can be represented as the functional integral average of $\psi_k \psi_k^\dagger$,

$$G(k) = -\beta \frac{\int \mathcal{D}\{\psi\} e^{-S_{\text{mat}}\{\psi\}} \psi_k \psi_k^\dagger}{\int \mathcal{D}\{\psi\} e^{-S_{\text{mat}}\{\psi\}}} \quad . \tag{3.6}$$

In absence of interactions this reduces to

$$G_0(k) = -\beta \frac{\int \mathcal{D}\{\psi\} e^{-S_0\{\psi\}} \psi_k \psi_k^\dagger}{\int \mathcal{D}\{\psi\} e^{-S_0\{\psi\}}} = \frac{1}{i\tilde{\omega}_n - \xi_k} \quad , \tag{3.7}$$

in agreement with Eq.(2.10). From the Matsubara Green's function we can obtain the real space imaginary time Green's function via Fourier transformation,

[1] The coupled electron-phonon system will be discussed in detail in Chap. 8.

$$G(\boldsymbol{r},\tau) = \frac{1}{\beta V}\sum_{k} \mathrm{e}^{\mathrm{i}(\boldsymbol{k}\cdot\boldsymbol{r}-\tilde{\omega}_n\tau)} G(k) \quad . \tag{3.8}$$

Defining

$$\psi(\boldsymbol{r},\tau) = \frac{1}{\sqrt{V}}\sum_{k} \mathrm{e}^{\mathrm{i}(\boldsymbol{k}\cdot\boldsymbol{r}-\tilde{\omega}_n\tau)} \psi_k \quad , \tag{3.9}$$

we can also write

$$G(\boldsymbol{r}-\boldsymbol{r}',\tau-\tau') = -\frac{\int \mathcal{D}\{\psi\}\,\mathrm{e}^{-S_{\mathrm{mat}}\{\psi\}}\psi(\boldsymbol{r},\tau)\psi^{\dagger}(\boldsymbol{r}',\tau')}{\int \mathcal{D}\{\psi\}\,\mathrm{e}^{-S_{\mathrm{mat}}\{\psi\}}} \quad . \tag{3.10}$$

Two-particle Green's functions can also be represented as functional integral averages. The density-density correlation function defined in Eq.(2.38) can be written as

$$\Pi(q) = \frac{\beta}{V}\frac{\int \mathcal{D}\{\psi\}\,\mathrm{e}^{-S_{\mathrm{mat}}\{\psi\}}\rho_q\rho_{-q}}{\int \mathcal{D}\{\psi\}\,\mathrm{e}^{-S_{\mathrm{mat}}\{\psi\}}} \quad , \tag{3.11}$$

where the composite Grassmann field corresponding to the Fourier components of the total density is (see Eq.(2.39))

$$\rho_q = \sum_{\alpha}\rho_q^{\alpha} = \sum_{k}\psi_k^{\dagger}\psi_{k+q} \quad . \tag{3.12}$$

Using Eq.(2.57) we may also write

$$\Pi(q) = \sum_{\alpha\alpha'}\Pi^{\alpha\alpha'}(q) \quad , \tag{3.13}$$

where for $q \neq 0$

$$\begin{aligned} \Pi^{\alpha\alpha'}(q) &= \frac{1}{\beta V}\int_0^{\beta}\mathrm{d}\tau\int_0^{\beta}\mathrm{d}\tau'\mathrm{e}^{-\mathrm{i}\omega_m(\tau-\tau')}\langle\mathcal{T}\left[\hat{\rho}_{\boldsymbol{q}}^{\alpha}(\tau)\hat{\rho}_{-\boldsymbol{q}}^{\alpha'}(\tau')\right]\rangle \\ &= \frac{\beta}{V}\frac{\int \mathcal{D}\{\psi\}\,\mathrm{e}^{-S_{\mathrm{mat}}\{\psi\}}\rho_q^{\alpha}\rho_{-q}^{\alpha'}}{\int \mathcal{D}\{\psi\}\,\mathrm{e}^{-S_{\mathrm{mat}}\{\psi\}}} \quad . \end{aligned} \tag{3.14}$$

We shall refer to $\Pi(q)$ as the *global* or *total* density-density correlation function, and to $\Pi^{\alpha\alpha'}(q)$ as the *local* or *sector* density-density correlation function. In the non-interacting limit Eq.(3.14) reduces to

$$\Pi_0^{\alpha\alpha'}(q) = -\frac{1}{V}\sum_{k}\Theta^{\alpha}(\boldsymbol{k})\Theta^{\alpha'}(\boldsymbol{k}+\boldsymbol{q})\frac{f(\xi_{\boldsymbol{k}+\boldsymbol{q}})-f(\xi_{\boldsymbol{k}})}{\xi_{\boldsymbol{k}+\boldsymbol{q}}-\xi_{\boldsymbol{k}}-\mathrm{i}\omega_m} \quad . \tag{3.15}$$

By relabeling $\boldsymbol{k}+\boldsymbol{q}\to\boldsymbol{k}$ it is easy to see that

$$\Pi_0^{\alpha\alpha'}(q) = \Pi_0^{\alpha'\alpha}(-q) \quad . \tag{3.16}$$

Substituting Eq.(3.15) into Eq.(3.13) and using $\sum_{\alpha}\Theta^{\alpha}(\boldsymbol{k}) = 1$, we recover the non-interacting Lindhard function given in Eq.(2.41). We would like to emphasize that in the above functional integral representations of the correlation functions the precise normalization for the integration measure $\mathcal{D}\{\psi\}$ is irrelevant, because the measure appears always in the numerator as well as in the denominator.

3.2 The first Hubbard-Stratonovich transformation

We decouple the two-body interaction between the fermions with the help of a Hubbard-Stratonovich field ϕ^α. After integrating over the Fermi fields, the single-particle Green's function can then be written as a quenched average with probability distribution given by the effective action of the ϕ^α-field.

3.2.1 Decoupling of the interaction

The generalized Landau parameters $f_q^{\alpha\alpha'}$ in Eq.(3.4) have units of energy × volume. Because we would like to work with dimensionless Hubbard-Stratonovich fields, it is useful to introduce the dimensionless Landau parameters

$$\tilde{f}_q^{\alpha\alpha'} = \frac{\beta}{V} f_q^{\alpha\alpha'} \quad . \tag{3.17}$$

The interaction part of our Grassmannian action can then be written as

$$S_{\text{int}}\{\psi\} = \frac{1}{2} \sum_q \sum_{\alpha\alpha'} \tilde{f}_q^{\alpha\alpha'} \rho_{-q}^\alpha \rho_q^{\alpha'} \quad . \tag{3.18}$$

Using the invariance of the sum in Eq.(3.18) under simultaneous relabelling $\alpha \leftrightarrow \alpha'$ and $q \to -q$, it is easy to see that, without loss of generality, we may assume that

$$\tilde{f}_q^{\alpha\alpha'} = \tilde{f}_{-q}^{\alpha'\alpha} \quad , \tag{3.19}$$

which is analogous to Eq.(3.16). We now decouple this action by means of the following Hubbard-Stratonovich transformation involving a dimensionless bosonic auxiliary field ϕ_q^α,

$$\exp\left[-S_{\text{int}}\{\psi\}\right] \equiv \exp\left[-\frac{1}{2}\sum_q \sum_{\alpha\alpha'} [\underline{\tilde{f}}_q]^{\alpha\alpha'} \rho_{-q}^\alpha \rho_q^{\alpha'}\right]$$

$$= \frac{\int \mathcal{D}\{\phi^\alpha\} \exp\left[-\frac{1}{2}\sum_q \sum_{\alpha\alpha'} [\underline{\tilde{f}}_q^{-1}]^{\alpha\alpha'} \phi_{-q}^\alpha \phi_q^{\alpha'} - \mathrm{i}\sum_q \sum_\alpha \phi_{-q}^\alpha \rho_q^\alpha\right]}{\int \mathcal{D}\{\phi^\alpha\} \exp\left[-\frac{1}{2}\sum_q \sum_{\alpha\alpha'} [\underline{\tilde{f}}_q^{-1}]^{\alpha\alpha'} \phi_{-q}^\alpha \phi_q^{\alpha'}\right]} \quad . \tag{3.20}$$

Here $\underline{\tilde{f}}_q$ is a matrix in the patch indices, with matrix elements given by

$$[\underline{\tilde{f}}_q]^{\alpha\alpha'} = \tilde{f}_q^{\alpha\alpha'} = \frac{\beta}{V} f_q^{\alpha\alpha'} \quad . \tag{3.21}$$

Throughout this work we shall use the convention that all underlined quantities are matrices in the patch indices. Eq.(3.20) is easily proved by shifting the ϕ^α-field in the numerator of the right-hand side according to

$$\phi_q^\alpha \to \phi_q^\alpha - \mathrm{i} \sum_{\alpha'} [\underline{\tilde{f}}_q]^{\alpha\alpha'} \rho_q^{\alpha'} \quad , \tag{3.22}$$

and using Eq.(3.19). For later convenience, let us fix the measure for the ϕ^α-integration such that

$$\int \mathcal{D}\{\phi^\alpha\} \exp\left[-\frac{1}{2}\sum_q \sum_{\alpha\alpha'} [\tilde{\underline{f}}_q^{-1}]^{\alpha\alpha'} \phi^\alpha_{-q}\phi^{\alpha'}_q\right] = \prod_q \det(\tilde{\underline{f}}_q) , \tag{3.23}$$

where det denotes the determinant with respect to the patch indices. Note that our complex auxiliary field satisfies $\phi^\alpha_{-q} = (\phi^\alpha_q)^*$, because it couples to the Fourier components of the density, which have also this symmetry. Of course, mathematically the ϕ^α-integrals in Eq.(3.20) and (3.23) are only well defined if the matrix $\tilde{\underline{f}}_q$ is positive definite. However, Eqs.(3.22) and (3.19) are sufficient to proof Eq.(3.20) as an algebraic identity, so that we shall use this transformation for intermediate algebraic manipulations even if the matrix $\tilde{\underline{f}}_q$ is not positive definite. Possible infinities due to vanishing (or even negative) eigenvalues of the matrix $\tilde{\underline{f}}_q$ cancel between the denominator and numerator of Eq.(3.20). For example, if all matrix elements of a $M \times M$-matrix have the same (non-zero) value, then $M-1$ of its eigenvalues are equal to zero, so that for constant matrices $\tilde{\underline{f}}_q$ we implicitly assume that the Gaussian integrations in Eq.(3.20) have been regularized in some convenient way. Note also that the appearance of $\tilde{\underline{f}}_q^{-1}$ is only an intermediate step in our calculation. The final expressions for physical correlation functions can be written entirely in terms of $\tilde{\underline{f}}_q$, and remain finite even if this matrix is not positive definite. Such a rather loose use of mathematics is quite common in statistical field theory, although for mathematicians it is certainly not acceptable. Formally, the appearance of $\tilde{\underline{f}}_q^{-1}$ at intermediate steps can be avoided with the help of the two-field Hubbard-Stratonovich transformation discussed in Sect. 3.3, see Eq.(3.42) below[2].

3.2.2 Transformation of the single-particle Green's function

Applying the Hubbard-Stratonovich transformation (3.20) to the functional integral representation (3.6) of the single-particle Green's function, we obtain

$$G(k) = -\beta \frac{\int \mathcal{D}\{\psi\}\mathcal{D}\{\phi^\alpha\} \mathrm{e}^{-S\{\psi,\phi^\alpha\}}\psi_k\psi_k^\dagger}{\int \mathcal{D}\{\psi\}\mathcal{D}\{\phi^\alpha\} \mathrm{e}^{-S\{\psi,\phi^\alpha\}}} , \tag{3.24}$$

where the decoupled action is given by

[2] Other formal ways to avoid this problem are briefly discussed in the books by Amit [3.2, p. 24], and by Itzykson and Drouffe [3.3, p. 153]. On the other hand, Zinn-Justin mentions this problem [2.1, p. 518], but does not hesitate to perform a transformation of the form (3.20) for a general matrix $\tilde{\underline{f}}_q$. Moreover, in the book by Negele and Orland [2.6, p. 198] as well as in Parisi's book [3.4, p. 209] this transformation is used without further comment. I would like to thank Kurt Schönhammer for giving me a copy of his notes with a summary and discussion of the relevant references.

$$S\{\psi, \phi^\alpha\} = S_0\{\psi\} + S_1\{\psi, \phi^\alpha\} + S_2\{\phi^\alpha\} \quad , \tag{3.25}$$

with

$$S_1\{\psi, \phi^\alpha\} = \sum_q \sum_\alpha \mathrm{i}\rho_q^\alpha \phi_{-q}^\alpha \quad , \tag{3.26}$$

$$S_2\{\phi^\alpha\} = \frac{1}{2} \sum_q \sum_{\alpha\alpha'} [\underline{\tilde{f}}_q^{-1}]^{\alpha\alpha'} \phi_{-q}^\alpha \phi_q^{\alpha'} \quad . \tag{3.27}$$

Thus, the fermionic two-body interaction has disappeared. Instead, we have the problem of a coupled field theory in which a dynamic bosonic field ϕ^α is coupled linearly to the fermionic density. The ϕ^α-field mediates the interaction between the fermionic matter in the sense that integration over the ϕ^α-field (i.e. undoing the Hubbard-Stratonovich transformation) generates an effective fermionic two-body interaction. In fact, because all interactions in nature can be viewed as the result of the exchange of some sort of particles, it is more general and fundamental to define the problem of interacting fermions in terms of an action that does not contain the fermionic two-body interaction explicitly, but involves the linear coupling of the fermionic density to another bosonic field. This point of view has been emphasized by Feynman and Hibbs [1.44]. We shall come back to the physical meaning of the Hubbard-Stratonovich field ϕ^α in Chap. 10.1.1, where we shall show that for the Maxwell action the ϕ^α-field can be identified physically with the scalar potential of electromagnetism.

In a functional integral we have the freedom of performing the integrations in any convenient order. Let us now perform the fermionic integration over the ψ-field in Eq.(3.24) before integrating over the ϕ^α-field. To do this, we write

$$S_0\{\psi\} + S_1\{\psi, \phi^\alpha\} = -\beta \sum_{kk'} \psi_k^\dagger [\hat{G}^{-1}]_{kk'} \psi_{k'} \quad , \tag{3.28}$$

where $\hat{G}^{-1}$ is an infinite matrix in momentum and frequency space, with matrix elements given by the formal Dyson equation

$$[\hat{G}^{-1}]_{kk'} = [\hat{G}_0^{-1}]_{kk'} - [\hat{V}]_{kk'} \quad . \tag{3.29}$$

Here $\hat{G}_0$ is the non-interacting Matsubara Green's function matrix,

$$[\hat{G}_0]_{kk'} = \delta_{kk'} G_0(k) \quad , \quad G_0(k) = \frac{1}{\mathrm{i}\tilde{\omega}_n - \xi_{\boldsymbol{k}}} \quad , \tag{3.30}$$

and the self-energy matrix $\hat{V}$ is defined by

$$[\hat{V}]_{kk'} = \sum_\alpha \Theta^\alpha(k) V_{k-k'}^\alpha \quad , \quad V_q^\alpha = \frac{\mathrm{i}}{\beta} \phi_q^\alpha \quad . \tag{3.31}$$

Recall that k denotes wave-vector and frequency, so that $\delta_{kk'} = \delta_{\boldsymbol{k}\boldsymbol{k}'} \delta_{nn'}$. Choosing the normalization of the integration measure $\mathcal{D}\{\psi\}$ suitably, the "trace-log" formula [2.4] yields

$$\int \mathcal{D}\{\psi\} \exp[-S_0\{\psi\} - S_1\{\psi, \phi^\alpha\}] = \det \hat{G}^{-1}$$
$$= \mathrm{e}^{\mathrm{Tr}\ln \hat{G}^{-1}} = \mathrm{e}^{\mathrm{Tr}\ln \hat{G}_0^{-1}} \mathrm{e}^{\mathrm{Tr}\ln[1-\hat{G}_0\hat{V}]} \quad , \tag{3.32}$$

$$-\beta \int \mathcal{D}\{\psi\} \psi_k \psi_k^\dagger \exp[-S_0\{\psi\} - S_1\{\psi, \phi^\alpha\}]$$
$$= [\hat{G}]_{kk} \mathrm{e}^{\mathrm{Tr}\ln \hat{G}_0^{-1}} \mathrm{e}^{\mathrm{Tr}\ln[1-\hat{G}_0\hat{V}]} \quad . \tag{3.33}$$

Hence, after integration over the fermions the exact interacting Green's function (3.24) can be written as a quenched average of the diagonal element $[\hat{G}]_{kk}$,

$$G(k) = \int \mathcal{D}\{\phi^\alpha\} \mathcal{P}\{\phi^\alpha\} [\hat{G}]_{kk} \equiv \left\langle [\hat{G}]_{kk} \right\rangle_{S_{\mathrm{eff}}} \quad . \tag{3.34}$$

Note that $[\hat{G}]_{kk}$ is in general a very complicated functional of the field ϕ^α. The normalized probability distribution $\mathcal{P}\{\phi^\alpha\}$ is

$$\mathcal{P}\{\phi^\alpha\} = \frac{\mathrm{e}^{-S_{\mathrm{eff}}\{\phi^\alpha\}}}{\int \mathcal{D}\{\phi^\alpha\} \mathrm{e}^{-S_{\mathrm{eff}}\{\phi^\alpha\}}} \quad , \tag{3.35}$$

where the effective action for the ϕ^α-field contains, in addition to the action $S_2\{\phi^\alpha\}$ defined in Eq.(3.27), a contribution due to the coupling to the electronic degrees of freedom,

$$S_{\mathrm{eff}}\{\phi^\alpha\} = S_2\{\phi^\alpha\} + S_{\mathrm{kin}}\{\phi^\alpha\} \quad , \tag{3.36}$$

with

$$S_{\mathrm{kin}}\{\phi^\alpha\} = -\mathrm{Tr}\ln[1 - \hat{G}_0\hat{V}] \quad . \tag{3.37}$$

Notc that in Eq.(3.34) one first calculates the Green's function for a frozen configuration of the ϕ^α-field, and then averages the resulting expression over all configurations this field, with the probability distribution given in Eq.(3.35). Such a procedure closely resembles the *background field method*, which is well-known in the field theory literature [3.5]. Following this terminology, we shall also refer to our auxiliary field ϕ_q^α as the *background field.*

The above transformations are exact. Of course, in practice it is impossible to calculate the interacting Green's function from Eq.(3.34), because (a) the matrix $\hat{G}^{-1}$ cannot be inverted exactly, (b) the kinetic energy contribution $S_{\mathrm{kin}}\{\phi^\alpha\}$ to the effective action of the ϕ^α-field can only be calculated perturbatively, and (c) the probability distribution $\mathcal{P}\{\phi^\alpha\}$ in Eq.(3.35) is not Gaussian, so that the averaging procedure cannot be carried out exactly. *The amazing fact is now that there exists a physically interesting limit where the difficulties (a), (b) and (c) can all be overcome.* The above method leads then to a new and non-perturbative approach to the fermionic many-body problem. The detailed description of this method and its application to physical problems is the central topic of this book. The highly non-perturbative character of this approach is evident from the fact that in $d = 1$ the well-known bosonization result for the Green's function of the Tomonaga-Luttinger model [1.19] can be obtained with this method [1.42]. This will be explicitly shown in Chap. 6.3.

3.3 The second Hubbard-Stratonovich transformation

In order to introduce collective bosonic density fields, we perform another change of variables in the functional integral by means of a second Hubbard-Stratonovich transformation. In this way we arrive at the general definition of the bosonized kinetic energy.

From Eq.(3.26) we see that after the first Hubbard-Stratonovich transformation the composite Grassmann field ρ^{α} couples linearly to the ϕ^{α}-field. Evidently the ϕ^{α}-field is related to the ρ^{α}-field in a very similar fashion as the chemical potential is related to the particle number. In other words, the ϕ^{α}-field is the ***conjugate field*** to the sector density ρ^{α}. We now use a second Hubbard-Stratonovich transformation to eliminate the composite Grassmann field ρ^{α} in favour of a collective bosonic field $\tilde{\rho}^{\alpha}$, which can then be identified physically with the bosonized density fluctuation. This additional transformation is useful for the calculation of quantities that can be written in terms of collective density fluctuations, such as the density-density correlation function or the bosonized Hamiltonian. On the other hand, for the calculation of the ***single-particle Green's function*** the first Hubbard-Stratonovich transformation introduced in the previous section is sufficient.

3.3.1 Transformation of the density-density correlation function

Applying the first Hubbard-Stratonovich transformation (3.20) to the Grassmannian functional integral representation (3.14) of the sector density-density correlation function, we obtain

$$\Pi^{\alpha\alpha'}(q) = \frac{\beta}{V} \frac{\int \mathcal{D}\{\psi\} \mathcal{D}\{\phi^{\alpha}\} \rho^{\alpha}_{q} \rho^{\alpha'}_{-q} \exp[-S\{\psi, \phi^{\alpha}\}]}{\int \mathcal{D}\{\psi\} \mathcal{D}\{\phi^{\alpha}\} \exp[-S\{\psi, \phi^{\alpha}\}]} . \tag{3.38}$$

We now decouple the quadratic action $S_2\{\phi^{\alpha}\}$ in this expression by means of an integration over another bosonic field $\tilde{\rho}^{\alpha}_{q}$,

$$\exp[-S_2\{\phi^{\alpha}\}] \equiv \exp\left[-\frac{1}{2}\sum_{q}\sum_{\alpha\alpha'}[\underline{\tilde{f}}_{q}^{-1}]^{\alpha\alpha'} \phi^{\alpha}_{-q}\phi^{\alpha'}_{q}\right]$$

$$= \frac{\int \mathcal{D}\{\tilde{\rho}^{\alpha}\} \exp\left[-\frac{1}{2}\sum_{q}\sum_{\alpha\alpha'}[\underline{\tilde{f}}_{q}]^{\alpha\alpha'} \tilde{\rho}^{\alpha}_{-q}\tilde{\rho}^{\alpha'}_{q} + \mathrm{i}\sum_{q}\sum_{\alpha}\phi^{\alpha}_{-q}\tilde{\rho}^{\alpha}_{q}\right]}{\int \mathcal{D}\{\tilde{\rho}^{\alpha}\} \exp\left[-\frac{1}{2}\sum_{q}\sum_{\alpha\alpha'}[\underline{\tilde{f}}_{q}]^{\alpha\alpha'} \tilde{\rho}^{\alpha}_{-q}\tilde{\rho}^{\alpha'}_{q}\right]} . \tag{3.39}$$

It is convenient to define the integration measure for the $\tilde{\rho}^{\alpha}$-integral such that

$$\int \mathcal{D}\{\tilde{\rho}^{\alpha}\} \exp\left[-\frac{1}{2}\sum_{q}\sum_{\alpha\alpha'}[\tilde{\underline{f}}_{q}]^{\alpha\alpha'}\tilde{\rho}^{\alpha}_{-q}\tilde{\rho}^{\alpha'}_{q}\right] = \prod_{q}\left[\det(\tilde{\underline{f}}_{q})\right]^{-1} \quad , \tag{3.40}$$

so that with Eq.(3.23) we have

$$\int \mathcal{D}\{\tilde{\rho}^{\alpha}\} \int \mathcal{D}\{\phi^{\alpha}\} \exp\Bigg[-\frac{1}{2}\sum_{q}\sum_{\alpha\alpha'}[\tilde{\underline{f}}_{q}]^{\alpha\alpha'}\tilde{\rho}^{\alpha}_{-q}\tilde{\rho}^{\alpha'}_{q}$$
$$-\frac{1}{2}\sum_{q}\sum_{\alpha\alpha'}[\tilde{\underline{f}}_{q}^{-1}]^{\alpha\alpha'}\phi^{\alpha}_{-q}\phi^{\alpha'}_{q}\Bigg] = 1 \quad . \tag{3.41}$$

Then our two-field decoupling of the original fermionic two-body interaction reads

$$\exp\left[-S_{\mathrm{int}}\{\psi\}\right] \equiv \exp\left[-\frac{1}{2}\sum_{q}\sum_{\alpha\alpha'}[\tilde{\underline{f}}_{q}]^{\alpha\alpha'}\rho^{\alpha}_{-q}\rho^{\alpha'}_{q}\right] = \int \mathcal{D}\{\tilde{\rho}^{\alpha}\} \int \mathcal{D}\{\phi^{\alpha}\}$$
$$\times \exp\left[-\frac{1}{2}\sum_{q}\sum_{\alpha\alpha'}[\tilde{\underline{f}}_{q}]^{\alpha\alpha'}\tilde{\rho}^{\alpha}_{-q}\tilde{\rho}^{\alpha'}_{q} + \mathrm{i}\sum_{q}\sum_{\alpha}\left[\tilde{\rho}^{\alpha}_{q} - \rho^{\alpha}_{q}\right]\phi^{\alpha}_{-q}\right] \quad . \tag{3.42}$$

Note that $\rho^{\alpha}_{q} = \sum_{\boldsymbol{k}} \Theta^{\alpha}(\boldsymbol{k})\psi^{\dagger}_{k}\psi_{k+q}$ on the left-hand side of this equation is a composite Grassmann field, while $\tilde{\rho}^{\alpha}_{q}$ on the right-hand side is a complex collective bosonic field. Eq.(3.42) can be viewed as a functional generalization of the elementary identity

$$\mathrm{e}^{-x^2} = \int_{-\infty}^{\infty} \mathrm{d}y \int_{-\infty}^{\infty} \frac{\mathrm{d}\phi}{2\pi} \mathrm{e}^{-y^2+\mathrm{i}(y-x)\phi} \quad . \tag{3.43}$$

Let us also point out that the two-field Hubbard-Stratonovich transformation (3.42) does not involve the inverse of the matrix $\tilde{\underline{f}}_{q}$, so that it is perfectly well defined for matrices with constant elements.

Applying the Hubbard-Stratonovich transformation (3.39) to the denominator in Eq.(3.38) and integrating over the fermionic ψ-field, we obtain with the help of the "trace-log" formula (3.32),

$$\int \mathcal{D}\{\psi\}\mathcal{D}\{\phi^{\alpha}\} \exp\left[-S\{\psi,\phi^{\alpha}\}\right] = \frac{\mathrm{e}^{\mathrm{Tr}\ln \hat{G}_0^{-1}}}{\int \mathcal{D}\{\tilde{\rho}^{\alpha}\}\mathrm{e}^{-\tilde{S}_2\{\tilde{\rho}^{\alpha}\}}}$$
$$\times \int \mathcal{D}\{\tilde{\rho}^{\alpha}\}\mathcal{D}\{\phi^{\alpha}\} \exp\left[-\tilde{S}_2\{\tilde{\rho}^{\alpha}\} + \mathrm{i}\sum_{q\alpha}\phi^{\alpha}_{-q}\tilde{\rho}^{\alpha}_{q} - S_{\mathrm{kin}}\{\phi^{\alpha}\}\right] \quad , \tag{3.44}$$

where the interaction contribution to the effective action of the collective $\tilde{\rho}^{\alpha}$-field is

$$\tilde{S}_2\{\tilde{\rho}^{\alpha}\} = \frac{1}{2}\sum_{q}\sum_{\alpha\alpha'}[\tilde{\underline{f}}_{q}]^{\alpha\alpha'}\tilde{\rho}^{\alpha}_{-q}\tilde{\rho}^{\alpha'}_{q} \quad , \tag{3.45}$$

and the action $S_{\mathrm{kin}}\{\phi^{\alpha}\}$ is defined in Eq.(3.37). The relation analogous to Eq.(3.44) for the numerator in Eq.(3.38) is

$$\int \mathcal{D}\{\psi\}\mathcal{D}\{\phi^\alpha\}\rho_q^\alpha \rho_{-q}^{\alpha'} \exp\left[-S\{\psi,\phi^\alpha\}\right] = \frac{\mathrm{e}^{\mathrm{Tr}\ln \hat{G}_0^{-1}}}{\int \mathcal{D}\{\tilde{\rho}^\alpha\}\mathrm{e}^{-\tilde{S}_2\{\tilde{\rho}^\alpha\}}}$$

$$\times \int \mathcal{D}\{\tilde{\rho}^\alpha\}\mathcal{D}\{\phi^\alpha\}\tilde{\rho}_q^\alpha \tilde{\rho}_{-q}^{\alpha'} \exp\left[-\tilde{S}_2\{\tilde{\rho}^\alpha\} + \mathrm{i}\sum_{q\alpha}\phi_{-q}^\alpha \tilde{\rho}_q^\alpha - S_{\mathrm{kin}}\{\phi^\alpha\}\right] . \quad (3.46)$$

To proof Eq.(3.46), we introduce the generating functional

$$\mathcal{F}\{\tilde{\phi}^\alpha\} = \int \mathcal{D}\{\psi\}\mathcal{D}\{\phi^\alpha\} \exp\left[-S\{\psi,\phi^\alpha\} + \mathrm{i}\sum_{q\alpha}\tilde{\phi}_{-q}^\alpha \rho_q^\alpha\right] , \quad (3.47)$$

which depends on external bosonic fields $\tilde{\phi}^\alpha$ and generates via differentiation (up to a constant factor) the left-hand side of Eq.(3.46),

$$\left.\frac{\partial^2 \mathcal{F}\{\tilde{\phi}^\alpha\}}{\partial\tilde{\phi}_{-q}^\alpha \partial\tilde{\phi}_q^{\alpha'}}\right|_{\tilde{\phi}^\alpha=0} = \mathrm{i}^2 \int \mathcal{D}\{\psi\}\mathcal{D}\{\phi^\alpha\}\rho_q^\alpha \rho_{-q}^{\alpha'} \exp\left[-S\{\psi,\phi^\alpha\}\right] . \quad (3.48)$$

Applying the Hubbard-Stratonovich transformation (3.39) to our generating functional, we obtain

$$\mathcal{F}\{\tilde{\phi}^\alpha\} = \frac{1}{\int \mathcal{D}\{\tilde{\rho}^\alpha\}\mathrm{e}^{-\tilde{S}_2\{\tilde{\rho}^\alpha\}}} \int \mathcal{D}\{\tilde{\rho}^\alpha\}\mathcal{D}\{\phi^\alpha\}\mathrm{e}^{-\tilde{S}_2\{\tilde{\rho}^\alpha\}}$$

$$\times \int \mathcal{D}\{\psi\} \exp\left[-S_0\{\psi\} + \mathrm{i}\sum_{q\alpha}[\phi_{-q}^\alpha \tilde{\rho}_q^\alpha - (\phi_{-q}^\alpha - \tilde{\phi}_{-q}^\alpha)\rho_q^\alpha]\right] . \quad (3.49)$$

Shifting the integration over the ϕ^α-field according to $\phi_q^\alpha \to \phi_q^\alpha + \tilde{\phi}_q^\alpha$, we replace in the last term of the exponential in Eq.(3.49)

$$\phi_{-q}^\alpha \tilde{\rho}_q^\alpha - (\phi_{-q}^\alpha - \tilde{\phi}_{-q}^\alpha)\rho_q^\alpha \to \tilde{\phi}_{-q}^\alpha \tilde{\rho}_q^\alpha + \phi_{-q}^\alpha(\tilde{\rho}_q^\alpha - \rho_q^\alpha) , \quad (3.50)$$

so that after the shift the derivatives with respect to the external $\tilde{\phi}^\alpha$-field generate factors of the collective bosonic density field $\tilde{\rho}^\alpha$. Performing now the fermionic integration and taking two derivatives with respect to the external field, we conclude from Eq.(3.49) that

$$\left.\frac{\partial^2 \mathcal{F}\{\tilde{\phi}^\alpha\}}{\partial\tilde{\phi}_{-q}^\alpha \partial\tilde{\phi}_q^{\alpha'}}\right|_{\tilde{\phi}^\alpha=0} = \mathrm{i}^2 \frac{\mathrm{e}^{\mathrm{Tr}\ln \hat{G}_0^{-1}}}{\int \mathcal{D}\{\tilde{\rho}^\alpha\}\mathrm{e}^{-\tilde{S}_2\{\tilde{\rho}^\alpha\}}} \int \mathcal{D}\{\tilde{\rho}^\alpha\}\mathcal{D}\{\phi^\alpha\}$$

$$\times \tilde{\rho}_q^\alpha \tilde{\rho}_{-q}^{\alpha'} \exp\left[-\tilde{S}_2\{\tilde{\rho}^\alpha\} + \mathrm{i}\sum_{q\alpha}\phi_{-q}^\alpha \tilde{\rho}_q^\alpha - S_{\mathrm{kin}}\{\phi^\alpha\}\right] . \quad (3.51)$$

Comparing the right-hand sides of Eqs.(3.48) and (3.51), the validity of Eq.(3.46) is evident. In summary, with the help of Eqs.(3.44), (3.46) and (3.38) the sector density-density correlation function (3.14) can be represented as

$$\Pi^{\alpha\alpha'}(q) =$$

$$\frac{\beta}{V}\frac{\int \mathcal{D}\{\tilde{\rho}^{\alpha}\}\,\mathrm{e}^{-\tilde{S}_2\{\tilde{\rho}^{\alpha}\}}\tilde{\rho}^{\alpha}_{q}\tilde{\rho}^{\alpha'}_{-q}\int \mathcal{D}\{\phi^{\alpha}\}\exp\left[\mathrm{i}\sum_{q\alpha}\phi^{\alpha}_{-q}\tilde{\rho}^{\alpha}_{q}-S_{\mathrm{kin}}\{\phi^{\alpha}\}\right]}{\int \mathcal{D}\{\tilde{\rho}^{\alpha}\}\,\mathrm{e}^{-\tilde{S}_2\{\tilde{\rho}^{\alpha}\}}\int \mathcal{D}\{\phi^{\alpha}\}\exp\left[\mathrm{i}\sum_{q\alpha}\phi^{\alpha}_{-q}\tilde{\rho}^{\alpha}_{q}-S_{\mathrm{kin}}\{\phi^{\alpha}\}\right]} \; . \tag{3.52}$$

3.3.2 Definition of the bosonized kinetic energy

In complete analogy with Eqs.(3.34)–(3.36), let us rewrite Eq.(3.52) as

$$\Pi^{\alpha\alpha'}(q)=\frac{\beta}{V}\int \mathcal{D}\{\tilde{\rho}^{\alpha}\}\tilde{\mathcal{P}}\{\tilde{\rho}^{\alpha}\}\tilde{\rho}^{\alpha}_{q}\tilde{\rho}^{\alpha'}_{-q}\equiv\frac{\beta}{V}\left\langle\tilde{\rho}^{\alpha}_{q}\tilde{\rho}^{\alpha'}_{-q}\right\rangle_{\tilde{S}_{\mathrm{eff}}} \quad , \tag{3.53}$$

where the normalized probability distribution $\tilde{\mathcal{P}}\{\tilde{\rho}^{\alpha}\}$ for the collective density field $\tilde{\rho}^{\alpha}$ is

$$\tilde{\mathcal{P}}\{\tilde{\rho}^{\alpha}\}=\frac{\mathrm{e}^{-\tilde{S}_{\mathrm{eff}}\{\tilde{\rho}^{\alpha}\}}}{\int \mathcal{D}\{\tilde{\rho}^{\alpha}\}\,\mathrm{e}^{-\tilde{S}_{\mathrm{eff}}\{\tilde{\rho}^{\alpha}\}}} \quad . \tag{3.54}$$

The effective action of the $\tilde{\rho}^{\alpha}$-field has again two contributions,

$$\tilde{S}_{\mathrm{eff}}\{\tilde{\rho}^{\alpha}\}=\tilde{S}_2\{\tilde{\rho}^{\alpha}\}+\tilde{S}_{\mathrm{kin}}\{\tilde{\rho}^{\alpha}\} \quad , \tag{3.55}$$

with $\tilde{S}_2\{\tilde{\rho}^{\alpha}\}$ given in Eq.(3.45), and

$$\tilde{S}_{\mathrm{kin}}\{\tilde{\rho}^{\alpha}\}=-\ln\left(\int \mathcal{D}\{\phi^{\alpha}\}\exp\left[\mathrm{i}\sum_{q\alpha}\phi^{\alpha}_{-q}\tilde{\rho}^{\alpha}_{q}-S_{\mathrm{kin}}\{\phi^{\alpha}\}\right]\right) \quad . \tag{3.56}$$

Note that $\tilde{S}_{\mathrm{kin}}\{\tilde{\rho}^{\alpha}\}$ is related to $S_{\mathrm{kin}}\{\phi^{\alpha}\}$ via a *functional Fourier transformation*, while the quadratic action $\tilde{S}_2\{\tilde{\rho}^{\alpha}\}$ is simply obtained from $S_{\mathrm{int}}\{\psi\}$ in Eq.(3.4) by replacing the composite Grassmann field ρ^{α} by the collective bosonic field $\tilde{\rho}^{\alpha}$. In this way the effect of the electron-electron interaction is taken into account exactly, while the contribution $\tilde{S}_{\mathrm{kin}}\{\tilde{\rho}^{\alpha}\}$ due to the kinetic energy can in general only be calculated approximately. In the next chapter we shall show that in the limit of long wavelengths and low energies the effective action $\tilde{S}_{\mathrm{eff}}\{\tilde{\rho}^{\alpha}\}$ in Eq.(3.55) is equivalent with the bosonized Hamiltonian of the interacting Fermi system. Obviously $\tilde{S}_{\mathrm{eff}}\{\tilde{\rho}^{\alpha}\}$ is in general *not* quadratic, so that the equivalent bosonized Hamiltonian contains terms describing interactions between the bosons. However, under certain conditions, which will be described in detail in Chap. 4.1, $\tilde{S}_{\mathrm{eff}}\{\tilde{\rho}^{\alpha}\}$ can be approximated by a quadratic form. In this case bosonization enormously simplifies the many-body problem. In a sense, the collective density fields $\tilde{\rho}^{\alpha}$ are the "correct coordinates" to parameterize the low-energy excitations of the system.

3.4 Summary and outlook

In this chapter we have used well-known representations of fermionic correlation functions as Grassmannian functional integrals and Hubbard-Stratonovich transformations to eliminate the fermionic degrees of freedom in favour of bosonic ones. The only new feature of these transformations is that our Hubbard-Stratonovich fields carry not only a momentum-frequency label q, but also a label α that refers to the sectors $K^{\alpha}_{\Lambda,\lambda}$ defined in Chaps. 2.4 and 2.5. Although our manipulations are formally exact, at this point the reader is perhaps rather skeptical whether they will turn out to be useful to obtain truly non-perturbative information about the interacting many-body system[3]. After all, the use of Hubbard-Stratonovich transformation is a well-known technique in the theory of strongly correlated systems [3.6–3.10], and in practice it is very difficult to go beyond the saddle point approximation. An important exception is a beautiful paper by Hertz [3.7], which has inspired the development of our functional bosonization approach. Hertz used a Hubbard-Stratonovich transformation to derive quantum Landau-Ginzburg-Wilson functionals for interacting Fermi systems, which form then the basis for a renormalization group analysis. Our fields ϕ^{α}_{q} are closely related to the Hubbard-Stratonovich fields introduced by Hertz; the only difference is that our fields carry an extra patch index α. As will be shown in Chap. 5, in this book we shall be able to *treat the full quantum dynamics of the Hubbard-Stratonovich field non-perturbatively* – we shall neither rely on saddle point approximations, nor on the naive perturbative calculation of fluctuation corrections around saddle points!

[3] As already mentioned, in $d = 1$ we have the ambitious goal to reproduce the exact solution of the Tomonaga-Luttinger model.

4. Bosonization of the Hamiltonian and the density-density correlation function

We use our functional integral formalism to bosonize the Hamiltonian of an interacting Fermi system with two-body density-density interactions. At the level of the Gaussian approximation the problem of deriving the bosonic representation of the Hamiltonian is closely related to the problem of calculating the density-density correlation function within the RPA. We develop a general formalism for obtaining corrections to the Gaussian approximation, and show that these are nothing but the local-field corrections to the RPA. Some of the results presented in this chapter has been published in [1.36].

In order to obtain the bosonized effective action $\tilde{S}_{\text{eff}}\{\tilde{\rho}^{\alpha}\}$ defined in Eq.(3.55), it is necessary to calculate first the effective action $S_{\text{eff}}\{\phi^{\alpha}\}$ of the ϕ^{α}-field given in Eq.(3.36). Note that the electron-electron interaction is taken into account exactly via $S_2\{\phi^{\alpha}\}$, so that the difficulty lies in the calculation of the kinetic energy contribution $S_{\text{kin}}\{\phi^{\alpha}\}$. Similarly, the interaction part $\tilde{S}_2\{\tilde{\rho}^{\alpha}\}$ of the effective action $\tilde{S}_{\text{eff}}\{\tilde{\rho}^{\alpha}\}$ for the collective density field can be obtained trivially by replacing $\rho_q^{\alpha} \rightarrow \tilde{\rho}_q^{\alpha}$ in the Grassmannian action $S_{\text{int}}\{\psi\}$ defined in Eq.(3.4). On the other hand, to obtain the bosonized kinetic energy $\tilde{S}_{\text{kin}}\{\tilde{\rho}^{\alpha}\}$ it is necessary to perform the functional Fourier transformation of $\exp[-S_{\text{kin}}\{\phi^{\alpha}\}]$ in Eq.(3.56).

Of course, in general the above kinetic energy contributions can only be calculated perturbatively by expanding

$$S_{\text{kin}}\{\phi^{\alpha}\} \equiv -\text{Tr}\ln[1-\hat{G}_0\hat{V}] = \sum_{n=1}^{\infty}\frac{1}{n}\text{Tr}\left[\hat{G}_0\hat{V}\right]^n \equiv \sum_{n=1}^{\infty} S_{\text{kin},n}\{\phi^{\alpha}\}\ , \quad (4.1)$$

and truncating the expansion at some finite order. The functional Fourier transformation in Eq.(3.56) should then also be performed perturbatively to this order. Within the *Gaussian approximation* all terms with $n \geq 3$ in Eq.(4.1) are neglected, so that one sets

$$S_{\text{kin}}\{\phi^{\alpha}\} \approx \text{Tr}\left[\hat{G}_0\hat{V}\right] + \frac{1}{2}\text{Tr}\left[\hat{G}_0\hat{V}\right]^2 \quad . \quad (4.2)$$

Because within this approximation $S_{\text{kin}}\{\phi^\alpha\}$ is a quadratic functional of the ϕ^α-field[1], the functional Fourier transformation (3.56) reduces to a trivial Gaussian integration. Evidently the effective action $\tilde{S}_{\text{eff}}\{\tilde{\rho}^\alpha\}$ of the collective density field is then also quadratic. Note that in the work by Houghton *et al.* [1.31,1.33] and Castro Neto and Fradkin [1.32,1.34] it is ***implicitly assumed that the Gaussian approximation is justified.*** However, in none of these works the corrections to the Gaussian approximation have been considered, so that the small parameter which actually controls the accuracy of the Gaussian approximation has not been determined.

On the other hand, in the exactly solvable one-dimensional Tomonaga-Luttinger model [1.17, 1.18] the bosonized Hamiltonian is known to be quadratic, so that the expansion in Eq.(4.1) truncates at the second order. In this case we have *exactly*

$$-\text{Tr}\ln[1-\hat{G}_0\hat{V}] = \text{Tr}\left[\hat{G}_0\hat{V}\right] + \frac{1}{2}\text{Tr}\left[\hat{G}_0\hat{V}\right]^2 \quad . \tag{4.3}$$

All higher order terms vanish identically due to a large scale cancellation between self-energy and vertex corrections, which has been discovered by Dzyaloshinskii and Larkin [4.1]. A few years later T. Bohr gave a much more readable proof of this cancellation [4.2], and formulated it as a theorem, which he called the *closed loop theorem.* In $d=1$ there are certainly alternative (but equivalent) approaches to the bosonization problem, which do not explicitly make use of the closed loop theorem [1.41, 1.42]. However, we find it advantageous to start from the closed loop theorem, because then it is very easy to see that the cancellations responsible for the validity of Eq.(4.3) in $d=1$ ***exist also in higher dimensions***, and control in the limit of high densities and small momentum-transfers the accuracy of the Gaussian approximation in arbitrary d. Following the terminology coined by T. Bohr [4.2], we shall describe the mechanism responsible for this cancellation in terms of a theorem, which we call the *generalized closed loop theorem.*

4.1 The generalized closed loop theorem

This is the fundamental reason why bosonization works.

Graphically, the traces $\text{Tr}[\hat{G}_0\hat{V}]^n$ in Eq.(4.1) can be represented as closed fermion loops with n external ϕ^α-fields, as shown in Fig. 4.1. Performing the trace of nth order term in Eq.(4.1), we obtain

[1] As shown in Eq.(4.20) below, the term $\text{Tr}[\hat{G}_0\hat{V}]$ in Eq.(4.2) gives rise to a contribution that is proportional to the $q=0$ component of the ϕ^α-field, which renormalizes the $q=0$ component of the collective density field $\tilde{\rho}^\alpha_q$ (see Eqs.(4.38) and (4.39)). In this work we shall restrict ourselves to the calculation of zero temperature correlation functions at finite q, in which case possible subtleties associated with these $q=0$ components of the Hubbard-Stratonovich fields can be ignored.

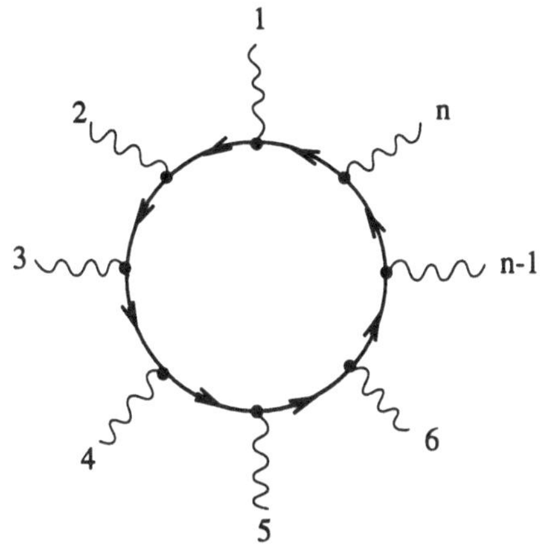

Fig. 4.1. Feynman diagram representing $\mathrm{Tr}[\hat{G}_0\hat{V}]^n$ with $n = 8$, see Eqs.(4.1) and (4.4). The lines with arrows denote non-interacting fermionic Green's functions, the dots represent the bare vertex, and the wavy lines denote external ϕ^α-fields.

$$S_{\mathrm{kin},n}\{\phi^\alpha\} = \frac{1}{n} \sum_{q_1 \ldots q_n} \sum_{\alpha_1 \ldots \alpha_n} U_n(q_1\alpha_1 \ldots q_n\alpha_n)\phi^{\alpha_1}_{q_1} \cdots \phi^{\alpha_n}_{q_n} \quad , \tag{4.4}$$

where the dimensionless vertices U_n are given by [4.3]

$$U_n(q_1\alpha_1 \ldots q_n\alpha_n) = \delta_{q_1+\ldots+q_n,0} \left(\frac{\mathrm{i}}{\beta}\right)^n \frac{1}{n!} \sum_{P(1\ldots n)} \sum_{k} \Theta^{\alpha_{P_1}}(\boldsymbol{k})$$
$$\times \Theta^{\alpha_{P_2}}(\boldsymbol{k} + \boldsymbol{q}_{P_2}) \cdots \Theta^{\alpha_{P_n}}(\boldsymbol{k} + \boldsymbol{q}_{P_2} + \ldots + \boldsymbol{q}_{P_n})$$
$$\times G_0(k)G_0(k + q_{P_2}) \cdots G_0(k + q_{P_2} + \ldots + q_{P_n}) \,. \tag{4.5}$$

Here $\delta_{q_1+\ldots+q_n,0}$ denotes a Kronecker-δ in wave-vector and frequency space. We have used the invariance of $S_{\mathrm{kin},n}\{\phi^\alpha\}$ under relabeling of the fields to symmetrize the vertices U_n with respect to the interchange of any two labels. The sum $\sum_{P(1\ldots n)}$ is over the $n!$ permutations of n integers, and P_i denotes the image of i under the permutation. Note that the vertices U_n are uniquely determined by the energy dispersion $\epsilon_{\boldsymbol{k}} - \mu$. The amazing fact is now that there exists a physically interesting limit where all higher order vertices U_n with $n \geq 3$ vanish. This limit is characterized by the requirement that the following two approximations ($A1$) and ($A2$) become accurate:

(A1): Diagonal-patch approximation

Let us assume that there exists a cutoff $q_{\mathrm{c}} \ll k_{\mathrm{F}}$ such that the contribution from fields ϕ^α_q with $|\boldsymbol{q}| \gtrsim q_{\mathrm{c}}$ to physical observables becomes negligibly small[2]. Because the fields ϕ^α_q mediate the interaction between the fermions, this condition is equivalent with the requirement that the nature of the *bare*

For the calculation of the free energy a more careful treatment of these terms is certainly necessary.

[2] As already mentioned in Chap. 2.4.3, in the case of the long-range part of the Coulomb interaction q_{c} can be identified with the Thomas-Fermi wave-vector κ, which is small compared with k_{F} at high densities.

interaction $\tilde{f}_{\underline{q}}$ should be such that the resulting *effective screened interaction* (which takes into account the modification of the bare interaction between two particles due to the presence of all other particles) is negligibly small for $|\boldsymbol{q}| \gtrsim q_c$. If this condition is satisfied, we may approximate in Eq.(4.5)

$$\Theta^{\alpha_{P_1}}(\boldsymbol{k})\Theta^{\alpha_{P_2}}(\boldsymbol{k}+\boldsymbol{q}_{P_2})\cdots\Theta^{\alpha_{P_n}}(\boldsymbol{k}+\boldsymbol{q}_{P_2}+\ldots+\boldsymbol{q}_{P_n})$$
$$\approx \delta^{\alpha_{P_1}\alpha_{P_2}}\delta^{\alpha_{P_1}\alpha_{P_3}}\ldots\delta^{\alpha_{P_1}\alpha_{P_n}}\Theta^{\alpha_{P_1}}(\boldsymbol{k}) \quad , \tag{4.6}$$

because the $\boldsymbol{k}$-sum in Eq.(4.5) is dominated by wave-vectors of the order of k_F. This approximation is correct to leading order in q_c/k_F, and becomes exact in the limit $q_c/k_F \rightarrow 0$. Note that this limit is approached either at high densities, where $k_F \rightarrow \infty$ at constant q_c, or in the limit that the range q_c of the effective interaction in momentum space approaches zero while k_F is held constant. It follows that, up to higher order corrections in q_c/k_F, the vertex $U_n(q_1\alpha_1\ldots q_n\alpha_n)$ is diagonal in all patch labels,

$$U_n(q_1\alpha_1\ldots q_n\alpha_n) = \delta^{\alpha_1\alpha_2}\ldots\delta^{\alpha_1\alpha_n}U_n^{\alpha_1}(q_1\ldots q_n) \quad , \tag{4.7}$$

with

$$U_n^{\alpha}(q_1\ldots q_n) = \delta_{q_1+\ldots+q_n,0}\left(\frac{\mathrm{i}}{\beta}\right)^n\frac{1}{n!}\sum_{P(1\ldots n)}\sum_{\boldsymbol{k}}\Theta^{\alpha}(\boldsymbol{k})$$
$$\times\, G_0(k)G_0(k+q_{P_2})\cdots G_0(k+q_{P_2}+\ldots+q_{P_n}) \quad . \tag{4.8}$$

Below we shall refer to the approximation (4.6) as the *diagonal-patch approximation.* It is important to note that at finite q_c/k_F this approximation can only become exact in $d=1$, because in this case the Fermi surface consists of two widely separated points. Except for special cases (see Chap. 7), in higher dimensions the covering of the Fermi surface involves always some adjacent patches, which can be connected by arbitrarily small momentum-transfers $\boldsymbol{q}$. These around-the-corner processes are ignored within the diagonal-patch approximation (*A*1). As discussed in detail in Chap. 2.4.3, this is only justified if the sector cutoffs Λ and λ are chosen large compared with q_c.

(A2): Local linearization of the energy dispersion

Suppose we put the origins $\boldsymbol{k}^{\alpha}$ of our local coordinate systems on the Fermi surface (so that $\epsilon_{\boldsymbol{k}^{\alpha}} = \mu$), and locally linearize the energy dispersion, $\xi_{\boldsymbol{q}}^{\alpha} \equiv \epsilon_{\boldsymbol{k}^{\alpha}+\boldsymbol{q}} - \mu \approx \boldsymbol{v}^{\alpha}\cdot\boldsymbol{q}$ (see Eq.(2.16)). Inserting unity in the form (2.60) into the non-interacting matter action $S_0\{\psi\}$ defined in Eq.(3.3), we see that the linearization amounts to replacing

$$S_0\{\psi\} \approx \beta\sum_{k}\sum_{\alpha}\Theta^{\alpha}(\boldsymbol{k})[-\mathrm{i}\tilde{\omega}_n + \boldsymbol{v}^{\alpha}\cdot(\boldsymbol{k}-\boldsymbol{k}^{\alpha})]\psi_k^{\dagger}\psi_k \quad . \tag{4.9}$$

Thus, the Fermi surface is approximated by a collection of *flat* $d-1$-dimensional hyper-surfaces, i.e. planes in $d=3$ and straight lines in $d=2$. The corresponding non-interacting Green's function is then approximated by

$$G_0(\boldsymbol{k}^\alpha + \boldsymbol{q}, \mathrm{i}\tilde{\omega}_n) \equiv G_0^\alpha(\tilde{q}) \approx \frac{1}{\mathrm{i}\tilde{\omega}_n - \boldsymbol{v}^\alpha \cdot \boldsymbol{q}} \quad . \tag{4.10}$$

Shifting the summation wave-vector in Eq.(4.8) according to $\boldsymbol{k} = \boldsymbol{k}^\alpha + \boldsymbol{q}$, we obtain

$$U_n^\alpha(q_1 \ldots q_n) = \delta_{q_1+\ldots+q_n,0} \left(\frac{\mathrm{i}}{\beta}\right)^n \frac{1}{n!} \sum_{P(1\ldots n)} \sum_{\tilde{q}} \Theta^\alpha(\boldsymbol{k}^\alpha + \boldsymbol{q})$$
$$\times\, G_0^\alpha(\tilde{q}) G_0^\alpha(\tilde{q} + q_{P_2}) \cdots G_0^\alpha(\tilde{q} + q_{P_2} + \ldots + q_{P_n}) \quad . \tag{4.11}$$

Recall that we have introduced the convention that $\tilde{q} = [\boldsymbol{q}, \mathrm{i}\tilde{\omega}_n]$ labels *fermionic* Matsubara frequencies, while $q = [\boldsymbol{q}, \mathrm{i}\omega_m]$ labels *bosonic* ones. Because the sum of a bosonic and a fermionic Matsubara frequency is a fermionic one, the external labels $q_1, \ldots, q_n$ in Eq.(4.11) depend on bosonic frequencies.

Having made the approximations $(A1)$ and $(A2)$, we are now ready to show that *in arbitrary dimensions* the vertices $U_n^\alpha(q_1 \ldots q_n)$ with $n \geq 3$ vanish in the limit $q_c \to 0$, so that in this limit *the Gaussian approximation becomes exact!* As already mentioned, in the context of the Tomonaga-Luttinger model the vanishing of the U_n for $n \geq 3$ has been called *closed loop theorem*, and is discussed and proved in unpublished lecture notes by T. Bohr [4.2]. Under the assumptions $(A1)$ and $(A2)$ the proof goes through in any dimension without changes. Note that the validity of $(A1)$ and $(A2)$ is implicitly built into the Tomonaga-Luttinger model by definition. The vanishing of U_n for $n \geq 3$ is equivalent with the statement that the RPA for the density-density correlation function is exact in this model. This is due to a complete cancellation between self-energy and vertex corrections [4.1]. In [4.2] the proof is formulated in the space-time domain, but for our purpose it is more convenient to work in momentum space, because here the Fermi surface and the patching construction are defined. The following two properties of our linearized non-interacting Green's function in Eq.(4.10) are essential,

$$G_0^\alpha(-\tilde{q}) = -G_0^\alpha(\tilde{q}) \quad , \tag{4.12}$$

$$G_0^\alpha(\tilde{q}) G_0^\alpha(\tilde{q} + q') = G_0^\alpha(q') \left[G_0^\alpha(\tilde{q}) - G_0^\alpha(\tilde{q} + q')\right] \quad . \tag{4.13}$$

Note that Eq.(4.12) follows trivially from the definition (4.10), while Eq.(4.13) is nothing but the partial fraction decomposition of the product of two rational functions. To show that the odd vertices $U_3, U_5, \ldots$ vanish, we only need Eq.(4.12) and the fact that the sector $K_{\Lambda,\lambda}^\alpha$ in Eq.(4.11) has inversion symmetry with respect to $\boldsymbol{k}^\alpha$, so that the domain for the $\boldsymbol{q}$-sum is invariant under $\boldsymbol{q} \to -\boldsymbol{q}$. Then it is easy to see that the contribution from a given permutation $(P_1 P_2 \ldots P_{n-1} P_n)$ is exactly cancelled by the contribution from the permutation $(P_n P_{n-1} \ldots P_2 P_1)$ in which the loop is traversed in the opposite direction. As already pointed out by T. Bohr [4.2], the vanishing of the odd vertices is a direct consequence of Furry's theorem [4.4]. To show that the even vertices U_n, $n = 4, 6, \ldots$ vanish, we use Eq.(4.13) n-times for the pairs

$$
\begin{aligned}
&G_0^\alpha(\tilde{q})G_0^\alpha(\tilde{q}+q_{P_2}) \ , \\
&G_0^\alpha(\tilde{q}+q_{P_2})G_0^\alpha(\tilde{q}+q_{P_2}+q_{P_3}) \ , \\
&\ldots \ , \\
&G_0^\alpha(\tilde{q}+q_{P_2}+\ldots+q_{P_{n-1}})G_0^\alpha(\tilde{q}+q_{P_2}+\ldots+q_{P_n}) \ , \\
&G_0^\alpha(\tilde{q}+q_{P_2}+\ldots+q_{P_n})G_0^\alpha(\tilde{q}) \ ,
\end{aligned}
\tag{4.14}
$$

and take into account that we may replace $q_{P_2}+\ldots+q_{P_n}=-q_{P_1}$ because of overall energy-momentum conservation. Using the fact that in Eq.(4.11) we sum over all permutations, it is easy to show that under the summation sign the second line in Eq.(4.11) can be replaced by

$$
\begin{aligned}
&G_0^\alpha(\tilde{q})G_0^\alpha(\tilde{q}+q_{P_2})\cdots G_0^\alpha(\tilde{q}+q_{P_2}+\ldots+q_{P_n}) \\
&\to \frac{1}{n}\left[G_0^\alpha(q_{P_1})-G_0^\alpha(q_{P_2})\right]\left\{G_0^\alpha(\tilde{q}+q_{P_2})\cdots G_0^\alpha(\tilde{q}+q_{P_2}+\ldots+q_{P_n})\right\} \ .
\end{aligned}
\tag{4.15}
$$

Substituting Eq.(4.15) in Eq.(4.11), noting that after the shift $\tilde{q}\to\tilde{q}-q_{P_2}+q_{P_1}$ of the summation label the factor in the curly braces in Eq.(4.15) can be replaced by the symmetrized (with respect to $q_{P_1}\leftrightarrow q_{P_2}$) expression

$$
\frac{1}{2}\left\{G_0^\alpha(\tilde{q}+q_{P_2})\cdots G_0^\alpha(\tilde{q}+q_{P_2}+\ldots+q_{P_n})+[q_{P_1}\leftrightarrow q_{P_2}]\right\} \ , \tag{4.16}
$$

and finally using again the fact that we may rename $q_{P_1}\leftrightarrow q_{P_2}$ because we sum over all permutations, it is easy to see that the resulting expression vanishes due to the antisymmetry of the first factor on the right-hand side of Eq.(4.15). This argument is not valid for $n=2$, because in this case $G_0^\alpha(q_{P_1})-G_0^\alpha(q_{P_2})=2G_0^\alpha(q_{P_1})$ due to energy-momentum conservation. We shall discuss the vertex U_2 in detail in Sect. 4.2.1. Note that the shift $\tilde{q}\to\tilde{q}-q_{P_2}+q_{P_1}$ affects also the patch cutoff, $\Theta^\alpha(\boldsymbol{k}^\alpha+\boldsymbol{q})\to\Theta^\alpha(\boldsymbol{k}^\alpha+\boldsymbol{q}-\boldsymbol{q}_{P_2}+\boldsymbol{q}_{P_1})$, but this leads to corrections of higher order in q_c. Because we have already ignored higher order terms in q_c by making the diagonal-patch approximation ($A1$), it is consistent to ignore this shift. We would like to encourage the reader to explicitly verify the above manipulations for the simplest non-trivial case $n=4$.

In fermionic language, the vanishing of the higher order vertices is due to a *complete cancellation between self-energy and vertex corrections.* This cancellation is automatically incorporated in our bosonic formulation via the symmetrization of the vertices U_n. We would like to emphasize again that this remarkable cancellation happens not only in $d=1$ [4.1,4.2] but in arbitrary dimensions[3]. The existence of these cancellations in the perturbative calculation of the dielectric function of the homogeneous electron gas in $d=3$ has already been noticed by Geldart and Taylor more than 20 years ago [4.6], although the origin for this cancellation has not been identified.

[3] It should be mentioned that recently W. Metzner has independently given an alternative proof of the generalized closed loop theorem in $d>1$ [4.5]. His approach is based on operator identities for the sector density operators $\hat{\rho}_q^\alpha$ defined in Eq.(2.55), and the resulting consequences for time-ordered expectation values of products of these operators.

The generalized closed loop theorem discussed here gives a clear mathematical explanation for this cancellation *to all orders in perturbation theory.* It is important to stress that the cancellation does not depend on the nature of the external fields that enter the closed loop; in particular, it occurs also in models where the fermionic current density is coupled to transverse gauge fields (see Chap. 10). The one-loop corrections to the RPA for the gauge invariant two-particle Green's functions of electrons interacting with transverse gauge fields have recently been calculated by Kim *et al.* [4.7]. They found that at long wavelengths and low frequencies the leading self-energy and vertex corrections cancel. In the light of the generalized closed loop theorem this cancellation is not surprising. However, the generalized closed loop theorem is a much stronger statement, because it implies a cancellation between the leading self-energy and vertex corrections to all orders in perturbation theory.

4.2 The Gaussian approximation

We now calculate the density-density correlation function and the bosonized Hamiltonian within the Gaussian approximation. We also show that at long wavelengths the resulting bosonized Hamiltonian agrees with the corresponding Hamiltonian derived via the conventional operator approach [1.31, 1.34].

4.2.1 The effective action for the ϕ^α-field

Within the Gaussian approximation the expansion for the kinetic energy contribution $S_{\mathrm{kin}}\{\phi^\alpha\}$ to the effective action for the ϕ^α-field in Eq.(4.1) is truncated at the second order (see Eq.(4.2)), so that the effective action (3.36) is approximated by

$$
\begin{aligned}
S_{\mathrm{eff}}\{\phi^\alpha\} &\approx S_2\{\phi^\alpha\} + S_{\mathrm{kin},1}\{\phi^\alpha\} + S_{\mathrm{kin},2}\{\phi^\alpha\} \\
&= \frac{1}{2}\sum_q \sum_{\alpha\alpha'} [\underline{\tilde{f}}_q^{-1}]^{\alpha\alpha'} \phi^\alpha_{-q}\phi^{\alpha'}_q \\
&\quad + \sum_q \sum_\alpha U_1(q\alpha)\phi^\alpha_q + \frac{1}{2}\sum_{q_1 q_2}\sum_{\alpha\alpha'} U_2(q_1\alpha, q_2\alpha')\phi^\alpha_{q_1}\phi^{\alpha'}_{q_2} \quad . \qquad (4.17)
\end{aligned}
$$

The generalized closed loop theorem implies that the Gaussian approximation is justified in a parameter regime where the approximations ($A1$) and ($A2$) discussed in Sect. 4.1 are accurate.

We now calculate the vertices U_1 and U_2. From Eq.(4.5) we obtain

$$
U_1(q\alpha) = \delta_{q,0}\frac{\mathrm{i}}{\beta}\sum_k \Theta^\alpha(k)\frac{1}{\mathrm{i}\tilde{\omega}_n - \xi_k} = \mathrm{i}\delta_{q,0}N_0^\alpha \quad , \qquad (4.18)
$$

where

$$N_0^\alpha = \sum_{\boldsymbol{k}} \Theta^\alpha(\boldsymbol{k}) f(\xi_{\boldsymbol{k}}) \tag{4.19}$$

is the number of occupied states in sector $K^\alpha_{\Lambda,\lambda}$ in the non-interacting limit. Thus,

$$S_{\text{kin},1}\{\phi^\alpha\} = \mathrm{i} \sum_\alpha \phi_0^\alpha N_0^\alpha \quad . \tag{4.20}$$

The second-order vertex is given by

$$U_2(q_1\alpha, q_2\alpha') = -\delta_{q_1+q_2,0} \frac{1}{2\beta^2} \sum_{k} \Big[\Theta^\alpha(\boldsymbol{k}) \Theta^{\alpha'}(\boldsymbol{k}+\boldsymbol{q}_2) G_0(k) G_0(k+q_2) + \Theta^{\alpha'}(\boldsymbol{k}) \Theta^\alpha(\boldsymbol{k}+\boldsymbol{q}_1) G_0(k) G_0(k+q_1) \Big] \quad . \tag{4.21}$$

Performing the frequency sum we obtain

$$U_2(-q\alpha, q\alpha') = \frac{V}{\beta} \Pi_0^{\alpha\alpha'}(q) \equiv \tilde{\Pi}_0^{\alpha\alpha'}(q) \quad , \tag{4.22}$$

where $\Pi_0^{\alpha\alpha'}(q)$ is the non-interacting sector polarization, see Eq.(3.15). We conclude that $S_{\text{kin},2}\{\phi^\alpha\}$ is given by

$$S_{\text{kin},2}\{\phi^\alpha\} = \frac{1}{2} \sum_q \sum_{\alpha\alpha'} \tilde{\Pi}_0^{\alpha\alpha'}(q) \phi_{-q}^\alpha \phi_q^{\alpha'} \quad . \tag{4.23}$$

For $|\boldsymbol{q}| \ll k_{\mathrm{F}}$ the diagonal-patch approximation ($A1$) is justified, so that we may replace $\Theta^\alpha(\boldsymbol{k}) \Theta^{\alpha'}(\boldsymbol{k}+\boldsymbol{q}) \approx \delta^{\alpha\alpha'} \Theta^\alpha(\boldsymbol{k})$. To leading order in $|\boldsymbol{q}|/k_{\mathrm{F}}$ we have therefore in any dimension

$$\Pi_0^{\alpha\alpha'}(q) \approx \delta^{\alpha\alpha'} \Pi_0^\alpha(q) \quad , \quad \Pi_0^\alpha(q) = \nu^\alpha \frac{\boldsymbol{v}^\alpha \cdot \boldsymbol{q}}{\boldsymbol{v}^\alpha \cdot \boldsymbol{q} - \mathrm{i}\omega_m} \quad , \tag{4.24}$$

where

$$\nu^\alpha = \frac{1}{V} \frac{\partial N_0^\alpha}{\partial \mu} = \frac{1}{V} \sum_{\boldsymbol{k}} \Theta^\alpha(\boldsymbol{k}) \left[-\frac{\partial f(\xi_{\boldsymbol{k}})}{\partial \xi_{\boldsymbol{k}}} \right] \tag{4.25}$$

is the *local* (or patch) density of states associated with sector $K^\alpha_{\Lambda,\lambda}$, and $\boldsymbol{v}^\alpha$ is the local Fermi velocity (see Eq.(2.17)). Note that the approximation (4.24) is valid for small $|\boldsymbol{q}|/k_{\mathrm{F}}$ but for *arbitrary* frequencies. The patch density of states ν^α is proportional to Λ^{d-1}, i.e. in dimensions $d > 1$ it is a cutoff-dependent quantity. To see this more clearly, we take the limit $\beta \to \infty$, $V \to \infty$ in Eq.(4.25) and convert the volume integral over the δ-function into a surface integral in the usual way,

$$\nu^\alpha = \int_{K^\alpha_{\Lambda,\lambda}} \frac{\mathrm{d}\boldsymbol{k}}{(2\pi)^d} \delta(\xi_{\boldsymbol{k}}) = \int_{P^\alpha_\Lambda} \frac{\mathrm{d}S_{\boldsymbol{k}}}{(2\pi)^d} \frac{1}{|\nabla_{\boldsymbol{k}} \xi_{\boldsymbol{k}}|} \quad , \tag{4.26}$$

where the $d-1$-dimensional surface integral is over the patch P^α_Λ, i.e. the intersection of the sector $K^\alpha_{\Lambda,\lambda}$ with the Fermi surface. Using now the fact that $|\nabla_{\boldsymbol{k}}\xi_{\boldsymbol{k}}| \approx |\boldsymbol{v}^{\alpha}|$ for $\boldsymbol{k} \in P^{\alpha}_{\Lambda}$, and that for linearized energy dispersion the area of P^{α}_{Λ} is by construction given by Λ^{d-1}, we have in d dimensions

$$\nu^{\alpha} \approx \frac{\Lambda^{d-1}}{(2\pi)^d |\boldsymbol{v}^{\alpha}|} \quad . \tag{4.27}$$

On the other hand, we shall show in this work that physical quantities depend only on the global density of states (or some weighted average of the ν^{α}),

$$\nu = \sum_{\alpha} \nu^{\alpha} = \frac{1}{V} \sum_{\boldsymbol{k}} \left[-\frac{\partial f(\xi_{\boldsymbol{k}})}{\partial \xi_{\boldsymbol{k}}} \right] \quad , \tag{4.28}$$

which is manifestly cutoff-independent.

In summary, within the Gaussian approximation the effective action of the ϕ^{α}-field is given by

$$S_{\mathrm{eff}}\{\phi^{\alpha}\} \approx \mathrm{i} \sum_{\alpha} \phi^{\alpha}_0 N^{\alpha}_0 + S_{\mathrm{eff},2}\{\phi^{\alpha}\} \quad , \tag{4.29}$$

with

$$\begin{aligned} S_{\mathrm{eff},2}\{\phi^{\alpha}\} &\equiv S_2\{\phi^{\alpha}\} + S_{\mathrm{kin},2}\{\phi^{\alpha}\} \\ &= \frac{1}{2} \sum_{q} \sum_{\alpha\alpha'} [\underline{\tilde{f}}_q^{-1} + \underline{\tilde{\Pi}}_0(q)]^{\alpha\alpha'} \phi^{\alpha}_{-q} \phi^{\alpha'}_{q} \quad , \end{aligned} \tag{4.30}$$

where the elements of the matrix $\underline{\tilde{\Pi}}_0(q)$ are defined by $[\underline{\tilde{\Pi}}_0(q)]^{\alpha\alpha'} = \tilde{\Pi}^{\alpha\alpha'}_0(q)$, with $\tilde{\Pi}^{\alpha\alpha'}_0(q)$ given in Eq.(4.22).

4.2.2 The Gaussian propagator of the ϕ^{α}-field

... which is also known under the name RPA interaction.

Within the Gaussian approximation the propagator of the ϕ^{α}-field is simply given by

$$\left\langle \phi^{\alpha}_{q} \phi^{\alpha'}_{-q} \right\rangle_{S_{\mathrm{eff},2}} = \left[[\underline{\tilde{f}}_q^{-1} + \underline{\tilde{\Pi}}_0(q)]^{-1} \right]^{\alpha\alpha'} \quad , \tag{4.31}$$

where the averaging $< \ldots >_{S_{\mathrm{eff},2}}$ is defined as in Eqs.(3.34) and (3.35), with $S_{\mathrm{eff}}\{\phi^{\alpha}\}$ approximated by $S_{\mathrm{eff},2}\{\phi^{\alpha}\}$. As already mentioned in the footnote after Eq.(4.2), the first term in Eq.(4.29) involving the $q = 0$ component of the ϕ^{α}-field does not contribute to correlation functions at finite q. From Eqs.(3.21) and (4.22) we have $\underline{\tilde{f}}_q = \frac{\beta}{V} \underline{f}_q$ and $\underline{\tilde{\Pi}}_0(q) = \frac{V}{\beta} \underline{\Pi}_0(q)$, so that Eq.(4.31) implies

$$\left\langle \phi^{\alpha}_{q} \phi^{\alpha'}_{-q} \right\rangle_{S_{\mathrm{eff},2}} = \frac{\beta}{V} [\underline{f}_q^{\mathrm{RPA}}]^{\alpha\alpha'} \quad , \tag{4.32}$$

where the RPA interaction matrix $\underline{f}_q^{\mathrm{RPA}}$ is defined via

$$\underline{f}_q^{\mathrm{RPA}} = \left[\underline{f}_q^{-1} + \underline{\Pi}_0(q)\right]^{-1} = \underline{f}_q \left[1 + \underline{\Pi}_0(q)\underline{f}_q\right]^{-1} \quad . \tag{4.33}$$

Thus, the Gaussian propagator of the ϕ^α-field is (up to a factor of β/V) given by the RPA interaction matrix $\underline{f}_q^{\mathrm{RPA}}$. In the special case that all matrix elements of the bare interaction are identical, $[\underline{f}_q]^{\alpha\alpha'} = f_q$, the matrix elements $[\underline{f}_q^{\mathrm{RPA}}]^{\alpha\alpha'}$ are also independent of the patch indices, and can be identified with the usual RPA interaction. To see this, we expand Eq.(4.33) as a Neumann series

$$\underline{f}_q^{\mathrm{RPA}} = \underline{f}_q - \underline{f}_q\underline{\Pi}_0(q)\underline{f}_q + \underline{f}_q\underline{\Pi}_0(q)\underline{f}_q\underline{\Pi}_0(q)\underline{f}_q - \dots \quad , \tag{4.34}$$

and then take matrix elements term by term. Using the fact that all matrix elements of $\underline{f}_q$ are identically given by f_q, we may sum the series again and obtain the usual RPA interaction,

$$[\underline{f}_q^{\mathrm{RPA}}]^{\alpha\alpha'} = f_q^{\mathrm{RPA}} \equiv \frac{f_q}{1 + f_q \Pi_0(q)} \quad , \tag{4.35}$$

where

$$\Pi_0(q) = \sum_{\alpha\alpha'} \Pi_0^{\alpha\alpha'}(q) \tag{4.36}$$

is the *total* non-interacting polarization (see Eq.(3.13)).

4.2.3 The effective action for the $\tilde{\rho}^\alpha$-field

According to Eq.(3.56) the kinetic energy contribution to the effective action for the collective density field is within the Gaussian approximation given by

$$\tilde{S}_{\mathrm{kin}}\{\tilde{\rho}^\alpha\} \approx$$
$$-\ln\left[\int \mathcal{D}\{\phi^\alpha\} \exp\left(\mathrm{i}\sum_{q\alpha} \phi^\alpha_{-q}\tilde{\rho}^\alpha_q - S_{\mathrm{kin},1}\{\phi^\alpha\} - S_{\mathrm{kin},2}\{\phi^\alpha\}\right)\right] \quad . \tag{4.37}$$

Using Eq.(4.20), the first two terms in the exponent can be combined as follows,

$$\mathrm{i}\sum_q\sum_\alpha \phi^\alpha_{-q}\tilde{\rho}^\alpha_q - S_{\mathrm{kin},1}\{\phi^\alpha\} = \mathrm{i}\sum_q\sum_\alpha \phi^\alpha_{-q}\left[\tilde{\rho}^\alpha_q - \delta_{q,0}N_0^\alpha\right] \quad , \tag{4.38}$$

so that it is obvious that the first order term $S_{\mathrm{kin},1}\{\phi^\alpha\}$ simply shifts the collective density field $\tilde{\rho}^\alpha$ according to

$$\tilde{\rho}^\alpha_q \to \tilde{\rho}^\alpha_q - \delta_{q,0}N_0^\alpha \quad , \tag{4.39}$$

i.e. the uniform component is shifted. Hence, $\tilde{S}_{\mathrm{kin}}\{\tilde{\rho}^\alpha\}$ in Eqs.(3.56) and (4.37) is actually a functional of the shifted field. For simplicity we shall from now on redefine the collective density field according to Eq.(4.39). Note that the $q = 0$ term in the interaction part $\tilde{S}_2\{\tilde{\rho}^\alpha\}$ given in Eq.(3.45) is usually excluded due to charge neutrality, so that the effective action $\tilde{S}_{\text{eff}}\{\tilde{\rho}^\alpha\}$ depends exclusively on the shifted field. The integration in Eq.(4.37) yields the usual Debye-Waller factor, so that within the Gaussian approximation

$$\tilde{S}_{\text{kin}}\{\tilde{\rho}^\alpha\} \approx \tilde{S}^{(0)}_{\text{kin},0} + \tilde{S}^{(0)}_{\text{kin},2}\{\tilde{\rho}^\alpha\} \quad , \tag{4.40}$$

where

$$\tilde{S}^{(0)}_{\text{kin},0} = -\ln\left[\int \mathcal{D}\{\phi^\alpha\}\,\mathrm{e}^{-S_{\text{kin},2}\{\phi^\alpha\}}\right] \tag{4.41}$$

is a constant independent of the $\tilde{\rho}^\alpha$-field, and

$$\tilde{S}^{(0)}_{\text{kin},2}\{\tilde{\rho}^\alpha\} = \frac{1}{2}\sum_q\sum_{\alpha\alpha'} \Gamma^{\alpha\alpha'}(q)\tilde{\rho}^\alpha_{-q}\tilde{\rho}^{\alpha'}_q \quad . \tag{4.42}$$

Here $\Gamma^{\alpha\alpha'}(q)$ is the propagator of the ϕ^α-field with respect to the quadratic action $S_{\text{kin},2}\{\phi^\alpha\}$ defined in Eq.(4.23), i.e.

$$\begin{aligned}\Gamma^{\alpha\alpha'}(q) &= \frac{\int \mathcal{D}\{\phi^\alpha\}\mathrm{e}^{-S_{\text{kin},2}\{\phi^\alpha\}}\phi^\alpha_q\phi^{\alpha'}_{-q}}{\int \mathcal{D}\{\phi^\alpha\}\mathrm{e}^{-S_{\text{kin},2}\{\phi^\alpha\}}} \equiv \left\langle \phi^\alpha_q\phi^{\alpha'}_{-q}\right\rangle_{S_{\text{kin},2}} \\ &= [\underline{\tilde{\Pi}}_0^{-1}(q)]^{\alpha\alpha'} \quad .\end{aligned} \tag{4.43}$$

Note that $\Gamma^{\alpha\alpha'}_q(q)$ is (up to a factor of β/V) given by the matrix inverse of the non-interacting sector polarization $\Pi_0^{\alpha\alpha'}(q)$. In summary, within the Gaussian approximation the effective action of the $\tilde{\rho}^\alpha$-field is given by

$$\tilde{S}_{\text{eff}}\{\tilde{\rho}^\alpha\} \approx \tilde{S}^{(0)}_{\text{kin},0} + \tilde{S}^{(0)}_{\text{eff},2}\{\tilde{\rho}^\alpha\} \quad , \tag{4.44}$$

with

$$\begin{aligned}\tilde{S}^{(0)}_{\text{eff},2}\{\tilde{\rho}^\alpha\} &\equiv \tilde{S}_2\{\tilde{\rho}^\alpha\} + \tilde{S}^{(0)}_{\text{kin},2}\{\tilde{\rho}^\alpha\} \\ &= \frac{1}{2}\sum_q\sum_{\alpha\alpha'}[\underline{\tilde{f}}_q + \underline{\Gamma}(q)]^{\alpha\alpha'}\tilde{\rho}^\alpha_{-q}\tilde{\rho}^{\alpha'}_q \quad ,\end{aligned} \tag{4.45}$$

where $\underline{\Gamma}(q) = \underline{\tilde{\Pi}}_0^{-1}(q)$. In contrast to $S_{\text{eff},2}\{\phi^\alpha\}$, the corresponding Gaussian action of the collective density field $\tilde{S}^{(0)}_{\text{eff},2}\{\tilde{\rho}^\alpha\}$ carries an extra superscript $^{(0)}$, which indicates that higher order corrections will renormalize the parameters of $\tilde{S}^{(0)}_{\text{eff},2}\{\tilde{\rho}^\alpha\}$. In the case of $S_{\text{eff},2}\{\phi^\alpha\}$ corrections of this type do not exist. In Sect. 4.3 we shall explicitly calculate the leading correction to the Gaussian approximation.

4.2.4 The Gaussian propagator of the $\tilde{\rho}^\alpha$-field

...which is nothing but the RPA polarization.

Having determined the effective action for the collective density field, we may calculate the density-density correlation function from Eq.(3.53) by performing the bosonic integration over the $\tilde{\rho}^\alpha$-field. Because within the Gaussian approximation $\tilde{S}_{\text{eff}}\{\tilde{\rho}^\alpha\}$ is quadratic, the integration can be carried out trivially, and we obtain

$$\left\langle \tilde{\rho}_q^\alpha \tilde{\rho}_{-q}^{\alpha'} \right\rangle_{\tilde{S}^{(0)}_{\text{eff},2}} = \left[\left[\underline{\tilde{f}}_q + \underline{\Gamma}(q) \right]^{-1} \right]^{\alpha\alpha'} . \tag{4.46}$$

Using again $\underline{\tilde{f}}_q = \frac{\beta}{V} \underline{f}_q$ and $\underline{\Gamma}(q) = \frac{\beta}{V} \underline{\Pi}_0^{-1}(q)$, we conclude that within the Gaussian approximation the sector density-density correlation function is approximated by

$$\Pi^{\alpha\alpha'}(q) \approx [\underline{\Pi}_{\text{RPA}}(q)]^{\alpha\alpha'} , \tag{4.47}$$

where the matrix $\underline{\Pi}_{\text{RPA}}(q)$ is given by

$$\underline{\Pi}_{\text{RPA}}(q) = \left[\underline{\Pi}_0^{-1}(q) + \underline{f}_q \right]^{-1} = \underline{\Pi}_0(q) \left[1 + \underline{f}_q \underline{\Pi}_0(q) \right]^{-1} . \tag{4.48}$$

Eq.(4.48) is nothing but the RPA for the sector density-density correlation function. Thus, the Gaussian propagator of the $\tilde{\rho}^\alpha$-field is simply given by the RPA polarization matrix $\underline{\Pi}_{\text{RPA}}(q)$.

To obtain the standard RPA result for the total density-density correlation function, we should sum Eq.(4.47) over both patch labels,

$$\Pi_{\text{RPA}}(q) = \sum_{\alpha\alpha'} \left[\left[\underline{\Pi}_0^{-1}(q) + \underline{f}_q \right]^{-1} \right]^{\alpha\alpha'} , \tag{4.49}$$

see Eq.(3.13). For simplicity let us assume that $[\underline{f}_q]^{\alpha\alpha'} = f_q$ is independent of the patch indices. Expanding

$$[\underline{\Pi}_0^{-1}(q) + \underline{f}_q]^{-1} = \underline{\Pi}_0(q) - \underline{\Pi}_0(q) \underline{f}_q \underline{\Pi}_0(q) + \dots , \tag{4.50}$$

and taking matrix elements, we see that Eqs.(4.49) and (4.50) reduce to the usual RPA result (2.47),

$$\Pi_{\text{RPA}}(q) = \frac{\Pi_0(q)}{1 + f_q \Pi_0(q)} , \tag{4.51}$$

where the total non-interacting polarization $\Pi_0(q)$ is given in Eq.(4.36). We would like to emphasize that up to this point we have not linearized the energy dispersion, so that Eq.(4.51) is the exact RPA result for all wave-vectors, including the short-wavelength regime.

4.2.5 The bosonized Hamiltonian

To make contact with the operator approach to bosonization [1.31, 1.34], let us now derive a bosonic Hamiltonian that at long wavelengths is equivalent with our Gaussian action $\tilde{S}^{(0)}_{\text{eff},2}\{\tilde{\rho}^\alpha\}$ in Eq.(4.45). The key observation is that, ***in the limit of high densities and long wavelengths*** (i.e. in the limit where the ***diagonal-patch approximation*** (*A*1) is correct), the sector polarization is diagonal in the sector indices, and is to leading order given in Eq.(4.24). It follows that the matrix elements of $\underline{\Gamma}(q)$ (which according to Eqs.(4.22) and (4.43) is proportional to the inverse non-interacting polarization) are in the above limit given by

$$\Gamma^{\alpha\alpha'}(q) \approx \delta^{\alpha\alpha'} \frac{\beta}{V\nu^{\alpha}} \frac{\boldsymbol{v}^{\alpha} \cdot \boldsymbol{q} - \mathrm{i}\omega_m}{\boldsymbol{v}^{\alpha} \cdot \boldsymbol{q}} \quad . \tag{4.52}$$

Hence the Gaussian action (4.45) can be written as

$$\tilde{S}^{(0)}_{\mathrm{eff},2}\{\tilde{\rho}^{\alpha}\} = \frac{\beta}{2V} \sum_{q} \sum_{\alpha\alpha'} \left[f_{\boldsymbol{q}}^{\alpha\alpha'} + \delta^{\alpha\alpha'} \frac{\boldsymbol{v}^{\alpha} \cdot \boldsymbol{q} - \mathrm{i}\omega_m}{\nu^{\alpha} \boldsymbol{v}^{\alpha} \cdot \boldsymbol{q}} \right] \tilde{\rho}^{\alpha}_{-q} \tilde{\rho}^{\alpha'}_{q} \quad . \tag{4.53}$$

The term proportional to $\mathrm{i}\omega_m$ defines the dynamics of the $\tilde{\rho}^{\alpha}$-field. We now recall that in the functional integral for canonically quantized bosons *the coefficient of the term proportional to $-\mathrm{i}\omega_m$ should be precisely β.* Any other value of this coefficient would describe operators with non-canonical commutation relations [4.8]. In a different context such a rescaling has also been performed in [4.9]. Thus, to write our effective action in terms of a canonical boson field b^{α}_{q}, we should rescale the $\tilde{\rho}^{\alpha}$-field accordingly. This is achieved by substituting in Eq.(4.53)

$$\tilde{\rho}^{\alpha}_{q} = (V\nu^{\alpha} |\boldsymbol{v}^{\alpha} \cdot \boldsymbol{q}|)^{1/2} \left[\Theta(\boldsymbol{v}^{\alpha} \cdot \boldsymbol{q}) b^{\alpha}_{q} + \Theta(-\boldsymbol{v}^{\alpha} \cdot \boldsymbol{q}) b^{\dagger\alpha}_{-q} \right] \quad . \tag{4.54}$$

The Θ-functions are necessary to make the coefficient of $-\mathrm{i}\omega_m$ equal to β for all patches, because the sign of $\mathrm{i}\omega_m$ in Eq.(4.53) depends on the sign of $\boldsymbol{v}^{\alpha} \cdot \boldsymbol{q}$. Our final result for the bosonized action $\tilde{S}_{\mathrm{b}}\{b^{\alpha}\} \equiv \tilde{S}^{(0)}_{\mathrm{eff},2}\{\tilde{\rho}^{\alpha}(b^{\alpha})\}$ is

$$\begin{aligned} \tilde{S}_{\mathrm{b}}\{b^{\alpha}\} &= \beta \sum_{q} \sum_{\alpha} \Theta(\boldsymbol{v}^{\alpha} \cdot \boldsymbol{q})(-\mathrm{i}\omega_m) b^{\alpha\dagger}_{q} b^{\alpha}_{q} \\ &\quad + \beta \left[H_{\mathrm{b,kin}}\{b^{\alpha}\} + H_{\mathrm{b,int}}\{b^{\alpha}\} \right] \quad , \end{aligned} \tag{4.55}$$

$$H_{\mathrm{b,kin}}\{b^{\alpha}\} = \sum_{q} \sum_{\alpha} \Theta(\boldsymbol{v}^{\alpha} \cdot \boldsymbol{q}) \boldsymbol{v}^{\alpha} \cdot \boldsymbol{q} b^{\alpha\dagger}_{q} b^{\alpha}_{q} \quad , \tag{4.56}$$

$$\begin{aligned} H_{\mathrm{b,int}}\{b^{\alpha}\} &= \frac{1}{2} \sum_{q} \sum_{\alpha\alpha'} \Theta(\boldsymbol{v}^{\alpha} \cdot \boldsymbol{q}) \sqrt{|\boldsymbol{v}^{\alpha} \cdot \boldsymbol{q}||\boldsymbol{v}^{\alpha'} \cdot \boldsymbol{q}|} \\ &\times \Big[\Theta(\boldsymbol{v}^{\alpha'} \cdot \boldsymbol{q}) \left(F^{\alpha\alpha'}_{\boldsymbol{q}} b^{\alpha\dagger}_{q} b^{\alpha'}_{q} + F^{\alpha'\alpha}_{\boldsymbol{q}} b^{\alpha'\dagger}_{q} b^{\alpha}_{q} \right) \\ &\quad + \Theta(-\boldsymbol{v}^{\alpha'} \cdot \boldsymbol{q}) \left(F^{\alpha\alpha'}_{\boldsymbol{q}} b^{\alpha\dagger}_{q} b^{\alpha'\dagger}_{-q} + F^{\alpha'\alpha}_{\boldsymbol{q}} b^{\alpha'}_{-q} b^{\alpha}_{q} \right) \Big] \quad , \end{aligned} \tag{4.57}$$

where $F^{\alpha\alpha'}_{\boldsymbol{q}} = \sqrt{\nu^{\alpha}\nu^{\alpha'}} f^{\alpha\alpha'}_{\boldsymbol{q}}$ are dimensionless couplings, and we have assumed that the bare interaction depends only on $\boldsymbol{q}$. For frequency-dependent bare interactions it is not possible to write down a conventional Hamiltonian that is equivalent to the effective action in Eq.(4.53). The functional integral for the b^{α}-field is now formally identical with a standard bosonic functional integral. The corresponding second-quantized bosonic Hamiltonian is therefore $\hat{H}_{\mathrm{b}} =$

$\hat{H}_{\mathrm{b,kin}} + \hat{H}_{\mathrm{b,int}}$, where $\hat{H}_{\mathrm{b,kin}}$ and $\hat{H}_{\mathrm{b,int}}$ are simply obtained by replacing the bosonic fields b_q^α in Eqs.(4.56) and (4.57) by operators $\hat{b}_q^\alpha$ satisfying $[\hat{b}_q^\alpha, \hat{b}_{q'}^{\alpha'\dagger}] = \delta^{\alpha\alpha'}\delta_{qq'}$. The resulting $\hat{H}_{\mathrm{b}}$ agrees with the bosonized Hamiltonian derived in [1.31, 1.34] by means of an operator approach.

Note, however, that the above identification with a canonical bosonic Hamiltonian is only possible in the limit of long wavelengths and high densities, so that our parameterization (4.45) of the effective Gaussian action is more general. Moreover, for practical calculations the substitution (4.54) is not very useful, because it maps the very simple form (4.45) of $\tilde{S}^{(0)}_{\mathrm{eff},2}\{\tilde{\rho}^\alpha\}$ onto the complicated effective action $\tilde{S}_{\mathrm{b}}\{b^\alpha\}$ in Eqs.(4.55)–(4.57) without containing new information.

4.3 Beyond the Gaussian approximation

We develop a systematic method for calculating the corrections to the Gaussian approximation, and then explicitly evaluate the one-loop correction. In this way we determine the hidden small parameter which determines the range of validity of the Gaussian approximation. We also show that bosonization leads to a new method for calculating the density-density correlation function beyond the RPA.

4.3.1 General expansion of the bosonized kinetic energy

The bosonized kinetic energy $\tilde{S}_{\mathrm{kin}}\{\tilde{\rho}^\alpha\}$ is calculated via a linked cluster expansion of the functional Fourier transformation in Eq.(3.56).

Defining $S'_{\mathrm{kin}}\{\phi^\alpha\}$ to be the sum of all non-Gaussian terms in the expansion (4.1) of $S_{\mathrm{kin}}\{\phi^\alpha\}$,

$$S'_{\mathrm{kin}}\{\phi^\alpha\} = \sum_{n=3}^{\infty} S_{\mathrm{kin},n}\{\phi^\alpha\} = \sum_{n=3}^{\infty} \frac{1}{n} \mathrm{Tr}\left[\hat{G}_0 \hat{V}\right]^n \quad , \tag{4.58}$$

we may write

$$\mathrm{e}^{-\tilde{S}_{\mathrm{kin}}\{\tilde{\rho}^\alpha\}} = \exp\left[-\tilde{S}^{(0)}_{\mathrm{kin},0} - \tilde{S}^{(0)}_{\mathrm{kin},2}\{\tilde{\rho}^\alpha\}\right] \left\langle \mathrm{e}^{-S'_{\mathrm{kin}}\{\phi^\alpha\}} \right\rangle^{\tilde{\rho}}_{S_{\mathrm{kin},2}} \quad , \tag{4.59}$$

where according to Eqs.(4.41) and (4.42),

$$\exp\left[-\tilde{S}^{(0)}_{\mathrm{kin},0} - \tilde{S}^{(0)}_{\mathrm{kin},2}\{\tilde{\rho}^\alpha\}\right] =$$

$$\int \mathcal{D}\{\phi^\alpha\} \exp\left[\mathrm{i}\sum_{q\alpha} \phi^\alpha_{-q}\tilde{\rho}^\alpha_q - S_{\mathrm{kin},2}\{\phi^\alpha\}\right] \quad , \tag{4.60}$$

and for any functional $\mathcal{F}\{\phi^\alpha\}$ the averaging in Eq.(4.59) is defined as follows,

$$\langle \mathcal{F}\{\phi^\alpha\}\rangle^{\tilde{\rho}}_{S_{\mathrm{kin},2}} = \frac{\int \mathcal{D}\{\phi^\alpha\}\mathcal{F}\{\phi^\alpha\}\exp\left[\mathrm{i}\sum_{q\alpha}\phi^\alpha_{-q}\tilde{\rho}^\alpha_q - S_{\mathrm{kin},2}\{\phi^\alpha\}\right]}{\int \mathcal{D}\{\phi^\alpha\}\exp\left[\mathrm{i}\sum_{q\alpha}\phi^\alpha_{-q}\tilde{\rho}^\alpha_q - S_{\mathrm{kin},2}\{\phi^\alpha\}\right]} \ . \quad (4.61)$$

Performing in this expression the shift transformation

$$\phi^\alpha_q \to \phi^\alpha_q + \mathrm{i}\sum_{\alpha'} \Gamma^{\alpha\alpha'}(q)\tilde{\rho}^{\alpha'}_q \ , \quad (4.62)$$

it is easy to see that

$$\begin{aligned}\langle \mathcal{F}\{\phi^\alpha\}\rangle^{\tilde{\rho}}_{S_{\mathrm{kin},2}} &= \frac{\int \mathcal{D}\{\phi^\alpha\}\mathcal{F}\{\phi^\alpha + \mathrm{i}\sum_{\alpha'}\Gamma^{\alpha\alpha'}\tilde{\rho}^{\alpha'}\}\exp\left[-S_{\mathrm{kin},2}\{\phi^\alpha\}\right]}{\int \mathcal{D}\{\phi^\alpha\}\exp\left[-S_{\mathrm{kin},2}\{\phi^\alpha\}\right]} \\ &\equiv \left\langle \mathcal{F}\{\phi^\alpha + \mathrm{i}\sum_{\alpha'}\Gamma^{\alpha\alpha'}\tilde{\rho}^{\alpha'}\}\right\rangle_{S_{\mathrm{kin},2}} \ . \end{aligned} \quad (4.63)$$

In our case we have to calculate

$$\left\langle \mathrm{e}^{-S'_{\mathrm{kin}}\{\phi^\alpha\}}\right\rangle^{\tilde{\rho}}_{S_{\mathrm{kin},2}} = \left\langle \mathrm{e}^{-S'_{\mathrm{kin}}\{\phi^\alpha + \mathrm{i}\sum_{\alpha'}\Gamma^{\alpha\alpha'}\tilde{\rho}^{\alpha'}\}}\right\rangle_{S_{\mathrm{kin},2}} \ . \quad (4.64)$$

Consider first the term of order $(\phi^\alpha)^n$ in the expansion (4.58) of $S'_{\mathrm{kin}}\{\phi^\alpha\}$. Clearly the substitution $\phi^\alpha \to \phi^\alpha + \mathrm{i}\sum_{\alpha'}\Gamma^{\alpha\alpha'}\tilde{\rho}^{\alpha'}$ generates (among many other terms) a term of order $(\tilde{\rho}^\alpha)^n$, which does not depend on the ϕ^α-field and can be pulled out of the average in Eq.(4.64). Let us denote this contribution by $\tilde{S}^{(0)}_{\mathrm{kin},n}\{\tilde{\rho}^\alpha\}$. From Eq.(4.4) it is easy to see that $\tilde{S}^{(0)}_{\mathrm{kin},n}\{\tilde{\rho}^\alpha\}$ is obtained by replacing $\phi^\alpha_q \to \mathrm{i}\sum_{\alpha'}\Gamma^{\alpha\alpha'}(q)\tilde{\rho}^{\alpha'}_q$ in $S_{\mathrm{kin},n}\{\phi^\alpha\}$, so that it is given by

$$\begin{aligned}\tilde{S}^{(0)}_{\mathrm{kin},n}\{\tilde{\rho}^\alpha\} &= S_{\mathrm{kin},n}\{\mathrm{i}\sum_{\alpha'}\Gamma^{\alpha\alpha'}(q)\tilde{\rho}^{\alpha'}_q\} \\ &= \frac{1}{n}\sum_{q_1\ldots q_n}\sum_{\alpha_1\ldots\alpha_n}\Gamma^{(0)}_n(q_1\alpha_1\ldots q_n\alpha_n)\tilde{\rho}^{\alpha_1}_{q_1}\cdots\tilde{\rho}^{\alpha_n}_{q_n} \ , \end{aligned} \quad (4.65)$$

where for $n \geq 3$ the vertices $\Gamma^{(0)}_n$ are

$$\begin{aligned}\Gamma^{(0)}_n(q_1\alpha_1\ldots q_n\alpha_n) = \mathrm{i}^n \sum_{\alpha'_1\ldots\alpha'_n} & U_n(q_1\alpha'_1\ldots q_n\alpha'_n) \\ & \times\Gamma^{\alpha'_1\alpha_1}(q_1)\ldots\Gamma^{\alpha'_n\alpha_n}(q_n) \ . \end{aligned} \quad (4.66)$$

Recall that $\Gamma^{\alpha\alpha'}(q)$ is according to Eq.(4.43) proportional to the matrix inverse of the non-interacting sector polarization $\Pi^{\alpha\alpha'}_0(q)$. Obviously the Gaussian action $\tilde{S}^{(0)}_{\mathrm{kin},2}\{\tilde{\rho}^\alpha\}$ in Eq.(4.42) is also of the form (4.65), with

$$\Gamma^{(0)}_2(q_1\alpha_1 q_2\alpha_2) = \delta_{q_1+q_2,0}\Gamma^{\alpha_1\alpha_2}(q_2) \ . \quad (4.67)$$

The vertex U_1 has been absorbed into the redefinition of $\tilde{\rho}^\alpha_q$ (see Eq.(4.39)), so that $\tilde{S}^{(0)}_{\mathrm{kin},1}\{\tilde{\rho}^\alpha_q\} = 0$. Defining

$$\tilde{S}^{(0)}_{\mathrm{kin}}\{\tilde{\rho}^{\alpha}\} = \tilde{S}^{(0)}_{\mathrm{kin},0} + \sum_{n=2}^{\infty} \tilde{S}^{(0)}_{\mathrm{kin},n}\{\tilde{\rho}^{\alpha}\} \quad , \tag{4.68}$$

$$S''_{\mathrm{kin}}\{\phi^{\alpha},\tilde{\rho}^{\alpha}\} = S'_{\mathrm{kin}}\{\phi^{\alpha} + \mathrm{i}\sum_{\alpha'} \Gamma^{\alpha\alpha'}\tilde{\rho}^{\alpha'}\} - S'_{\mathrm{kin}}\{\mathrm{i}\sum_{\alpha'} \Gamma^{\alpha\alpha'}\tilde{\rho}^{\alpha'}\} \quad , \tag{4.69}$$

the general perturbative expansion for $\tilde{S}_{\mathrm{kin}}\{\tilde{\rho}^{\alpha}\}$ is

$$\tilde{S}_{\mathrm{kin}}\{\tilde{\rho}^{\alpha}\} = \tilde{S}^{(0)}_{\mathrm{kin}}\{\tilde{\rho}^{\alpha}\} - \ln\left[1 + \sum_{n=1}^{\infty} \frac{(-1)^n}{n!} \left\langle [S''_{\mathrm{kin}}\{\phi^{\alpha},\tilde{\rho}^{\alpha}\}]^n \right\rangle_{S_{\mathrm{kin},2}}\right] . \tag{4.70}$$

According to the linked cluster theorem [2.10] the logarithm eliminates all disconnected diagrams, so that Eq.(4.70) can also be written as

$$\tilde{S}_{\mathrm{kin}}\{\tilde{\rho}^{\alpha}\} = \tilde{S}^{(0)}_{\mathrm{kin}}\{\tilde{\rho}^{\alpha}\} - \sum_{n=1}^{\infty} \frac{(-1)^n}{n} \left\langle [S''_{\mathrm{kin}}\{\phi^{\alpha},\tilde{\rho}^{\alpha}\}]^n \right\rangle^{\mathrm{con}}_{S_{\mathrm{kin},2}} \quad , \tag{4.71}$$

where the superscript ^{con} means that all different connected diagrams should be retained [2.10]. From this expression it is easy to see that $\tilde{S}_{\mathrm{kin}}\{\tilde{\rho}^{\alpha}\}$ is in general of the following form

$$\tilde{S}_{\mathrm{kin}}\{\tilde{\rho}^{\alpha}\} = \tilde{S}_{\mathrm{kin},0} + \sum_{n=1}^{\infty} \tilde{S}_{\mathrm{kin},n}\{\tilde{\rho}^{\alpha}\} \quad , \tag{4.72}$$

where $\tilde{S}_{\mathrm{kin},0}$ is a constant independent of the fields that cancels in the calculation of correlation functions, and for $n \geq 1$

$$\tilde{S}_{\mathrm{kin},n}\{\tilde{\rho}^{\alpha}\} = \frac{1}{n} \sum_{q_1 \ldots q_n} \sum_{\alpha_1 \ldots \alpha_n} \Gamma_n(q_1\alpha_1 \ldots q_n\alpha_n) \tilde{\rho}^{\alpha_1}_{q_1} \cdots \tilde{\rho}^{\alpha_n}_{q_n} \quad , \tag{4.73}$$

where the vertices Γ_n have an expansion of the form

$$\Gamma_n(q_1\alpha_1 \ldots q_n\alpha_n) = \sum_{m=0}^{\infty} \Gamma^{(m)}_n(q_1\alpha_1 \ldots q_n\alpha_n) \quad . \tag{4.74}$$

Here $\Gamma^{(m)}_n$ is the interaction vertex between n collective density fields $\tilde{\rho}^{\alpha}$, that is generated from all diagrams in the linked cluster expansion (4.71) containing m internal loops of the ϕ^{α}-field. Note that the vertices $\Gamma^{(0)}_n$ in Eq.(4.66) are the tree-approximation for the exact vertices Γ_n, because they do not involve any internal ϕ^{α}-loops. Each internal ϕ^{α}-loop attached to a vertex U_n reduces the number of external ϕ^{α}-fields by 2, so that for $m \geq 1$ the vertices $\Gamma^{(m)}_n$ can only by determined by vertices $U_{n'}$ with $n' > n$. Within the Gaussian approximation all U_n with $n \geq 3$ are set equal to zero, while the contribution from U_1 can be absorbed into the redefinition of $\tilde{\rho}^{\alpha}_0$, see Eq.(4.39). Hence the Gaussian approximation amounts to setting

$$\Gamma_2(-q\alpha, q\alpha') \approx \Gamma^{(0)}_2(-q\alpha, q\alpha') = \Gamma^{\alpha\alpha'}(q) \quad , \tag{4.75}$$

$$\Gamma^{(m)}_n = 0 \quad , \quad \text{for } n > 2 \text{ or } m > 0 \quad , \tag{4.76}$$

where $\Gamma^{\alpha\alpha'}(q)$ is defined in Eq.(4.43). Although $\Gamma_1 = 0$ within the Gaussian approximation, the higher order terms will in general lead to a finite value of Γ_1, which describes the fluctuations of the total number of occupied states in the sectors $K^{\alpha}_{\Lambda,\lambda}$. As already pointed out in the footnote after Eq.(4.2), at zero temperature these terms do not contribute to correlation functions at finite q, but they are certainly important for the calculation of the free energy.

4.3.2 The leading correction to the effective action

We now show that our formalism can indeed be used in practice for a systematic calculation of the corrections to the non-interacting boson approximation.

The leading correction to the Gaussian approximation is obtained from the one-loop approximation for our effective bosonic theory, which amounts to a two-loop calculation at the fermionic level. Note that we have mapped the problem of calculating a two-particle Green's function of the original fermionic model onto the problem of calculating a one-particle Green's function of an effective bosonic model. The latter is conceptually simpler, because the symmetrized vertices U_n and $\Gamma_n^{(m)}$ automatically contain the relevant self-energy and vertex corrections of the underlying fermionic problem. This will become evident below.

At one-loop order, it is sufficient to truncate the expansion of the interaction part $S'_{\mathrm{kin}}\{\phi^{\alpha}\}$ of the effective action (4.58) of the ϕ^{α}-field at the fourth order,

$$
\begin{aligned}
S'_{\mathrm{kin}}\{\phi^{\alpha}\} &\approx S_{\mathrm{kin},3}\{\phi^{\alpha}\} + S_{\mathrm{kin},4}\{\phi^{\alpha}\} \\
&= \frac{1}{3}\sum_{q_1 q_2 q_3}\sum_{\alpha_1\alpha_2\alpha_3} U_3(q_1\alpha_1 q_2\alpha_2 q_3\alpha_3)\phi^{\alpha_1}_{q_1}\phi^{\alpha_2}_{q_2}\phi^{\alpha_3}_{q_3} \\
&+\frac{1}{4}\sum_{q_1 q_2 q_3 q_4}\sum_{\alpha_1\alpha_2\alpha_3\alpha_4} U_4(q_1\alpha_1 q_2\alpha_2 q_3\alpha_3 q_4\alpha_4)\phi^{\alpha_1}_{q_1}\phi^{\alpha_2}_{q_2}\phi^{\alpha_3}_{q_3}\phi^{\alpha_4}_{q_4} \quad ,
\end{aligned}
\tag{4.77}
$$

where the vertices U_3 and U_4 are defined in Eq.(4.5). According to the general formalism outlined above, the bosonized kinetic energy $\tilde{S}_{\mathrm{kin}}\{\tilde{\rho}^{\alpha}\}$ is obtained by calculating the functional Fourier transform of $S_{\mathrm{kin}}\{\phi^{\alpha}\}$. Within the one-loop approximation it is sufficient to retain only the term $n = 1$ in the linked cluster expansion (4.71), so that

$$
\tilde{S}_{\mathrm{kin}}\{\tilde{\rho}^{\alpha}\} \approx \tilde{S}^{(0)}_{\mathrm{kin}}\{\tilde{\rho}^{\alpha}\} + \langle S''_{\mathrm{kin}}\{\phi^{\alpha},\tilde{\rho}^{\alpha}\}\rangle^{\mathrm{con}}_{S_{\mathrm{kin},2}} \quad ,
\tag{4.78}
$$

where

$$
\begin{aligned}
\tilde{S}^{(0)}_{\mathrm{kin}}\{\tilde{\rho}^{\alpha}\} &\approx \tilde{S}^{(0)}_{\mathrm{kin},0} + \frac{1}{2}\sum_{q}\sum_{\alpha\alpha'}\Gamma^{\alpha\alpha'}(q)\tilde{\rho}^{\alpha}_{-q}\tilde{\rho}^{\alpha'}_{q} \\
&+\frac{1}{3}\sum_{q_1 q_2 q_3}\sum_{\alpha_1\alpha_2\alpha_3}\Gamma^{(0)}_3(q_1\alpha_1 q_2\alpha_2 q_3\alpha_3)\tilde{\rho}^{\alpha_1}_{q_1}\tilde{\rho}^{\alpha_2}_{q_2}\tilde{\rho}^{\alpha_3}_{q_3}
\end{aligned}
$$

$$+\frac{1}{4}\sum_{q_1q_2q_3q_4}\sum_{\alpha_1\alpha_2\alpha_3\alpha_4}\Gamma_4^{(0)}(q_1\alpha_1q_2\alpha_2q_3\alpha_3q_4\alpha_4)\tilde{\rho}_{q_1}^{\alpha_1}\tilde{\rho}_{q_2}^{\alpha_2}\tilde{\rho}_{q_3}^{\alpha_3}\tilde{\rho}_{q_4}^{\alpha_4} \quad , \tag{4.79}$$

with

$$\Gamma_3^{(0)}(q_1\alpha_1q_2\alpha_2q_3\alpha_3) = -\mathrm{i}\sum_{\alpha_1'\alpha_2'\alpha_3'}U_3(q_1\alpha_1'q_2\alpha_2'q_3\alpha_3') \times\Gamma^{\alpha_1'\alpha_1}(q_1)\Gamma^{\alpha_2'\alpha_2}(q_2)\Gamma^{\alpha_3'\alpha_3}(q_3) \quad , \tag{4.80}$$

$$\Gamma_4^{(0)}(q_1\alpha_1q_2\alpha_2q_3\alpha_3q_4\alpha_4) = \sum_{\alpha_1'\alpha_2'\alpha_3'\alpha_4'}U_4(q_1\alpha_1'q_2\alpha_2'q_3\alpha_3'q_4\alpha_4') \times\Gamma^{\alpha_1'\alpha_1}(q_1)\Gamma^{\alpha_2'\alpha_2}(q_2)\Gamma^{\alpha_3'\alpha_3}(q_3)\Gamma^{\alpha_4'\alpha_4}(q_4) \quad . \tag{4.81}$$

The correction term due to one internal ϕ^α-loop is

$$\langle S''_{\mathrm{kin}}\{\phi^\alpha,\tilde{\rho}^\alpha\}\rangle^{\mathrm{con}}_{S_{\mathrm{kin},2}} = \tilde{S}^{(1)}_{\mathrm{kin},0} + \tilde{S}^{(1)}_{\mathrm{kin},1}\{\tilde{\rho}^\alpha\} + \tilde{S}^{(1)}_{\mathrm{kin},2}\{\tilde{\rho}^\alpha\} \quad , \tag{4.82}$$

where

$$\tilde{S}^{(1)}_{\mathrm{kin},0} = \frac{3}{2}\sum_{qq'}\sum_{\alpha_1\alpha_2\alpha_3\alpha_4}U_4(-q\alpha_1,q\alpha_2,-q'\alpha_3,q'\alpha_4) \times\Gamma^{\alpha_2\alpha_1}(q)\Gamma^{\alpha_4\alpha_3}(q') \quad , \tag{4.83}$$

$$\tilde{S}^{(1)}_{\mathrm{kin},1}\{\tilde{\rho}^\alpha\} = \sum_\alpha \Gamma_1^{(1)}(0\alpha)\tilde{\rho}_0^\alpha \quad , \tag{4.84}$$

$$\tilde{S}^{(1)}_{\mathrm{kin},2}\{\tilde{\rho}^\alpha\} = \frac{1}{2}\sum_q\sum_{\alpha\alpha'}\Gamma_2^{(1)}(-q\alpha,q\alpha')\tilde{\rho}^\alpha_{-q}\tilde{\rho}^{\alpha'}_q \quad , \tag{4.85}$$

with

$$\Gamma_1^{(1)}(0\alpha) = \mathrm{i}\sum_q\sum_{\alpha_1\alpha_2\alpha_3}U_3(-q\alpha_1,q\alpha_2,0\alpha_3) \times\Gamma^{\alpha_2\alpha_1}(q)\Gamma^{\alpha\alpha_3}(0) \quad , \tag{4.86}$$

$$\Gamma_2^{(1)}(-q\alpha,q\alpha') = -3\sum_{q'}\sum_{\alpha_1\alpha_2\alpha_3\alpha_4}U_4(-q\alpha_1,q\alpha_2,-q'\alpha_3,q'\alpha_4) \times\Gamma^{\alpha\alpha_1}(q)\Gamma^{\alpha_2\alpha'}(q)\Gamma^{\alpha_4\alpha_3}(q') \quad . \tag{4.87}$$

Recall that the superscript $^{(1)}$ indicates that these terms contain one internal bosonic loop. Thus, within the one-loop approximation the constant in Eq.(4.72) is $\tilde{S}_{\mathrm{kin},0} = \tilde{S}^{(0)}_{\mathrm{kin},0} + \tilde{S}^{(1)}_{\mathrm{kin},0}$ (see Eqs.(4.41) and (4.83)), and the vertices Γ_n in Eq.(4.74) are approximated by

$$\Gamma_1(q\alpha) = \Gamma_1^{(1)}(q\alpha) \quad , \tag{4.88}$$

$$\Gamma_2(-q\alpha,q\alpha') = \Gamma^{\alpha\alpha'}(q) + \Gamma_2^{(1)}(-q\alpha,q\alpha') \quad , \tag{4.89}$$

$$\Gamma_3(q_1\alpha_1q_2\alpha_2q_3\alpha_3) = \Gamma_3^{(0)}(q_1\alpha_1q_2\alpha_2q_3\alpha_3) \quad , \tag{4.90}$$

$$\Gamma_4(q_1\alpha_1q_2\alpha_2q_3\alpha_3q_4\alpha_4) = \Gamma_4^{(0)}(q_1\alpha_1q_2\alpha_2q_3\alpha_3q_4\alpha_4) \quad , \tag{4.91}$$

and all Γ_n with $n \geq 5$ are set equal to zero. The term with Γ_1 can again be ignored for a calculation of correlation functions at finite q, because it involves only the $q = 0$ component of the density fields. Furthermore, for our one-loop calculation we may also ignore the vertex Γ_3, because the Gaussian expectation value of a product of three $\tilde{\rho}^{\alpha}$-fields vanishes. Combining the relevant contributions from the kinetic energy with the interaction contribution, we finally arrive at the effective action

$$\tilde{S}_{\mathrm{eff}}\{\tilde{\rho}^{\alpha}\} \approx \frac{1}{2}\sum_{q}\sum_{\alpha\alpha'}\left[[\underline{\tilde{f}}_q]^{\alpha\alpha'} + \Gamma^{\alpha\alpha'}(q)\right]\tilde{\rho}^{\alpha}_{-q}\tilde{\rho}^{\alpha'}_{q}$$
$$+\frac{1}{2}\sum_{q}\sum_{\alpha\alpha'}\Gamma_2^{(1)}(-q\alpha, q\alpha')\tilde{\rho}^{\alpha}_{-q}\tilde{\rho}^{\alpha'}_{q}$$
$$+\frac{1}{4}\sum_{q_1q_2q_3q_4}\sum_{\alpha_1\alpha_2\alpha_3\alpha_4}\Gamma_4^{(0)}(q_1\alpha_1 q_2\alpha_2 q_3\alpha_3 q_4\alpha_4)\tilde{\rho}^{\alpha_1}_{q_1}\tilde{\rho}^{\alpha_2}_{q_2}\tilde{\rho}^{\alpha_3}_{q_3}\tilde{\rho}^{\alpha_4}_{q_4} \quad , \tag{4.92}$$

which should be compared with the Gaussian action in Eq.(4.45). We emphasize that this effective action is only good for the purpose of calculating the one-loop corrections to the Gaussian approximation. At two-loop order one should also retain the terms with Γ_3 and Γ_6. The last two terms in Eq.(4.92) contain the one-loop corrections to the non-interacting boson approximation for the bosonized collective density fluctuations. In the limit of long wavelengths we may again write down an equivalent effective Hamiltonian of canonically quantized bosons by using the substitution (4.54). However, we shall not even bother writing down this complicated expression, because this mapping is only valid at long wavelengths and high densities, and does not lead to any simplification. For all practical purposes the parameterization in terms of the $\tilde{\rho}^{\alpha}$-field is superior. We shall now use this parameterization to calculate the leading correction to the free bosonic propagator, and in this way determine the hidden small parameter which controls the range of validity of the Gaussian approximation.

4.3.3 The leading correction to the bosonic propagator

The calculation in this section takes non-linearities in the energy dispersion as well as momentum-transfer between different patches (i.e. around-the-corner processes) into account.

Let us define a dimensionless proper self-energy matrix $\underline{\Sigma}_*(q)$ via

$$\left\langle \tilde{\rho}^{\alpha}_{q}\tilde{\rho}^{\alpha'}_{-q}\right\rangle_{\tilde{S}_{\mathrm{eff}}} = \left[\left[\underline{\tilde{f}}_q + \underline{\Gamma}(q) - \underline{\Sigma}_*(q)\right]^{-1}\right]^{\alpha\alpha'} \quad , \tag{4.93}$$

where the probability distribution for the average is determined by the exact effective action $\tilde{S}_{\mathrm{eff}}\{\tilde{\rho}^{\alpha}\}$, see Eqs.(3.53)–(3.56). From Eq.(4.46) it is clear that the self-energy $\underline{\Sigma}_*(q)$ contains by definition all corrections to the RPA. Introducing the exact proper polarization matrix $\underline{\Pi}_*(q)$ via

$$\underline{\Pi}_*^{-1}(q) = \underline{\Pi}_0^{-1}(q) - \underline{g}(q) \quad , \quad \underline{g}(q) = \frac{V}{\beta}\underline{\Sigma}_*(q) \quad , \tag{4.94}$$

the exact total density-density correlation function can be written as

$$\Pi(q) = \sum_{\alpha\alpha'} \left[\left[\underline{\Pi}_*^{-1}(q) + \underline{f}_q \right]^{-1} \right]^{\alpha\alpha'} \quad . \tag{4.95}$$

If all matrix elements of $\underline{f}_q$ are identical and equal to f_q, we may repeat the manipulations in Eqs.(4.49)–(4.51), so that Eq.(4.95) reduces to Eq.(2.51), with

$$\Pi_*(q) = \sum_{\alpha\alpha'} [\underline{\Pi}_*(q)]^{\alpha\alpha'} \quad . \tag{4.96}$$

Comparing Eq.(4.94) with Eq.(2.50), we see that the quantities $[\underline{\Sigma}_*(q)]^{\alpha\alpha'}$ can be identified physically with generalized local field corrections $[\underline{g}(q)]^{\alpha\alpha'}$, which differentiate between the contributions from the various sectors.

We now calculate the irreducible bosonic self-energy to first order in an expansion in the number of bosonic loops. To this order we simply have to add the two diagrams shown in Fig. 4.2. Note that according to Eq.(4.87) the shaded semi-circle vertex in the diagram (a) implicitly involves one internal loop summation. Hence, within the one-loop approximation, the diagram (a) should be added to the diagram (b), which explicitly contains a bosonic loop. Because we have symmetrized the vertices, the diagram (b) has a combina-

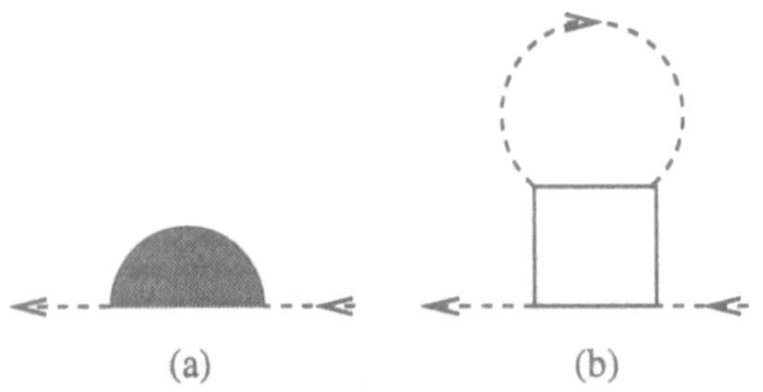

Fig. 4.2. Leading self-energy corrections to the Gaussian propagator of the collective $\tilde{\rho}^\alpha$-field. Diagram (a) represents the first term in Eq.(4.97), while diagram (b) represents the second term. Dashed arrows denote collective density fields $\tilde{\rho}^\alpha$, and the dashed loop is the Gaussian propagator of the $\tilde{\rho}^\alpha$-field, see Eq.(4.46). The vertex $\Gamma_2^{(1)}$ is represented by the shaded semi-circle. The shading indicates that this vertex involves an internal bosonic loop summation, see Eq.(4.87). The tree-level vertex $\Gamma_4^{(0)}$ given in Eq.(4.81) is represented by an empty square.

torial factor of three, so that at one-loop order we obtain $\underline{\Sigma}_*(q) \approx \underline{\Sigma}_*^{(1)}(q)$, with

$$[\underline{\Sigma}_*^{(1)}(q)]^{\alpha\alpha'} = -\Gamma_2^{(1)}(-q\alpha, q\alpha') - 3\sum_{q'}\sum_{\alpha_3\alpha_4} \Gamma_4^{(0)}(-q\alpha, q\alpha', -q'\alpha_3, q'\alpha_4) \times \left[\left[\underline{\tilde{f}}_{q'} + \underline{\Gamma}(q')\right]^{-1}\right]^{\alpha_4\alpha_3} . \tag{4.97}$$

Using the definitions of $\Gamma_2^{(1)}$ and $\Gamma_4^{(0)}$ (see Eqs.(4.87) and (4.81)), it is easy to show that Eq.(4.97) can also be written as

$$[\underline{\Sigma}_*^{(1)}(q)]^{\alpha\alpha'} = 3\sum_{q'}\sum_{\alpha_1\alpha_2\alpha_3\alpha_4} U_4(-q\alpha_1, q\alpha_2, -q'\alpha_3, q'\alpha_4)\Gamma^{\alpha\alpha_1}(q)\Gamma^{\alpha_2\alpha'}(q) \times \left[\underline{\Gamma}(q') - \underline{\Gamma}(q')\left[\underline{\tilde{f}}_{q'} + \underline{\Gamma}(q')\right]^{-1}\underline{\Gamma}(q')\right]^{\alpha_4\alpha_3} . \tag{4.98}$$

A simple manipulation shows that the matrix in the last line of Eq.(4.98) can be identified with $\frac{\beta}{V}\underline{f}_q^{\mathrm{RPA}}$, where $\underline{f}_q^{\mathrm{RPA}}$ is the RPA interaction matrix defined in Eq.(4.33). We conclude that

$$[\underline{\Sigma}_*^{(1)}(q)]^{\alpha\alpha'} = 3\frac{\beta}{V}\sum_{q'}\sum_{\alpha_1\alpha_2\alpha_3\alpha_4} \Gamma^{\alpha\alpha_1}(q)\Gamma^{\alpha_2\alpha'}(q) \times U_4(-q\alpha_1, q\alpha_2, -q'\alpha_3, q'\alpha_4)\left[\underline{f}_{q'}^{\mathrm{RPA}}\right]^{\alpha_4\alpha_3} . \tag{4.99}$$

Note that $\underline{\Sigma}_*^{(1)}(q)$ is proportional to the RPA screened interaction and vanishes in the non-interacting limit, as it should. Eq.(4.99) is the general result for the leading correction to the Gaussian approximation due to non-linearities in the energy dispersion and around-the-corner processes for arbitrary sectorizations and bare interaction matrices $\underline{f}_q$.

4.3.4 The hidden small parameter

We now neglect the around-the-corner processes, but keep the non-linearities in the energy dispersion.

To make further progress, we shall ignore from now on scattering processes that transfer momentum between different sectors, i.e. the around-the-corner processes. As discussed in Chap. 2.5, for non-linear energy dispersion we are free to choose rather large patches with finite curvature, so that the neglect of the around-the-corner processes is not a serious restriction. Moreover, this approximation is always justified if there exists a cutoff $q_{\mathrm{c}} \ll \Lambda, \lambda \ll k_{\mathrm{F}}$ such that for wave-vectors $|\boldsymbol{q}| \gtrsim q_{\mathrm{c}}$ the effective interaction $\underline{f}_q^{\mathrm{RPA}}$ becomes negligibly small. Choosing also the magnitude of the external wave-vector $\boldsymbol{q}$ in Eq.(4.99) small compared with the cutoffs Λ and λ, the diagonal-patch approximation ($A1$) is justified, so that $\underline{\Gamma}(q)$ and $U_4(-q\alpha_1, q\alpha_2, -q'\alpha_3, q'\alpha_4)$ are diagonal in the patch indices, see Eqs.(4.52) and (4.7). Then Eq.(4.99) reduces to

$$[\underline{\Sigma}_{*}^{(1)}(q)]^{\alpha\alpha'} = \delta^{\alpha\alpha'} \frac{\beta}{V\nu^{\alpha}} \left(\frac{\boldsymbol{v}^{\alpha} \cdot \boldsymbol{q} - \mathrm{i}\omega_m}{\boldsymbol{v}^{\alpha} \cdot \boldsymbol{q}} \right)^2 A_q^{\alpha} \quad , \tag{4.100}$$

where the dimensionless function A_q^{α} is given by

$$A_q^{\alpha} = \frac{3}{\nu^{\alpha}} \left(\frac{\beta}{V} \right)^2 \sum_{q'} U_4^{\alpha}(-q, q, -q', q') \left[\underline{f}_{q'}^{\mathrm{RPA}} \right]^{\alpha\alpha} \quad , \tag{4.101}$$

with U_4^{α} defined in Eq.(4.8). We thus obtain in the limit of high densities and long wavelengths to first order in the screened interaction

$$\Gamma^{\alpha\alpha'}(q) - [\underline{\Sigma}_{*}^{(1)}(q)]^{\alpha\alpha'} = \frac{\beta}{V} [\underline{\Pi}_{*}^{-1}(q)]^{\alpha\alpha'}$$

$$= \delta^{\alpha\alpha'} \frac{\beta}{V\nu^{\alpha}} \frac{(1 - A_q^{\alpha})\boldsymbol{v}^{\alpha} \cdot \boldsymbol{q} - (1 - 2A_q^{\alpha})\mathrm{i}\omega_m - A_q^{\alpha} \frac{(\mathrm{i}\omega_m)^2}{\boldsymbol{v}^{\alpha} \cdot \boldsymbol{q}}}{\boldsymbol{v}^{\alpha} \cdot \boldsymbol{q}} \quad . \tag{4.102}$$

Comparing this expression with Eq.(4.52), it is evident that the non-interacting boson approximation is quantitatively correct provided the condition $|A_q^{\alpha}| \ll 1$ is satisfied for all α, because then the corrections to the propagator of the collective density field $\tilde{\rho}^{\alpha}$ are small. Using the general definition of the vertices U_n in Eq.(4.5), it is easy to show that

$$A_q^{\alpha} = -\frac{1}{\nu^{\alpha}\beta V} \sum_{k} \Theta^{\alpha}(\boldsymbol{k}) \Big\{ G_0(k)\Sigma^{(1)}(k)G_0(k)[G_0(k+q) + G_0(k-q)]$$

$$+ \frac{1}{2} G_0(k)[\Lambda^{(1)}(k;q)G_0(k+q) + \Lambda^{(1)}(k;-q)G_0(k-q)] \Big\} \quad , \tag{4.103}$$

with

$$\Sigma^{(1)}(k) = -\frac{1}{\beta V} \sum_{q'} f_{q'}^{\mathrm{RPA}} G_0(k+q') \quad , \tag{4.104}$$

$$\Lambda^{(1)}(k;q) = -\frac{1}{\beta V} \sum_{q'} f_{q'}^{\mathrm{RPA}} G_0(k+q') G_0(k+q'+q) \quad . \tag{4.105}$$

Note that $A_{-q}^{\alpha} = A_q^{\alpha}$ due to the symmetrization of the vertex U_4 with respect to the interchange of any two labels. It is now obvious that the vertices of our effective bosonic action automatically contain the relevant self-energy and vertex corrections of the underlying fermionic problem [4.6]. The first term in Eq.(4.103) corresponds to the self-energy corrections to the non-interacting polarization bubble shown in Fig. 4.3 (a) and (b), while the last term is due to the vertex correction shown in Fig. 4.3 (c).

In order to determine the range of validity of the non-interacting boson approximation, we have to calculate the dependence of A_q^{α} on the various parameters in the problem. In the limit of long wavelengths and low energies, it is to leading order in $\boldsymbol{v}^{\alpha} \cdot \boldsymbol{q}$ and ω_m consistent to replace in Eq.(4.102) $A_q^{\alpha} \to A_0^{\alpha}$. Actually, the $q \to 0$ limit of A_q^{α} should be taken in such a way that the ratio $\mathrm{i}\omega_m/(\boldsymbol{v}^{\alpha} \cdot \boldsymbol{q})$ is held constant, because in this case we obtain the low-energy behavior of A_q^{α} close to the poles of the Gaussian propagator. However,

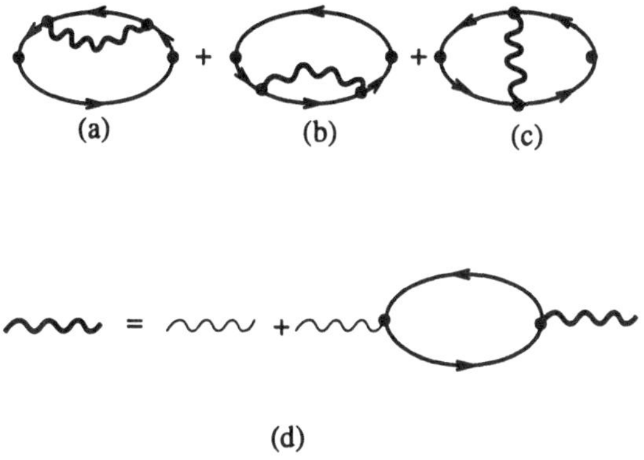

Fig. 4.3. Leading local field corrections to the non-interacting polarization. (a) and (b) are the leading self-energy corrections (see Eq.(4.104)), while (c) is the leading vertex correction (see Eq.(4.105)). The thick wavy line denotes the RPA interaction, as defined in (d). The thin wavy line represents the bare interaction.

since we are only interested in the order of magnitude of A_q^α for small ω_m and $\boldsymbol{q}$, it is sufficient to consider the "$\boldsymbol{q}$-limit" $A_0^\alpha = \lim_{\boldsymbol{q}\to 0}[\lim_{\omega_m \to 0} A_q^\alpha]$. For simplicity, let us assume that $[\underline{f}_q^{\mathrm{RPA}}]^{\alpha\alpha'} = f_q^{\mathrm{RPA}}$ is independent of the sector labels and depends only on the wave-vector. This amounts to the static approximation for the dielectric function, which seems reasonable to obtain the correct order of magnitude of A_0^α. The "$\boldsymbol{q}$-limit" is obtained by setting $q = 0$ under the summation sign and performing the Matsubara sums before doing the wave-vector integrations. For $\beta \to \infty$ we obtain

$$A_0^\alpha = \frac{1}{\nu^\alpha V^2} \sum_{\boldsymbol{kq}} \Theta^\alpha(\boldsymbol{k}) f_q^{\mathrm{RPA}} \left\{ f(\xi_{\boldsymbol{k}+\boldsymbol{q}}) \frac{\partial^2}{\partial \mu^2} f(\xi_{\boldsymbol{k}}) + \frac{\partial}{\partial \mu} f(\xi_{\boldsymbol{k}+\boldsymbol{q}}) \frac{\partial}{\partial \mu} f(\xi_{\boldsymbol{k}}) \right\} \quad . \tag{4.106}$$

Because the $\boldsymbol{k}$-sum extends over the entire sector $K_{\Lambda,\lambda}^\alpha$ and by assumption the $\boldsymbol{q}$-sum is cut off by the interaction at $q_c \ll \Lambda, \lambda$, we may set $\xi_{\boldsymbol{k}+\boldsymbol{q}} \approx \xi_{\boldsymbol{k}}$ in the Fermi functions of Eq.(4.106). Then the summations factorize, and we obtain

$$A_0^\alpha = \left[\frac{1}{V} \sum_{\boldsymbol{q}} f_q^{\mathrm{RPA}} \right] \frac{1}{\nu^\alpha} \int_{-\infty}^{\infty} \mathrm{d}\xi \nu^\alpha(\xi) \left[f(\xi) f''(\xi) + f'(\xi) f'(\xi) \right] \quad , \tag{4.107}$$

where $\nu^\alpha(\xi)$ is the energy-dependent patch density of states,

$$\nu^\alpha(\xi) = \frac{1}{V} \sum_{\boldsymbol{k}} \Theta^\alpha(\boldsymbol{k}) \delta(\xi - \xi_{\boldsymbol{k}}) \quad . \tag{4.108}$$

Note that from the definition (4.25) of ν^α it follows that

$$\nu^\alpha = \int_{-\infty}^{\infty} \mathrm{d}\xi \nu^\alpha(\xi) \left[-f'(\xi) \right] \quad . \tag{4.109}$$

Integrating by parts and taking the limit $\beta \to \infty$, the integral in Eq.(4.107) can be written as

$$\int_{-\infty}^{\infty} \mathrm{d}\xi \nu^\alpha(\xi) \frac{\partial}{\partial \xi} \left[f(\xi) f'(\xi) \right] = \frac{1}{2} \frac{\partial \nu^\alpha}{\partial \mu} \quad . \tag{4.110}$$

Because by assumption f_{q}^{RPA} becomes negligibly small for $|q| \gtrsim q_{\mathrm{c}}$, the first factor in Eq.(4.107) is for $V \to \infty$ given by

$$\frac{1}{V} \sum_{q} f_{q}^{\mathrm{RPA}} = q_{\mathrm{c}}^{d} \langle f^{\mathrm{RPA}} \rangle \quad , \tag{4.111}$$

where $\langle f^{\mathrm{RPA}} \rangle$ is some suitably defined measure for the average strength of the screened interaction. Ignoring a numerical factor of the order of unity, the final result for A_0^α can be written as

$$A_0^\alpha = \frac{q_{\mathrm{c}}^{d} \langle f^{\mathrm{RPA}} \rangle}{\mu} C^\alpha \quad , \tag{4.112}$$

where the dimensionless parameter

$$C^\alpha = \frac{\mu}{\nu^\alpha} \frac{\partial \nu^\alpha}{\partial \mu} = \frac{\mu \partial^2 N_0^\alpha / \partial \mu^2}{\partial N_0^\alpha / \partial \mu} \tag{4.113}$$

is for $d > 1$ a measure for the *local curvature of the Fermi surface in patch* P_Λ^α. Although the patch density of states ν^α is proportional to Λ^{d-1} (see Eq.(4.27)), the cutoff-dependence cancels in Eq.(4.113), because it appears in the numerator as well as in the denominator. Therefore C^α is a cutoff-independent quantity. In fact, writing ν^α as a surface integral over the curved patch P_Λ^α (see Eq.(4.26)), simple geometric considerations lead to the result

$$C^\alpha = \frac{\langle k_{\mathrm{F}} \rangle}{m^\alpha |\boldsymbol{v}^\alpha|} \quad , \tag{4.114}$$

where $\langle k_{\mathrm{F}} \rangle$ is some suitably defined average radius of the Fermi surface, and m^α is the effective mass close to $\mathbf{k}^\alpha$, see Eq.(2.66). Note that $\langle k_{\mathrm{F}} \rangle$ characterizes the *global* geometry of the Fermi surface, while m^α and $\boldsymbol{v}^\alpha$ depend on the *local* shape of the Fermi surface in patch P_Λ^α. Evidently C^α vanishes if we linearize the energy dispersion in patch P_Λ^α, because the linearization amounts to taking the limit $|m^\alpha| \to \infty$ while keeping $\langle k_{\mathrm{F}} \rangle$ finite. Then there is no correction to the Gaussian approximation. Of course, we already know from the closed loop theorem that the Gaussian approximation is exact if the energy dispersion is linearized and the around-the-corner processes are neglected.

As usual, we introduce the dimensionless interaction $\langle F^{\mathrm{RPA}} \rangle = \nu \langle f^{\mathrm{RPA}} \rangle$, which measures the strength of the potential energy relative to the kinetic energy. Using the fact that the global density of states is in d dimensions proportional to k_{F}^{d-2} (see Eq.(A.5)), we conclude that the Gaussian approximation is quantitatively accurate as long as for all patches P_Λ^α

$$|A_0^\alpha| \equiv \left(\frac{q_{\mathrm{c}}}{k_{\mathrm{F}}} \right)^d |\langle F^{\mathrm{RPA}} \rangle| |C^\alpha| \ll 1 \quad . \tag{4.115}$$

The appearance of three parameters that control the accuracy of the Gaussian approximation has a very simple intuitive interpretation. First of all, if everywhere on the Fermi surface the curvature is intrinsically small (i.e. $|C^\alpha| \ll 1$

for all α) then the corrections to the linearization of the energy dispersion are negligible, and hence the Gaussian approximation becomes accurate. Note that in the one-dimensional Tomonaga-Luttinger model $C^{\alpha} = 0$, because the energy dispersion is linear by definition. However, in $d > 1$ and for realistic energy dispersions of the form $\epsilon_{\boldsymbol{k}} = \boldsymbol{k}^2/(2m)$ the dimensionless curvature parameter C^{α} is of the order of unity. But even then the Gaussian approximation is accurate, provided the nature of the interaction is such that it involves only small momentum-transfers. This is also intuitively obvious, because in this case the scattering processes probe only a thin shell around the Fermi surface and do not feel the deviations from linearity. Finally, it is clear that also the strength of the effective interaction should determine the range of validity of Gaussian approximation, because in the limit that the strength of the interaction approaches zero all corrections to the Gaussian approximation vanish. We would like to emphasize, however, that we have not explicitly calculated the corrections to the Gaussian approximation due to around-the-corner processes, although our general result for the bosonic self-energy in Eq.(4.99) includes also these corrections. Nevertheless, the around-the-corner processes can to a large extent be eliminated by subdividing the Fermi surface into a small number of curved patches, as discussed in Chap. 2.5.

4.3.5 Calculating corrections to the RPA via bosonization

Here comes the first practical application of our formalism.

For simplicity let us assume that the diagonal-patch approximation ($A1$) is justified, so that Eq.(4.94) reduces to an equation for the diagonal elements,

$$\begin{aligned} \Pi_*^{\alpha}(q) &= \frac{\Pi_0^{\alpha}(q)}{1 - g^{\alpha}(q)\Pi_0^{\alpha}(q)} \\ &\approx \Pi_0^{\alpha}(q) + \Pi_0^{\alpha}(q) g^{\alpha}(q) \Pi_0^{\alpha}(q) + \ldots \quad . \end{aligned} \tag{4.116}$$

Here $\Pi_*^{\alpha}(q) = [\underline{\Pi}_*(q)]^{\alpha\alpha}$, $g^{\alpha}(q) = [\underline{g}(q)]^{\alpha\alpha}$, and $\Pi_0^{\alpha}(q)$ is at long wavelengths given in Eq.(4.24). Note that our approach is based on the perturbative calculation of the ***inverse*** proper polarization, while in the naive perturbative approach the corrections to the proper polarization are obtained by direct expansion of $\Pi_*(q)$ in powers of the interaction [2.15]. Such a procedure does not correspond to the perturbative calculation of the ***irreducible self-energy*** in our effective bosonic problem, but is equivalent to a direct expansion of the ***Green's function.*** As discussed in Chap. 1.1, close to the poles of the Green's function this expansion cannot be expected to be reliable. To first order, only the leading correction in the expansion of the Dyson equation (i.e. the second line in Eq.(4.116)) is kept in this method, so that the total proper polarization is approximated by

$$\Pi_*(q) \approx \sum_{\alpha} \left[\frac{\nu^{\alpha} \boldsymbol{v}^{\alpha} \cdot \boldsymbol{q}}{\boldsymbol{v}^{\alpha} \cdot \boldsymbol{q} - \mathrm{i}\omega_m} + \nu^{\alpha} A_q^{\alpha} \right]$$

$$= \Pi_0(q) - \frac{1}{\beta V}\sum_k \Big\{ G_0(k)\Sigma^{(1)}(k)G_0(k)[G_0(k+q)+G_0(k-q)]$$

$$+ \frac{1}{2}G_0(k)[\Lambda^{(1)}(k;q)G_0(k+q) + \Lambda^{(1)}(k;-q)G_0(k-q)]\Big\} \quad , \tag{4.117}$$

see Eqs.(4.103)–(4.105). For the Coulomb interaction in $d = 3$ the correction term in Eq.(4.117) has been discussed by Geldart and Taylor [4.6], as well as by Holas *et al.* [2.15]. Note, however, that these authors have evaluated the fermionic self-energy $\Sigma^{(1)}(k)$ and vertex correction $\Lambda^{(1)}(k;q)$ with the bare Coulomb interaction. Holas *et al.* [2.15] have also pointed out that the expansion (4.117) leads to unphysical singularities in the dielectric function close to the plasmon poles. The origin for these singularities is easy to understand within our bosonization approach. The crucial point is that the problem of calculating the corrections to the RPA can be completely mapped onto an effective bosonic problem: our functional bosonization method allows us to explicitly construct the *interacting* bosonic Hamiltonian. Once we accept the validity of this mapping, standard many-body theory tells us that the corrections to the propagator of this effective bosonic theory should be calculated by expanding its irreducible self-energy $\underline{\Sigma}_*(q)$ in the number of internal bosonic loops, and then resumming the perturbation series by means of the Dyson equation. A similar resummation has been suggested in [2.13–2.15], but it is not so easy to justify this procedure at the fermionic level. Our bosonization approach provides the natural justification for this resummation. The unphysical singularities [2.15] that are encountered in the naive perturbative approach are easy to understand from the point of view of bosonization: they are most likely due to the fact that one attempts to calculate a bosonic single-particle Green's function by direct expansion. This expansion is bound to fail close to the poles of the Green's function, i.e. close to the plasmon poles!

Based on the insights gained from our bosonization approach, we would like to suggest that corrections to the RPA should be calculated by expanding the generalized local field corrections $\underline{g}(q)$ in powers of the RPA interaction. We suspect that in this way unphysical singularities in the dielectric function can be avoided. From the first line in Eq.(4.116) we obtain in our method for the total proper polarization at long wave-lengths within the diagonal-patch approximation

$$\Pi_*(q) = \sum_\alpha \frac{\frac{\nu^\alpha}{1-A_q^\alpha}\boldsymbol{v}^\alpha\cdot\boldsymbol{q}}{\boldsymbol{v}^\alpha\cdot\boldsymbol{q} - \mathrm{i}\omega_m(1-B_q^\alpha) - \frac{(\mathrm{i}\omega_m)^2}{\boldsymbol{v}^\alpha\cdot\boldsymbol{q}}B_q^\alpha} \quad , \tag{4.118}$$

with $B_q^\alpha = A_q^\alpha/(1-A_q^\alpha)$. Recall that the around-the-corner processes have been neglected in the derivation of Eq.(4.118), so that it is expected to be accurate for sufficiently small $\boldsymbol{q}$ and for interactions that are dominated by small momentum-transfers.

4.4 Summary and outlook

In this chapter we have developed a general formalism which allows us to bosonize the Hamiltonian of fermions interacting with two-body density-density forces in arbitrary dimensions. We have also shown that the bosonization of the Hamiltonian is closely related to the problem of calculating the density-density correlation function. In general, the bosonized system is described by an effective action of collective density fields which contains also multiple-particle interactions between the bosons. However, the generalized closed loop theorem discussed in Sect. 4.1 guarantees that in certain parameter regimes the vertices describing the interactions are small. To leading order, the collective density fields can then be treated as non-interacting bosons. The relevant small parameter justifying this approximation has been explicitly calculated, and is given in Eqs.(4.114) and (4.115).

From the practical point of view, higher-dimensional bosonization might lead to a new systematic method for **calculating corrections to the RPA**. This is an old problem, which in the context of the homogeneous electron gas has been discussed thoroughly by Geldart and Taylor [4.6] long time ago. These authors already observed partial cancellations between the leading corrections to the RPA. We now know that these cancellations occur to all orders in perturbation theory, and are a direct consequence of the generalized closed loop theorem. The calculation of the local field corrections to the RPA is still an active area of research [4.10, 4.11], which could get some fresh momentum from the non-perturbative insights gained via higher-dimensional bosonization. Note that the corrections to the RPA describe the damping of the collective density oscillations. This and other effects can in principle be obtained from Eq.(4.118) and the resulting dielectric function $\epsilon(q) = 1 + f_q \Pi_*(q)$. This calculation requires a careful analysis of the analytic properties of the function A_q^α defined in Eq.(4.103), and still remains to be done. The possibility that higher-dimensional bosonization might lead to a new systematic method for calculating corrections to the RPA has also been suggested by Khveshchenko [1.48].

5. The single-particle Green's function

In this central chapter of this book we calculate the single-particle Green's function by means of the background field method outlined in Chap. 3.2. We carefully examine the approximations and limitations inherent in higher-dimensional bosonization, and develop a new systematic method for including the non-linear terms in the expansion of the energy dispersion close to the Fermi surface into the bosonization procedure. Short accounts of the results and methods developed in this chapter have been published in [1.35,1.37,1.38].

According to Eq.(3.34) the Matsubara Green's function $G(k) \equiv G(\boldsymbol{k}, \mathrm{i}\tilde{\omega}_n)$ can be *exactly* written as

$$G(k) = \int \mathcal{D}\{\phi^\alpha\}\mathcal{P}\{\phi^\alpha\}[\hat{G}]_{kk} \equiv \left\langle [\hat{G}]_{kk} \right\rangle_{S_{\mathrm{eff}}} \quad , \tag{5.1}$$

where the probability distribution $\mathcal{P}\{\phi^\alpha\}$ is defined in Eq.(3.35), and the matrix elements of the *inverse* of the infinite matrix $\hat{G}$ are given by

$$[\hat{G}^{-1}]_{kk'} = \sum_\alpha \Theta^\alpha(\boldsymbol{k}) \left[\delta_{kk'}(\mathrm{i}\tilde{\omega}_n - \epsilon_{\boldsymbol{k}} + \mu) - V^\alpha_{k-k'}\right] \quad , \tag{5.2}$$

with $V^\alpha_q = \frac{\mathrm{i}}{\beta}\phi^\alpha_q$, see Eq.(3.31). The cutoff function $\Theta^\alpha(\boldsymbol{k})$ refers either to the boxes intersecting the Fermi surface discussed in Chap. 2.4, or to the more general sectors introduced in Chap. 2.5, which by construction cover the entire momentum space. Note also that Eq.(5.2) includes the special case (discussed at the end of Chap. 2.5) that the entire momentum space is identified with a single sector. Then the α-sum contains only a single term, and by definition we may replace the cutoff-function $\Theta^\alpha(\boldsymbol{k})$ by unity.

5.1 The Gaussian approximation with linearized energy dispersion

We show how for linearized energy dispersion the calculation of the Green's function from Eq.(5.1) is carried out in practice. We first discuss the inversion problem of the infinite matrix $\hat{G}^{-1}$. The averaging of the diagonal

elements $[\hat{G}]_{kk}$ with respect to the Gaussian probability distribution $\mathcal{P}_2\{\phi^\alpha\}$ yields then a simple Debye-Waller factor.

In a parameter regime where the approximations ($A1$) and ($A2$) discussed in Chap. 4.1 are justified, the generalized closed loop theorem guarantees that the Gaussian approximation is very accurate. As shown in Chap. 4.2.1, the effective action $S_{\text{eff}}\{\phi^\alpha\}$ of the ϕ^α-field is then to a good approximation given by (see Eqs.(4.29) and (4.30)),

$$S_{\text{eff}}\{\phi^\alpha\} \approx \mathrm{i} \sum_\alpha \phi_0^\alpha N_0^\alpha + S_{\text{eff},2}\{\phi^\alpha\} \quad , \tag{5.3}$$

where the quadratic part is

$$S_{\text{eff},2}\{\phi^\alpha\} = \frac{V}{2\beta} \sum_q \sum_{\alpha\alpha'} [[\underline{f}_q^{-1}]^{\alpha\alpha'} + \delta^{\alpha\alpha'} \Pi_0^\alpha(q)] \phi_{-q}^\alpha \phi_q^{\alpha'} \quad , \tag{5.4}$$

with $\Pi_0^\alpha(q)$ given in Eq.(4.24). The probability distribution $\mathcal{P}\{\phi^\alpha\}$ associated with the ϕ^α-field in Eq.(3.35) is then Gaussian,

$$\mathcal{P}\{\phi^\alpha\} \approx \mathcal{P}_2\{\phi^\alpha\} \equiv \frac{\mathrm{e}^{-S_{\text{eff},2}\{\phi^\alpha\}}}{\int \mathcal{D}\{\phi^\alpha\} \mathrm{e}^{-S_{\text{eff},2}\{\phi^\alpha\}}} \quad . \tag{5.5}$$

The first term in Eq.(5.3) involving ϕ_0^α can be ignored for the calculation of correlation functions at $q \neq 0$. Although within the Gaussian approximation the density-density correlation function is given by the usual RPA result, the ***single-particle Green's function*** in Eq.(5.1) can exhibit a large variety of behaviors, which range from conventional Fermi liquids over Luttinger liquids to even more exotic quantum liquids. Which of these possibilities is realized depends crucially on the dimensionality of the system, on the nature of the interaction, and on the symmetry of the Fermi surface.

Of course, in general it is impossible to invert $\hat{G}^{-1}$ exactly, so that one usually has to use some sort of perturbation theory to calculate the matrix elements $[\hat{G}]_{kk}$. However, in the parameter regime where the conditions ($A1$) and ($A2$) are accurate, it is possible to *calculate the matrix elements $[\hat{G}]_{kk}$ exactly as functionals of the ϕ^α-field.* Note that the conditions ($A1$) and ($A2$) imply also the validity of the closed loop theorem, which in turn insures that the probability distribution $\mathcal{P}\{\phi^\alpha\}$ is Gaussian. In other words, the conditions under which $\mathcal{P}\{\phi^\alpha\}$ can be approximated by a Gaussian are also sufficient to guarantee that $\hat{G}^{-1}$ can be inverted exactly.

5.1.1 The Green's function for fixed background field

To invert $\hat{G}^{-1}$, we proceed in two steps. We first show that the condition ($A1$) discussed in Chap. 4.1 implies that $\hat{G}^{-1}$ is approximately block diagonal, with diagonal blocks $(\hat{G}^\alpha)^{-1}$ labelled by the sector (or patch) indices. Therefore the problem of inverting $\hat{G}^{-1}$ can be reduced to the problem of inverting each *diagonal block separately. We then show that, after linearization of the energy dispersion, each block $(\hat{G}^{\alpha})^{-1}$ can be inverted exactly.*

Block diagonalization

The quadratic form defining the matrix elements $[\hat{G}^{-1}]_{kk'}$ in Eq.(3.29) can be written as

$$S_0\{\psi\} + S_1\{\psi, \phi^{\alpha}\} = -\beta \sum_{kq} \psi^{\dagger}_{k+q} [\hat{G}^{-1}]_{k+q,k} \psi_k \quad , \tag{5.6}$$

with

$$[\hat{G}^{-1}]_{k+q,k} = \sum_{\alpha} \Theta^{\alpha}(\boldsymbol{k}) \left[\delta_{q,0}(\mathrm{i}\tilde{\omega}_n - \xi^{\alpha}_{k-k^{\alpha}} - \epsilon_{k^{\alpha}} + \mu) - V^{\alpha}_{q} \right] \quad , \tag{5.7}$$

where $\xi^{\alpha}_{q} = \epsilon_{k^{\alpha}+q} - \epsilon_{k^{\alpha}}$ is the excitation energy relative to the energy at $\boldsymbol{k}^{\alpha}$ (see Eq.(2.65)), and $V^{\alpha}_{q} = \frac{\mathrm{i}}{\beta}\phi^{\alpha}_{q}$ (see Eq.(3.31)). The cutoff function $\Theta^{\alpha}(\boldsymbol{k})$ groups the matrix elements of the infinite matrix $\hat{G}^{-1}$ into rows labelled by the patch index α. To see this more clearly, consider for simplicity a spherical

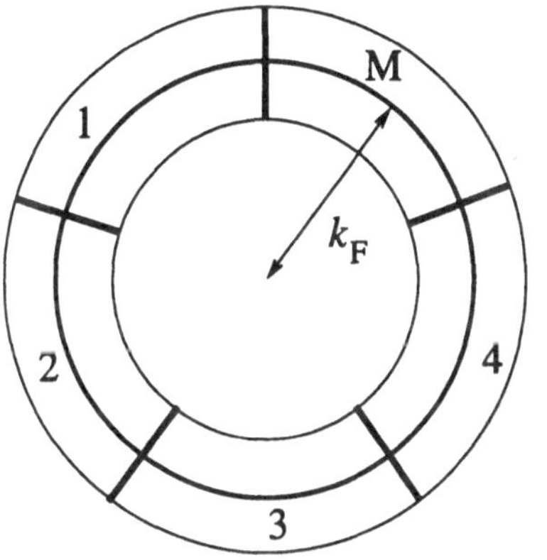

Fig. 5.1. Subdivision of momentum space close to a spherical Fermi surface in $d = 2$ into $M = 5$ sectors $K^{\alpha}_{\Lambda,\lambda}$, $\alpha = 1, \ldots, M$.

Fermi surface in $d = 2$. We partition the degrees of freedom in the vicinity of the Fermi surface into M sectors $K^{\alpha}_{\Lambda,\lambda}$, and label neighboring sectors in increasing order, as shown in Fig. 5.1. The group of matrix elements corresponding to a given label α in Eq.(5.7) can be found in the horizontal stripes in the schematic representation of the matrix $\hat{G}^{-1}$ shown in Fig. 5.2(a). The width of the diagonal band with non-zero matrix elements is determined by the range q_c of the interaction in momentum space, because the vanishing of the interaction $f^{\alpha\alpha'}_{q}$ for $|\boldsymbol{q}| \gtrsim q_c$ implies that the field V^{α}_{q} mediating this interaction must also vanish. But $q_c \ll k_F$ by assumption (A1) in Chap. 4.1, so that we have the freedom of choosing the sector cutoffs Λ and λ such that $q_c \ll \Lambda, \lambda \ll k_F$ (see Eqs.(2.61), (2.63), and (2.64)). As shown in Fig. 5.2(b), in this way $\hat{G}^{-1}$ is subdivided into block matrices associated with the sectors

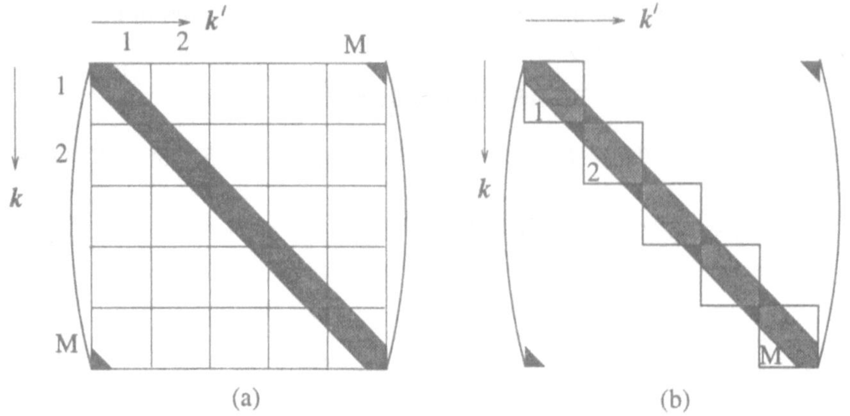

Fig. 5.2. (a) Schematic representation of the matrix $\hat{G}^{-1}$ defined in Eq.(5.7) for $d = 2$. Only the wave-vector index is shown, i.e. each matrix-element is an infinite matrix in frequency space. Regions with non-zero matrix elements are shaded. The triangles in the upper right and lower left corner represent scattering processes between sectors 1 and M. Because these sectors are adjacent, they can be connected by small momentum-transfers. (b) Diagonal blocks and around-the-corner processes (represented by black triangles).

such that $\hat{G}^{-1}$ is *approximately block diagonal.* The block diagonalization is not exact, because non-zero matrix elements are also located in the black triangles of Fig. 5.2(b). These matrix elements represent scattering processes that transfer momentum between different sectors (the around-the-corner processes mentioned in Chap. 2.4.3). The crucial approximation is now to neglect these processes. This is precisely the diagonal-patch approximation ($A1$) discussed in Chap. 4.1, which is also necessary to insure that the probability distribution $\mathcal{P}\{\phi^{\alpha}\}$ can be approximated by a Gaussian. The justification for this step is that the relative number of matrix elements representing momentum-transfer between different sectors is small as long as $q_c \ll \Lambda, \lambda$. In $d > 1$ dimensions the relative number of around-the-corner matrix elements associated with a given sector $K^{\alpha}_{\Lambda,\lambda}$ is of the order of $q_c^d/(\Lambda^{d-1}\lambda)$. Note that this approximation makesectors only sense *if the sector cutoffs are kept finite and large compared with the range of the interaction in momentum space.*

Although the relative number of matrix elements describing around-the-corner processes is small, we have to make one important caveat: Possible non-perturbative effects that depend on the *global topology* of the Fermi surface cannot be described within this approximation. For example, in $d = 2$ each sector has two neighbors, but the first and the last sector are adjacent, so that there exist also around-the-corner processes connecting the sectors 1 and M, which give rise to the off-diagonal triangles in the lower left and upper right corners of the matrix shown in Fig. 5.2. More generally, in higher dimensions the number of non-zero blocks corresponding to around-the-corner processes in each row or column of $\hat{G}^{-1}$ is equal to the coordination number

z_{d-1} of the patches P^α_Λ on the Fermi surface. For example, for hyper-cubic patches on a Fermi surface in d dimensions the coordination number is $z_{d-1} = 2(d-1)$. Hence, the total number of blocks corresponding to around-the-corner processes is $N_c = Mz_{d-1}$, where M is the total number of patches that cover the Fermi surface[1]. Note that $N_c = 0$ for $d = 1$, because around-the-corner processes are absent due to the widely separated Fermi points. On the other hand, in any dimension the number of diagonal blocks is equal to the number M of the patches, so that in $d > 2$ only a small number of the around-the-corner triangles can be found in the vicinity of the diagonal band. The case $d = 2$ shown in Fig. 5.2 is special, because there exist only two off-diagonal around-the-corner blocks, independently of the number M of the patches. In higher dimensions, however, the off-diagonal around-the corner blocks are distributed in a complicated manner over the matrix $\hat{G}^{-1}$. The precise position of the blocks depends on the way in which the patches are labelled on the Fermi surface. The effect of these sparsely distributed around-the-corner blocks is difficult to estimate, and we are assuming that they do not lead to qualitatively new effects. This is an important *assumption* which is implicitly made in all of the following calculations. In systems where the *topological structure* of the Fermi surface is crucial, this assumption may not be justified. We would like to emphasize that this assumption is implicitly also made in the operator bosonization approach [1.31, 1.32], as well as in the Ward identity approach by Castellani, Di Castro and Metzner [1.50–1.52].

Once we have disposed of the around-the-corner matrix elements, the matrix $\hat{G}^{-1}$ is a direct sum of diagonal blocks $(\hat{G}^\alpha)^{-1}$, $\alpha = 1, \ldots, M$. Hence,

$$[\hat{G}^{-1}]_{\boldsymbol{k}\boldsymbol{k}'} = \sum_\alpha \Theta^\alpha(\boldsymbol{k})\Theta^\alpha(\boldsymbol{k}')[(\hat{G}^\alpha)^{-1}]_{\boldsymbol{k}\boldsymbol{k}'} \quad , \tag{5.8}$$

where the matrix $(\hat{G}^\alpha)^{-1}$ is the diagonal block of $\hat{G}^{-1}$ associated with sector $K^\alpha_{\Lambda,\lambda}$, with matrix elements given by

$$[(\hat{G}^\alpha)^{-1}]_{\boldsymbol{k}\boldsymbol{k}'} = \delta_{\boldsymbol{k}\boldsymbol{k}'}[\mathrm{i}\tilde{\omega}_n - \xi^\alpha_{\boldsymbol{k}-\boldsymbol{k}^\alpha} - \epsilon_{\boldsymbol{k}^\alpha} + \mu] - V^\alpha_{\boldsymbol{k}-\boldsymbol{k}'} \quad . \tag{5.9}$$

Thus, *the problem of inverting $\hat{G}^{-1}$ is reduced to the problem of inverting each diagonal block separately.* The diagonal elements of $\hat{G}$ are then simply

$$[\hat{G}]_{\boldsymbol{k}\boldsymbol{k}} = \sum_\alpha \Theta^\alpha(\boldsymbol{k})[\hat{G}^\alpha]_{\boldsymbol{k}\boldsymbol{k}} \quad . \tag{5.10}$$

Note that $\hat{G}^\alpha$ is still an infinite matrix in frequency space, so that the quantum dynamics is fully taken into account.

Inversion of the diagonal blocks

Up to this point we have *not* linearized the energy dispersion, so that the above block diagonalization is valid for arbitrary dispersion $\xi^\alpha_{\boldsymbol{q}}$. The crucial

[1] In Fig. 5.2 we have $M = 5$, $z_1 = 2$ and $N_c = 10$

advantage of the subdivision of $\hat{G}^{-1}$ into blocks is that, to a first approximation, within a given block we may linearize the energy dispersion, $\xi_{\boldsymbol{q}}^{\alpha} \approx \boldsymbol{v}^{\alpha} \cdot \boldsymbol{q}$ (see Eqs.(2.16) and (2.17)). It is also convenient to choose the centers $\boldsymbol{k}^{\alpha}$ of the sectors such that $\epsilon_{\boldsymbol{k}^{\alpha}} = \mu$, so that the last two terms in the square brace of Eq.(5.9) cancel. In Chap. 2.4.2 we have argued[2] that the linearization is justified if the sectors are sufficiently small, so that within a given sector the variation of the local normal vector to the Fermi surface is small. On the other hand, as discussed in detail in Chap. 2.4.3, the cutoffs must be kept large compared with q_{c} in order to guarantee that the patching construction leads to an approximate block diagonalization of $\hat{G}^{-1}$.

Once the linearization has been made, it is possible to invert the diagonal blocks $(\hat{G}^{\alpha})^{-1}$ exactly. Note that $\hat{G}^{\alpha}$ is still an infinite matrix in frequency space. Shifting the wave-vector labels according to $\boldsymbol{k} = \boldsymbol{k}^{\alpha} + \boldsymbol{q}$ and $\boldsymbol{k}' = \boldsymbol{k}^{\alpha} + \boldsymbol{q}'$, the diagonal block $\hat{G}^{\alpha}$ is determined by the equation

$$\sum_{\tilde{q}'} \left[\delta_{\tilde{q}\tilde{q}'} [G_0^{\alpha}(\tilde{q})]^{-1} - V_{\tilde{q}-\tilde{q}'}^{\alpha} \right] [\hat{G}^{\alpha}]_{\tilde{q}'\tilde{q}''} = \delta_{\tilde{q}\tilde{q}''} \quad , \tag{5.11}$$

where $[G_0^{\alpha}(\tilde{q})]^{-1} = \mathrm{i}\tilde{\omega}_n - \boldsymbol{v}^{\alpha} \cdot \boldsymbol{q}$, see Eq.(4.10). The important point is now that Eq.(5.11) is first order and linear, and can therefore be solved exactly by means of a trivial generalization of a method due to Schwinger [5.1]. Defining

$$\mathcal{G}^{\alpha}(\boldsymbol{r}, \boldsymbol{r}', \tau, \tau') = \frac{1}{\beta V} \sum_{\tilde{q}\tilde{q}'} \mathrm{e}^{\mathrm{i}(\boldsymbol{q}\cdot\boldsymbol{r} - \tilde{\omega}_n \tau)} \mathrm{e}^{-\mathrm{i}(\boldsymbol{q}'\cdot\boldsymbol{r}' - \tilde{\omega}_{n'} \tau')} [\hat{G}^{\alpha}]_{\tilde{q}\tilde{q}'} \quad , \tag{5.12}$$

$$V^{\alpha}(\boldsymbol{r}, \tau) = \sum_{q} \mathrm{e}^{\mathrm{i}(\boldsymbol{q}\cdot\boldsymbol{r} - \omega_m \tau)} V_q^{\alpha} \quad , \tag{5.13}$$

it is easy to see that Eq.(5.11) is equivalent with

$$\left[-\partial_{\tau} + \mathrm{i}\boldsymbol{v}^{\alpha} \cdot \nabla_{\boldsymbol{r}} - V^{\alpha}(\boldsymbol{r}, \tau) \right] \mathcal{G}^{\alpha}(\boldsymbol{r}, \boldsymbol{r}', \tau, \tau') = \delta(\boldsymbol{r} - \boldsymbol{r}') \delta^{*}(\tau - \tau') \; , \tag{5.14}$$

where

$$\delta^{*}(\tau - \tau') = \frac{1}{\beta} \sum_{n} \mathrm{e}^{-\mathrm{i}\tilde{\omega}_n (\tau - \tau')} \tag{5.15}$$

is the antiperiodic δ-function. Note that the Fourier transformation in Eq.(5.12) involves fermionic Matsubara frequencies, because $\mathcal{G}^{\alpha}(\boldsymbol{r}, \boldsymbol{r}', \tau, \tau')$ has to satisfy the Kubo-Martin-Schwinger (KMS) boundary condition [1.2, 5.2]

$$\mathcal{G}^{\alpha}(\boldsymbol{r}, \boldsymbol{r}', \tau + \beta, \tau') = \mathcal{G}^{\alpha}(\boldsymbol{r}, \boldsymbol{r}', \tau, \tau' + \beta) = -\mathcal{G}^{\alpha}(\boldsymbol{r}, \boldsymbol{r}', \tau, \tau') \quad . \tag{5.16}$$

[2] As will be discussed in Sect. 5.2 and in more detail in Chap. 10, in $d > 1$ the linearization of the energy dispersion is not always a good approximation, because in $d > 1$ the condition $\boldsymbol{v}^{\alpha} \cdot \boldsymbol{q} = 0$ defines hyper-planes in momentum space on which the leading term in the expansion of $\xi_{\boldsymbol{q}}^{\alpha}$ is quadratic in $\boldsymbol{q}$. Linearization is only allowed if the contribution from these hyper-planes to the physical quantity of interest is negligible. Whether this is really the case depends also on the nature of the interaction. For example, in physically relevant models of transverse gauge fields that couple to the electronic current density (to be discussed in Chap. 10) the linearization is *not* allowed.

In contrast, $V^\alpha(\boldsymbol{r},\tau)$ is by definition a periodic function of τ, so that the sum in Eq.(5.13) involves bosonic Matsubara frequencies[3]. Following Schwinger [5.1], let us make the ansatz

$$\mathcal{G}^\alpha(\boldsymbol{r},\boldsymbol{r}',\tau,\tau') = G_0^\alpha(\boldsymbol{r}-\boldsymbol{r}',\tau-\tau')\mathrm{e}^{\Phi^\alpha(\boldsymbol{r},\tau)-\Phi^\alpha(\boldsymbol{r}',\tau')} \quad , \tag{5.17}$$

where $G_0^\alpha(\boldsymbol{r}-\boldsymbol{r}',\tau-\tau')$ satisfies

$$[-\partial_\tau + \mathrm{i}\boldsymbol{v}^\alpha\cdot\nabla_{\boldsymbol{r}}]\, G_0^\alpha(\boldsymbol{r}-\boldsymbol{r}',\tau-\tau') = \delta(\boldsymbol{r}-\boldsymbol{r}')\delta^*(\tau-\tau') \quad . \tag{5.18}$$

To take the KMS boundary condition (5.16) into account, we require that $G_0^\alpha(\boldsymbol{r}-\boldsymbol{r}',\tau-\tau')$ should be antiperiodic in τ and τ', while $\Phi^\alpha(\boldsymbol{r},\tau)$ should be a *periodic* function of τ,

$$\Phi^\alpha(\boldsymbol{r},\tau+\beta) = \Phi^\alpha(\boldsymbol{r},\tau) \quad . \tag{5.19}$$

Substituting Eq.(5.17) into Eq.(5.14), it is easy to show that

$$\begin{aligned}&[-\partial_\tau + \mathrm{i}\boldsymbol{v}^\alpha\cdot\nabla_{\boldsymbol{r}} - V^\alpha(\boldsymbol{r},\tau)]\,\mathcal{G}^\alpha(\boldsymbol{r},\boldsymbol{r}',\tau,\tau') = \delta(\boldsymbol{r}-\boldsymbol{r}')\delta^*(\tau-\tau')\\ &\quad + \mathcal{G}^\alpha(\boldsymbol{r},\boldsymbol{r}',\tau,\tau')\left\{[-\partial_\tau + \mathrm{i}\boldsymbol{v}^\alpha\cdot\nabla_{\boldsymbol{r}}]\,\Phi^\alpha(\boldsymbol{r},\tau) - V^\alpha(\boldsymbol{r},\tau)\right\} \quad .\end{aligned} \tag{5.20}$$

Comparing Eq.(5.20) with Eq.(5.14), we see that our ansatz is consistent provided $\Phi^\alpha(\boldsymbol{r},\tau)$ satisfies

$$[-\partial_\tau + \mathrm{i}\boldsymbol{v}^\alpha\cdot\nabla_{\boldsymbol{r}}]\,\Phi^\alpha(\boldsymbol{r},\tau) = V^\alpha(\boldsymbol{r},\tau) \quad . \tag{5.21}$$

Eqs.(5.18) and (5.21) are first order linear partial differential equations. The solution with the correct boundary condition is easily obtained via Fourier transformation,

$$G_0^\alpha(\boldsymbol{r},\tau) = \frac{1}{\beta V}\sum_{\tilde{q}} \frac{\mathrm{e}^{\mathrm{i}(\boldsymbol{q}\cdot\boldsymbol{r}-\tilde{\omega}_n\tau)}}{\mathrm{i}\tilde{\omega}_n - \boldsymbol{v}^\alpha\cdot\boldsymbol{q}} \quad , \tag{5.22}$$

$$\Phi^\alpha(\boldsymbol{r},\tau) = \sum_{q} \frac{\mathrm{e}^{\mathrm{i}(\boldsymbol{q}\cdot\boldsymbol{r}-\omega_m\tau)}}{\mathrm{i}\omega_m - \boldsymbol{v}^\alpha\cdot\boldsymbol{q}} V_q^\alpha \quad . \tag{5.23}$$

Let us emphasize again that the Matsubara sum in Eq.(5.23) involves bosonic frequencies because we have to satisfy the boundary condition (5.19). Having determined $G_0^\alpha(\boldsymbol{r},\tau)$ and $\Phi^\alpha(\boldsymbol{r},\tau)$, the diagonal blocks $(\hat{G}^\alpha)^{-1}$ have been inverted, so that $\hat{G}^\alpha$ is known as functional of the ϕ^α-field. The diagonal elements are explicitly given by

[3] The $q = 0$ component of the interaction requires a special treatment, and should be excluded from the q-sum in Eq.(5.13). Formally this condition can be taken into account by setting $\phi_{q=0}^\alpha = 0$, so that the $q = 0$ term in the sum (5.13) (as well as in all subsequent q-sums in this chapter) is automatically excluded. Note that this is equivalent with $\int \mathrm{d}\boldsymbol{r} \int_0^\beta \mathrm{d}\tau V^\alpha(\boldsymbol{r},\tau) = 0$. Any finite value of this integral can be absorbed into a redefinition of the chemical potential, which has disappeared in Eq.(5.14), because by assumption we have linearized the energy dispersion at the true chemical potential. See also the footnote after Eq.(4.2) in Chap. 4.

$$[\hat{G}^{\alpha}]_{kk} = \frac{1}{\beta V} \int \mathrm{d}\boldsymbol{r} \int \mathrm{d}\boldsymbol{r}' \int_0^{\beta} \mathrm{d}\tau \int_0^{\beta} \mathrm{d}\tau' \mathrm{e}^{-\mathrm{i}[(\boldsymbol{k}-\boldsymbol{k}^{\alpha})\cdot(\boldsymbol{r}-\boldsymbol{r}')-\tilde{\omega}_n(\tau-\tau')]}$$
$$\times G_0^{\alpha}(\boldsymbol{r}-\boldsymbol{r}', \tau-\tau') \exp\left[\frac{\mathrm{i}}{\beta}\sum_{\boldsymbol{q}} \frac{\mathrm{e}^{\mathrm{i}(\boldsymbol{q}\cdot\boldsymbol{r}-\omega_m\tau)} - \mathrm{e}^{\mathrm{i}(\boldsymbol{q}\cdot\boldsymbol{r}'-\omega_m\tau')}}{\mathrm{i}\omega_m - \boldsymbol{v}^{\alpha}\cdot\boldsymbol{q}} \phi_q^{\alpha}\right] . \quad (5.24)$$

5.1.2 Gaussian averaging: calculation of the Debye-Waller

This is the easy part of the calculation, because we have to average an exponential of ϕ^{α} with respect to a Gaussian probability distribution. This yields, of course, a Debye-Waller factor!

Combining Eqs.(5.1), (5.10) and (5.24), and using the fact that averaging restores translational invariance in space and time, we conclude that the interacting Matsubara Green's function is given by

$$G(k) = \sum_{\alpha} \Theta^{\alpha}(\boldsymbol{k}) \int \mathrm{d}\boldsymbol{r} \int_0^{\beta} \mathrm{d}\tau \mathrm{e}^{-\mathrm{i}[(\boldsymbol{k}-\boldsymbol{k}^{\alpha})\cdot\boldsymbol{r}-\tilde{\omega}_n\tau]}$$
$$\times G_0^{\alpha}(\boldsymbol{r}, \tau) \left\langle \mathrm{e}^{\Phi^{\alpha}(\boldsymbol{r},\tau)-\Phi^{\alpha}(0,0)} \right\rangle_{S_{\mathrm{eff},2}} . \quad (5.25)$$

Using Eqs.(3.31) and (5.23), we may write

$$\Phi^{\alpha}(\boldsymbol{r}, \tau) - \Phi^{\alpha}(0, 0) = \sum_{q} \mathcal{J}_{-q}^{\alpha}(\boldsymbol{r}, \tau)\phi_q^{\alpha} , \quad (5.26)$$

with

$$\mathcal{J}_q^{\alpha}(\boldsymbol{r}, \tau) = \frac{\mathrm{i}}{\beta}\left[\frac{1-\mathrm{e}^{-\mathrm{i}(\boldsymbol{q}\cdot\boldsymbol{r}-\omega_m\tau)}}{\mathrm{i}\omega_m - \boldsymbol{v}^{\alpha}\cdot\boldsymbol{q}}\right] . \quad (5.27)$$

The problem of calculating the interacting Green's function is now reduced to a trivial Gaussian integration, which simply yields the usual *Debye-Waller factor*,

$$\left\langle \mathrm{e}^{\Phi^{\alpha}(\boldsymbol{r},\tau)-\Phi^{\alpha}(0,0)} \right\rangle_{S_{\mathrm{eff},2}} = \left\langle \mathrm{e}^{\sum_q \mathcal{J}_{-q}^{\alpha}(\boldsymbol{r},\tau)\phi_q^{\alpha}} \right\rangle_{S_{\mathrm{eff},2}}$$
$$= \exp\left[\frac{1}{2}\sum_q \langle \phi_q^{\alpha}\phi_{-q}^{\alpha}\rangle_{S_{\mathrm{eff},2}} \mathcal{J}_{-q}^{\alpha}(\boldsymbol{r},\tau)\mathcal{J}_q^{\alpha}(\boldsymbol{r},\tau)\right] = \mathrm{e}^{Q^{\alpha}(\boldsymbol{r},\tau)} , \quad (5.28)$$

with

$$Q^{\alpha}(\boldsymbol{r}, \tau) = \frac{\beta}{2V}\sum_q [\underline{f}_q^{\mathrm{RPA}}]^{\alpha\alpha} \mathcal{J}_{-q}^{\alpha}(\boldsymbol{r},\tau)\mathcal{J}_q^{\alpha}(\boldsymbol{r},\tau) . \quad (5.29)$$

We have used the fact that the Gaussian propagator of the ϕ^{α}-field is according to Eq.(4.32) proportional to the RPA interaction. For consistency, in Eq.(5.29) the polarization contribution to $[\underline{f}_q^{\mathrm{RPA}}]^{\alpha\alpha}$ should be approximated by its leading long-wavelength limit given in Eq.(4.24), because in deriving Eq.(5.28) we have neglected momentum transfer between different sectors (i.e. the around-the-corner processes represented by the black triangles in Fig. 5.2(b)). Using

$$\mathcal{J}^{\alpha}_{-q}(\boldsymbol{r},\tau)\mathcal{J}^{\alpha}_{q}(\boldsymbol{r},\tau) = \frac{2}{\beta^2}\frac{1-\cos(\boldsymbol{q}\cdot\boldsymbol{r}-\omega_m\tau)}{(\mathrm{i}\omega_m - \boldsymbol{v}^{\alpha}\cdot\boldsymbol{q})^2} \quad , \tag{5.30}$$

we conclude that

$$Q^{\alpha}(\boldsymbol{r},\tau) = R^{\alpha} - S^{\alpha}(\boldsymbol{r},\tau) \quad , \tag{5.31}$$

with

$$R^{\alpha} = \frac{1}{\beta V}\sum_{q}\frac{f_q^{\mathrm{RPA},\alpha}}{(\mathrm{i}\omega_m - \boldsymbol{v}^{\alpha}\cdot\boldsymbol{q})^2} = S^{\alpha}(0,0) \quad , \tag{5.32}$$

$$S^{\alpha}(\boldsymbol{r},\tau) = \frac{1}{\beta V}\sum_{q}\frac{f_q^{\mathrm{RPA},\alpha}\cos(\boldsymbol{q}\cdot\boldsymbol{r}-\omega_m\tau)}{(\mathrm{i}\omega_m - \boldsymbol{v}^{\alpha}\cdot\boldsymbol{q})^2} \quad . \tag{5.33}$$

Here $f_q^{\mathrm{RPA},\alpha}$ is the diagonal element of the RPA interaction matrix defined in Eq.(4.33),

$$f_q^{\mathrm{RPA},\alpha} \equiv [\underline{f}_q^{\mathrm{RPA}}]^{\alpha\alpha} = \left[\underline{f}_q\left[1 + \underline{\Pi}_0(q)\underline{f}_q\right]^{-1}\right]^{\alpha\alpha} \quad . \tag{5.34}$$

An important special case is a patch-independent bare interaction, i.e. $[\underline{f}_q]^{\alpha\alpha'} = f_q$ for all α and α'. From Eq.(4.35) we know that in this case $f_q^{\mathrm{RPA},\alpha}$ can be identified with the usual RPA interaction,

$$f_q^{\mathrm{RPA},\alpha} = f_q^{\mathrm{RPA}} \equiv \frac{f_q}{1 + f_q \Pi_0(q)} \quad , \quad \text{if} \quad [\underline{f}_q]^{\alpha\alpha'} = f_q \quad . \tag{5.35}$$

In summary, the averaged diagonal blocks are given by

$$\left\langle [\hat{G}^{\alpha}]_{\boldsymbol{k}\boldsymbol{k}} \right\rangle_{S_{\mathrm{eff},2}} = \int \mathrm{d}\boldsymbol{r} \int_0^{\beta} \mathrm{d}\tau e^{-\mathrm{i}[(\boldsymbol{k}-\boldsymbol{k}^{\alpha})\cdot\boldsymbol{r} - \tilde{\omega}_n\tau]} G^{\alpha}(\boldsymbol{r},\tau) \quad , \tag{5.36}$$

with

$$G^{\alpha}(\boldsymbol{r},\tau) = G_0^{\alpha}(\boldsymbol{r},\tau) e^{Q^{\alpha}(\boldsymbol{r},\tau)} \quad . \tag{5.37}$$

From Eqs.(5.1) and (5.10) we finally obtain for the Matsubara Green's function of the interacting many-body system

$$G(k) = \sum_{\alpha} \Theta^{\alpha}(\boldsymbol{k}) G^{\alpha}(\boldsymbol{k}-\boldsymbol{k}^{\alpha}, \mathrm{i}\tilde{\omega}_n) \quad , \tag{5.38}$$

where

$$G^{\alpha}(\tilde{q}) \equiv G^{\alpha}(\boldsymbol{q},\mathrm{i}\tilde{\omega}_n) = \int \mathrm{d}\boldsymbol{r} \int_0^{\beta} \mathrm{d}\tau e^{-\mathrm{i}(\boldsymbol{q}\cdot\boldsymbol{r} - \tilde{\omega}_n\tau)} G^{\alpha}(\boldsymbol{r},\tau) \quad . \tag{5.39}$$

Shifting in Eq.(5.38) $\boldsymbol{k} = \boldsymbol{k}^{\alpha'} + \boldsymbol{q}$ and choosing $|\boldsymbol{q}|$ small compared with the cutoffs Λ and λ that determine the size of the sector $K^{\alpha}_{\Lambda,\lambda}$, it is easy to see that only the term $\alpha' = \alpha$ in the sum (5.38) contributes, so that (after renaming again $\alpha' \to \alpha$)

$$G(\boldsymbol{k}^{\alpha} + \boldsymbol{q}, \mathrm{i}\tilde{\omega}_n) = G^{\alpha}(\boldsymbol{q}, \mathrm{i}\tilde{\omega}_n) \quad , \quad |\boldsymbol{q}| \ll \Lambda, \lambda \quad . \tag{5.40}$$

5.1.3 The Green's function in real space

The real space Green's function $G(\boldsymbol{r}, \tau)$ should not be confused with the sector Green's function $G^{\alpha}(\boldsymbol{r}, \tau)$ in Eq.(5.37). Here we derive the precise relation between these functions.

Given the exact Matsubara Green's function $G(k)$, we can use Eq.(3.8) to reconstruct the exact real space imaginary time Green's function $G(\boldsymbol{r}, \tau)$ by inverse Fourier transformation. Substituting Eqs.(5.38) and (5.39) into Eq.(3.8), we obtain

$$G(\boldsymbol{r}, \tau) = \sum_{\alpha} \int \mathrm{d}\boldsymbol{r}' \mathrm{e}^{\mathrm{i}\boldsymbol{k}^{\alpha} \cdot \boldsymbol{r}'} \frac{1}{V} \sum_{\boldsymbol{k}} \Theta^{\alpha}(\boldsymbol{k}) \mathrm{e}^{\mathrm{i}\boldsymbol{k} \cdot (\boldsymbol{r} - \boldsymbol{r}')} G^{\alpha}(\boldsymbol{r}', \tau) \quad . \tag{5.41}$$

At distances large compared with the inverse sector cutoffs Λ^{-1} and λ^{-1} we may approximate

$$\frac{1}{V} \sum_{\boldsymbol{k}} \Theta^{\alpha}(\boldsymbol{k}) \mathrm{e}^{\mathrm{i}\boldsymbol{k} \cdot (\boldsymbol{r} - \boldsymbol{r}')} \approx \frac{1}{V} \sum_{\boldsymbol{k}} \mathrm{e}^{\mathrm{i}\boldsymbol{k} \cdot (\boldsymbol{r} - \boldsymbol{r}')} = \delta(\boldsymbol{r} - \boldsymbol{r}') \quad , \tag{5.42}$$

so that Eq.(5.41) reduces to

$$G(\boldsymbol{r}, \tau) = \sum_{\alpha} \mathrm{e}^{\mathrm{i}\boldsymbol{k}^{\alpha} \cdot \boldsymbol{r}} G^{\alpha}(\boldsymbol{r}, \tau) \quad , \tag{5.43}$$

which is the real space imaginary time version of Eq.(5.38).

To see the role of the cutoffs more clearly, it is instructive to calculate the non-interacting sector Green's function $G^{\alpha}_{0}(\boldsymbol{r}, \tau)$ defined in Eq.(5.22). Performing the fermionic Matsubara sum we obtain

$$G^{\alpha}_{0}(\boldsymbol{r}, \tau) = \frac{1}{V} \sum_{\boldsymbol{q}} \mathrm{e}^{\mathrm{i}\boldsymbol{q} \cdot \boldsymbol{r}} G^{\alpha}_{0}(\boldsymbol{q}, \tau) \quad , \tag{5.44}$$

with

$$G^{\alpha}_{0}(\boldsymbol{q}, \tau) = \mathrm{e}^{-\boldsymbol{v}^{\alpha} \cdot \boldsymbol{q} \tau} \left[f(\boldsymbol{v}^{\alpha} \cdot \boldsymbol{q}) \Theta(-\tau + 0^{+}) - f(-\boldsymbol{v}^{\alpha} \cdot \boldsymbol{q}) \Theta(\tau - 0^{+}) \right] \quad . \tag{5.45}$$

Because $G^{\alpha}_{0}(\boldsymbol{q}, \tau)$ depends on $\boldsymbol{q}$ only via the component $\boldsymbol{v}^{\alpha} \cdot \boldsymbol{q}$, it is convenient to choose the orientation of the local coordinate system attached to sector $K^{\alpha}_{\Lambda,\lambda}$ such that one of its axes matches the direction $\hat{\boldsymbol{v}}^{\alpha} = \boldsymbol{v}^{\alpha} / |\boldsymbol{v}^{\alpha}|$ of the local Fermi velocity. In this coordinate system we have the decomposition

$$q = q_{\parallel}^{\alpha} \hat{v}^{\alpha} + q_{\perp}^{\alpha} \; , \; q_{\parallel}^{\alpha} = q \cdot \hat{v}^{\alpha} \; , \; q_{\perp}^{\alpha} \cdot \hat{v}^{\alpha} = 0 \; , \tag{5.46}$$

$$r = r_{\parallel}^{\alpha} \hat{v}^{\alpha} + r_{\perp}^{\alpha} \; , \; r_{\parallel}^{\alpha} = r \cdot \hat{v}^{\alpha} \; , \; r_{\perp}^{\alpha} \cdot \hat{v}^{\alpha} = 0 \; . \tag{5.47}$$

For $\beta \to \infty$ and $V \to \infty$ we obtain then after a simple calculation

$$G_0^{\alpha}(r, \tau) = \delta^{(d-1)}(r_{\perp}^{\alpha}) \left(\frac{-\mathrm{i}}{2\pi} \right) \frac{1}{r_{\parallel}^{\alpha} + \mathrm{i}|v^{\alpha}|\tau} \; , \tag{5.48}$$

with

$$\delta^{(d-1)}(r_{\perp}^{\alpha}) = \int \frac{\mathrm{d}r_{\perp}^{\alpha}}{(2\pi)^{d-1}} \mathrm{e}^{\mathrm{i}q_{\perp}^{\alpha} \cdot r_{\perp}^{\alpha}} \; , \tag{5.49}$$

where the integral is over the $d-1$ components of r that are perpendicular to v^{α}. In deriving Eq.(5.48) we have not been very careful about cutoffs. In order not to over-count the degrees of freedom, the q-summations should be restricted to the sectors $K_{\Lambda,\lambda}^{\alpha}$. Hence, there is a hidden cutoff function $\Theta^{\alpha}(k^{\alpha} + q)$ in all q-sums, which we have not explicitly written out. However, we may ignore this cutoff function as long as we are interested in length scales

$$|r_{\perp}^{\alpha}| \gg \Lambda^{-1} \; , \; |r_{\parallel}^{\alpha}| \gg \lambda^{-1} \; , \tag{5.50}$$

because the oscillating exponential in Eq.(5.44) cuts off the q-summations before the boundaries of the sectors are reached. It should be kept in mind, however, that Eq.(5.48) is only correct if the conditions (5.50) are satisfied. More precisely, in a finite system of volume $V = L^d$ the δ-function in Eq.(5.48) should be replaced by the cutoff-dependent function

$$\delta_{\Lambda}^{(d-1)}(r_{\perp}^{\alpha}) = \frac{1}{L^{d-1}} \sum_{q_{\perp}^{\alpha}} \Theta^{\alpha}(k^{\alpha} + q_{\perp}^{\alpha}) \mathrm{e}^{\mathrm{i}q_{\perp}^{\alpha} \cdot r_{\perp}^{\alpha}} \; . \tag{5.51}$$

We conclude that for length scales $|r_{\perp}^{\alpha}| \gg \Lambda^{-1}$ we may replace $r \to r_{\parallel}^{\alpha} \hat{v}^{\alpha}$ in the argument of the Debye-Waller factor, so that Eq.(5.37) becomes

$$G^{\alpha}(r, \tau) = G_0^{\alpha}(r, \tau) \mathrm{e}^{Q^{\alpha}(r_{\parallel}^{\alpha} \hat{v}^{\alpha}, \tau)} \; . \tag{5.52}$$

From Eq.(5.43) we obtain then for the real space Green's function of the interacting system,

$$G(r, \tau) = \frac{-\mathrm{i}}{2\pi} \sum_{\alpha} \delta^{(d-1)}(r_{\perp}^{\alpha}) \frac{\exp\left[\mathrm{i}k^{\alpha} \cdot r + Q^{\alpha}(r_{\parallel}^{\alpha} \hat{v}^{\alpha}, \tau) \right]}{r_{\parallel}^{\alpha} + \mathrm{i}|v^{\alpha}|\tau} \; . \tag{5.53}$$

Note that this expression has units of V^{-1}, as expected from a real space single-particle Green's function in d dimensions.

5.1.4 The underlying asymptotic Ward identity

Our bosonization formula (5.37) for the Green's function is the result of an infinite resummation of the perturbation series, but it is not clear which type of diagrams have been summed. In this section we shall clarify this point. We first derive from Eq.(5.37) an integral equation, which is exactly equivalent with the integral equation derived by Castellani, Di Castro and Metzner [1.50]. We then combine the integral equation with the Dyson equation to obtain a Ward identity. In this way we see the relation between bosonization and diagrammatic perturbation theory.

The integral equation

Let us apply the differential operator $-\partial_\tau + \mathrm{i}\boldsymbol{v}^\alpha \cdot \nabla_{\boldsymbol{r}}$ to the bosonization result (5.37) for the sector Green's function $G^\alpha(\boldsymbol{r},\tau)$. Using the fact that according to Eqs.(5.18) and (5.15) the application of this operator to $G_0^\alpha(\boldsymbol{r},\tau)$ generates as δ-function, it is easy to show that

$$[-\partial_\tau + \mathrm{i}\boldsymbol{v}^\alpha \cdot \nabla_{\boldsymbol{r}} + X^\alpha(\boldsymbol{r}-\boldsymbol{r}',\tau-\tau')]\, G^\alpha(\boldsymbol{r}-\boldsymbol{r}',\tau-\tau') = \delta(\boldsymbol{r}-\boldsymbol{r}')\delta^*(\tau-\tau') \quad , \tag{5.54}$$

with

$$X^\alpha(\boldsymbol{r}-\boldsymbol{r}',\tau-\tau') = -[-\partial_\tau + \mathrm{i}\boldsymbol{v}^\alpha \cdot \nabla_{\boldsymbol{r}}] Q^\alpha(\boldsymbol{r}-\boldsymbol{r}',\tau-\tau') \quad . \tag{5.55}$$

From the explicit expression for $Q^\alpha(\boldsymbol{r},\tau)$ given in Eqs.(5.31)–(5.33) we find that the function $X^\alpha(\boldsymbol{r},\tau)$ has the Fourier expansion

$$X^\alpha(\boldsymbol{r},\tau) = \frac{1}{\beta V}\sum_q \mathrm{e}^{\mathrm{i}(\boldsymbol{q}\cdot\boldsymbol{r}-\omega_m\tau)} X_q^\alpha \quad , \tag{5.56}$$

with Fourier coefficients given by

$$X_q^\alpha = \frac{f_q^{\mathrm{RPA},\alpha}}{\mathrm{i}\omega_m - \boldsymbol{v}^\alpha\cdot\boldsymbol{q}} \quad . \tag{5.57}$$

In Fourier space Eq.(5.54) becomes

$$[\mathrm{i}\tilde{\omega}_n - \boldsymbol{v}^\alpha\cdot\boldsymbol{q}]G^\alpha(\tilde{q}) + \frac{1}{\beta V}\sum_{\tilde{q}'} X^\alpha_{\tilde{q}-\tilde{q}'} G^\alpha(\tilde{q}') = 1 \quad , \tag{5.58}$$

or equivalently

$$[\mathrm{i}\tilde{\omega}_n - \boldsymbol{v}^\alpha\cdot\boldsymbol{q}]G^\alpha(\boldsymbol{q},\mathrm{i}\tilde{\omega}_n) = 1 - \frac{1}{\beta V}\sum_{\boldsymbol{q}',n'} \frac{f^{\mathrm{RPA},\alpha}_{\boldsymbol{q}-\boldsymbol{q}',\mathrm{i}\omega_{n-n'}}}{\mathrm{i}\omega_{n-n'} - \boldsymbol{v}^\alpha\cdot(\boldsymbol{q}-\boldsymbol{q}')} G^\alpha(\boldsymbol{q}',\mathrm{i}\tilde{\omega}_{n'}) \quad . \tag{5.59}$$

Because the difference between two fermionic Matsubara frequencies is a bosonic one, the kernel $X^\alpha_{\tilde{q}-\tilde{q}'}$ in Eq.(5.58) depends on bosonic frequencies. In the limit $\beta \to \infty$ Eq.(5.59) is equivalent with the integral equation given in Eq.(13) of the work [1.50] by Castellani, Di Castro and Metzner. Our bosonization approach maps the solution of Eq.(5.59) onto the standard problem of solving a linear partial differential equation (Eq.(5.20)) and calculating a Debye-Waller factor in a Gaussian integral. The solution for the Green's function is given in Eqs.(5.37)–(5.39), with the Debye-Waller factor given in

Eqs.(5.31)–(5.33). On the other hand, Castellani, Di Castro and Metzner do not directly solve Eq.(5.59) but *first* perform an angular averaging operation on this integral equation and *then* solve the resulting averaged equation. Although in general the operations of averaging and solving integral equations do not commute (i.e. the solution of the averaged integral equation is not necessarily identical with the average of the solution of the integral equation), in the particular case of interest the final result seems to be equivalent, at least up to re-definitions of cutoffs.

The Ward identity

In modern many-body theory it is sometimes convenient [1.5, 1.6] to define so-called skeleton diagrams in order to exhibit the structure of the perturbation series more clearly. The skeleton diagram for the exact self-energy is shown in Fig. 5.3. In the Matsubara formalism this diagram represents the following

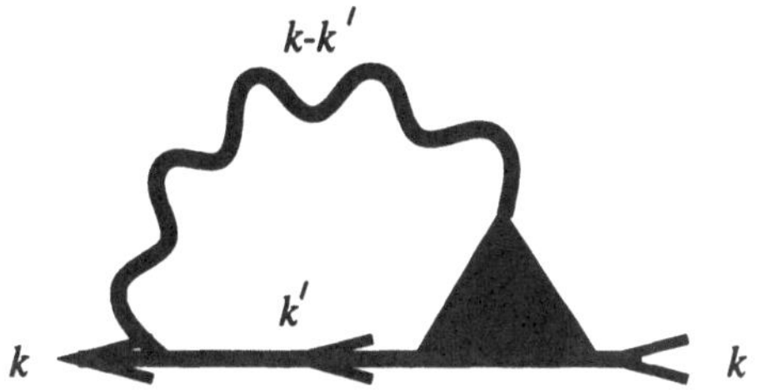

Fig. 5.3. Skeleton diagram for the irreducible self-energy. The thick wavy line denotes the exact screened effective interaction f^*_q, the shaded triangle is the exact three-legged vertex $\Lambda(k; k - k')$, and the solid line with arrow is the exact Green's function.

expression,

$$\Sigma(k) = -\frac{1}{\beta V} \sum_{k'} f^*_{k-k'} \Lambda(k; k - k') G(k') \quad . \tag{5.60}$$

The exact effective interaction f^*_q is related to the bare interaction f_q via the dielectric function, $f^*_q = f_q/\epsilon(q)$, which in turn can be expressed in terms of the exact proper polarization via $\epsilon(q) = 1 + f_q \Pi_*(q)$, see Eqs.(2.50)–(2.52). By definition, the vertex function $\Lambda(k; q)$ is the sum of all diagrams with three external ends corresponding to two solid lines and a single interaction line. To lowest non-trivial order we have $\Lambda(k; q) \approx 1 + \Lambda^{(1)}(k; q)$, with $\Lambda^{(1)}(k; q)$ given in Eq.(4.105). Because $G(k')$ on the right-hand side of Eq.(5.60) depends again on $\Sigma(k')$ via the Dyson equation, Eq.(5.60) is a complicated integral equation, which can only be solved approximately. Moreover, the formal kernel $f^*_{k-k'} \Lambda(k; k - k')$ of this integral equation is again a functional of $G(k)$, so that it cannot be calculated exactly unless the entire perturbation series has been summed. In practice one therefore replaces the effective interaction $f^*_{k-k'}$ and the vertex $\Lambda(k; k - k')$ by some "reasonable" approximation.

For better comparison with the self-energy calculated within our bosonization approach, let us shift again $\boldsymbol{k} = \boldsymbol{k}^{\alpha} + \boldsymbol{q}$ and $\boldsymbol{k}' = \boldsymbol{k}^{\alpha} + \boldsymbol{q}'$, so that wave-vectors are measured with respect to the local coordinate system with origin in sector $K^{\alpha}_{\Lambda,\lambda}$. Defining

$$G(\boldsymbol{k}^\alpha + \boldsymbol{q}, \mathrm{i}\tilde{\omega}_n) = G^\alpha(\tilde{q}) \quad , \tag{5.61}$$

$$\Sigma(\boldsymbol{k}^\alpha + \boldsymbol{q}, \mathrm{i}\tilde{\omega}_n) = \Sigma^\alpha(\tilde{q}) \quad , \tag{5.62}$$

$$\Lambda(\boldsymbol{k}^\alpha + \boldsymbol{q}, \mathrm{i}\tilde{\omega}_n; \boldsymbol{q} - \boldsymbol{q}', \mathrm{i}\omega_{n-n'}) = \Lambda^\alpha(\tilde{q}; \tilde{q} - \tilde{q}') \quad , \tag{5.63}$$

the skeleton equation (5.60) reads

$$\Sigma^\alpha(\tilde{q}) = -\frac{1}{\beta V} \sum_{\tilde{q}'} f^*_{\tilde{q}-\tilde{q}'} \Lambda^\alpha(\tilde{q}; \tilde{q} - \tilde{q}') G^\alpha(\tilde{q}') \quad , \tag{5.64}$$

while the Dyson equation can be written as

$$[G^\alpha(\tilde{q})]^{-1} = [G_0^\alpha(\tilde{q})]^{-1} - \Sigma^\alpha(\tilde{q}) \quad . \tag{5.65}$$

Let us now determine the skeleton parameters that correspond to our bosonization result for the Green's function. Starting point is the integral equation (5.58). Keeping in mind that after linearization we may write $\mathrm{i}\tilde{\omega}_n - \boldsymbol{v}^\alpha \cdot \boldsymbol{q} = [G_0^\alpha(\tilde{q})]^{-1}$ and dividing both sides of Eq.(5.58) by $G^\alpha(\tilde{q})$, we obtain

$$[G^\alpha(\tilde{q})]^{-1} = [G_0^\alpha(\tilde{q})]^{-1} + \frac{1}{\beta V} \sum_{\tilde{q}'} \frac{X^\alpha_{\tilde{q}-\tilde{q}'}}{G^\alpha(\tilde{q})} G^\alpha(\tilde{q}') \quad . \tag{5.66}$$

Comparing this with Eq.(5.65), we conclude that in our bosonization approach the self-energy satisfies

$$\Sigma^\alpha(\tilde{q}) = -\frac{1}{\beta V} \sum_{\tilde{q}'} \frac{X^\alpha_{\tilde{q}-\tilde{q}'}}{G^\alpha(\tilde{q})} G^\alpha(\tilde{q}') \quad . \tag{5.67}$$

From Eqs.(5.64) and (5.67) we finally obtain

$$f^*_{\tilde{q}-\tilde{q}'} \Lambda^\alpha(\tilde{q}; \tilde{q} - \tilde{q}') = \frac{X^\alpha_{\tilde{q}-\tilde{q}'}}{G^\alpha(\tilde{q})} = \frac{f^{\mathrm{RPA},\alpha}_{\tilde{q}-\tilde{q}'}}{[\mathrm{i}\omega_{n-n'} - \boldsymbol{v}^\alpha \cdot (\boldsymbol{q} - \boldsymbol{q}')] G^\alpha(\tilde{q})} \quad . \tag{5.68}$$

Note that in the skeleton equation (5.60) it is assumed that the bare interaction depends only on the momentum-transfer. From Eq.(4.35) we know that in this case $f_q^{\mathrm{RPA},\alpha} = f_q^{\mathrm{RPA}}$, the usual RPA interaction. Then we see from Eq.(5.68) that the approximations inherent in our bosonization approach amount to replacing the exact effective interaction f_q^* by the RPA interaction f_q^{RPA}, and setting the vertex function equal to

$$\Lambda^\alpha(\tilde{q}; \tilde{q} - \tilde{q}') = \frac{1}{[\mathrm{i}\omega_{n-n'} - \boldsymbol{v}^\alpha \cdot (\boldsymbol{q} - \boldsymbol{q}')] G^\alpha(\tilde{q})} \quad , \tag{5.69}$$

which is equivalent with

$$\left[\frac{1}{G_0^\alpha(\tilde{q})} - \frac{1}{G_0^\alpha(\tilde{q}')} \right] \Lambda^\alpha(\tilde{q}; \tilde{q} - \tilde{q}') = \frac{1}{G^\alpha(\tilde{q})} \quad , \tag{5.70}$$

or, after shifting $\tilde{q} - \tilde{q}' \to q' \equiv [\boldsymbol{q}', \mathrm{i}\omega_{m'}]$,

$$[\mathrm{i}\omega_{m'} - \boldsymbol{v}^\alpha \cdot \boldsymbol{q}'] \Lambda^\alpha(\tilde{q}; q') = [G^\alpha(\tilde{q})]^{-1} \quad . \tag{5.71}$$

In terms of the symmetrized vertex function

$$\tilde{\Lambda}^{\alpha}(\tilde{q};\tilde{q}') = \Lambda^{\alpha}(\tilde{q};\tilde{q}-\tilde{q}') + \Lambda^{\alpha}(\tilde{q}';\tilde{q}'-\tilde{q}) \quad , \tag{5.72}$$

Eq.(5.70) can also be rewritten in the more symmetric form

$$\left[\frac{1}{G_0^{\alpha}(\tilde{q})} - \frac{1}{G_0^{\alpha}(\tilde{q}')}\right]\tilde{\Lambda}^{\alpha}(\tilde{q};\tilde{q}') = \left[\frac{1}{G^{\alpha}(\tilde{q})} - \frac{1}{G^{\alpha}(\tilde{q}')}\right] \quad . \tag{5.73}$$

The important point is that the right-hand side of Eq.(5.73) depends again on the exact Green's function. Such a relation between a vertex function and a Green's function is called a *Ward identity*. In the limit $\beta \to \infty$ Eq.(5.73) is equivalent with the Ward identity derived in [1.50]. Of course, in $d > 1$ or for non-linear energy dispersion this Ward identity is not exact. In Sect. 5.2 we shall develop a powerful method for calculating in a controlled and quantitative way the corrections neglected in Eq.(5.73).

In summary, although within our bosonization approach the dielectric function is approximated by the RPA expression, bosonization does not simply reproduce the usual RPA self-energy, because it sums in addition infinitely many other diagrams by means of a non-trivial Ward identity for the vertex function. The analytic expressions for these diagrams can be generated order by order in the RPA interaction by iterating the integral equation (5.59).

Finally, let us compare the skeleton equation (5.60) with the dynamically screened exchange diagram, the so-called GW approximation for the self-energy [5.3]. In this approximation the effective interaction f_q^* is approximated by the RPA interaction, just like in our bosonization approach. The crucial difference with bosonization is that vertex corrections are completely ignored within the GW approximation, so that one sets $\Lambda(k, k-k') \to 1$. Then Eq.(5.64) reduces to the simpler integral equation

$$\Sigma^{\alpha}(\tilde{q}) = -\frac{1}{\beta V}\sum_{\tilde{q}'} f^{\mathrm{RPA}}_{\tilde{q}-\tilde{q}'} G^{\alpha}(\tilde{q}') \quad , \quad \text{GW approximation} \quad . \tag{5.74}$$

If we replace the interacting Green's function on the right-hand side of Eq.(5.74) by the non-interacting one, we recover the lowest order self-energy correction given in Eq.(4.104). The self-consistent solution of Eq.(5.74) contains infinite orders in perturbation theory. It seems, however, that the only reason for ignoring vertex corrections is that one is unable to calculate them in a controlled way. As recently pointed out by Farid [5.4], the errors due to the omission of vertex corrections seem to be partially cancelled by the replacement $G \to G_0$ on the right-hand side of Eq.(5.74). In other words, non-self-consistent GW can be better than self-consistent GW! Evidently such an approximation cannot be systematic. On the other hand, for interactions that are dominated by forward scattering the bosonization approach uses the small parameter q_c/k_F to sum the dominant terms of the entire perturbation series.

5.1.5 The Fermi liquid renormalization factors Z^α and Z^α_{m}

We show how in a Fermi liquid the quasi-particle residue Z^α and the effective mass renormalization Z^α_{m} can be obtained from the Debye-Waller factor $Q^\alpha(\boldsymbol{r},\tau)$.

The quasi-particle residue Z^α

As shown in Chap. 2.2.3 (see Eq.(2.25)), the quasi-particle residue Z^α of a Fermi liquid can be identified from the discontinuity $\delta n^\alpha_{\boldsymbol{q}}$ of the momentum distribution at the Fermi surface. Hence, in order to relate our Debye-Waller factor $Q^\alpha(\boldsymbol{r},\tau)$ given in Eq.(5.31) to the quasi-particle residue, we simply have to calculate $\delta n^\alpha_{\boldsymbol{q}}$ from Eq.(5.37). Substituting Eqs.(5.37), (5.39) and (5.40) into Eq.(2.13) we obtain

$$n_{\boldsymbol{k}^\alpha+\boldsymbol{q}} = \int \mathrm{d}\boldsymbol{r} \mathrm{e}^{-\mathrm{i}\boldsymbol{q}\cdot\boldsymbol{r}} G^\alpha_0(\boldsymbol{r},0) \mathrm{e}^{Q^\alpha(\boldsymbol{r},0)} \quad , \tag{5.75}$$

so that the change $\delta n^\alpha_{\boldsymbol{q}}$ of the momentum distribution defined in Eq.(2.25) is given by

$$\delta n^\alpha_{\boldsymbol{q}} = 2\mathrm{i} \int \mathrm{d}\boldsymbol{r} \sin(\boldsymbol{q}\cdot\boldsymbol{r}) G^\alpha_0(\boldsymbol{r},0) \mathrm{e}^{Q^\alpha(\boldsymbol{r},0)} \quad . \tag{5.76}$$

From Eq.(5.48) we obtain for the non-interacting sector Green's function at equal times

$$G^\alpha_0(\boldsymbol{r},0) = \delta^{(d-1)}(\boldsymbol{r}^\alpha_\perp) \frac{-\mathrm{i}}{2\pi r^\alpha_\parallel} \quad , \tag{5.77}$$

so that

$$\delta n^\alpha_{\boldsymbol{q}} = \frac{2}{\pi} \int_0^\infty \mathrm{d}x \frac{\sin(q^\alpha_\parallel x)}{x} \mathrm{e}^{Q^\alpha(x\hat{\boldsymbol{v}}^\alpha,0)} \quad , \tag{5.78}$$

where we have renamed $r^\alpha_\parallel = x$. As discussed in Chap. 2.4.3, bosonization should lead to cutoff-independent results if the interaction is dominated by wave-vectors $|\boldsymbol{q}| \lesssim q_{\mathrm{c}} \ll \Lambda, \lambda$. Hence, in real space the bosonization result for the Green's function is accurate at length scales $x \gg q_{\mathrm{c}}^{-1}$. We therefore separate from Eq.(5.78) the non-universal short-distance regime,

$$\delta n^\alpha_{\boldsymbol{q}} = \frac{2}{\pi} \left[\int_0^{q_{\mathrm{c}}^{-1}} \mathrm{d}x \frac{\sin(q^\alpha_\parallel x)}{x} \mathrm{e}^{Q^\alpha(x\hat{\boldsymbol{v}}^\alpha,0)} + \int_{q_{\mathrm{c}}^{-1}}^\infty \mathrm{d}x \frac{\sin(q^\alpha_\parallel x)}{x} \mathrm{e}^{Q^\alpha(x\hat{\boldsymbol{v}}^\alpha,0)} \right] . \tag{5.79}$$

For $|q^\alpha_\parallel| \ll q_{\mathrm{c}}$ it is allowed to expand the sine-function in the first term. Evidently, this yields an analytic contribution to $\delta n^\alpha_{\boldsymbol{q}}$, which to leading order is proportional to $q^\alpha_\parallel / q_{\mathrm{c}}$. In a Fermi liquid this term is negligible compared with the contribution from the second term in Eq.(5.79), so that we obtain to leading order

$$\delta n^\alpha_{\boldsymbol{q}} \sim \mathrm{sgn}(q^\alpha_\parallel) \frac{2}{\pi} \int_{|q^\alpha_\parallel|/q_{\mathrm{c}}}^\infty \mathrm{d}x' \frac{\sin(x')}{x'} \mathrm{e}^{Q^\alpha(x'\hat{\boldsymbol{v}}^\alpha/|q^\alpha_\parallel|,0)} \quad , \tag{5.80}$$

where we have rescaled $x = x'/|q_{\parallel}^{\alpha}|$. For $q_{\parallel}^{\alpha} \to 0$ we may set $|q_{\parallel}^{\alpha}|/q_{\mathrm{c}} = 0$ in the lower limit and replace $Q^{\alpha}(x'\hat{\boldsymbol{v}}^{\alpha}/|q_{\parallel}^{\alpha}|, 0)$ by its asymptotic expansion for large $x'/|q_{\parallel}^{\alpha}|$. Assuming that the limit

$$\lim_{r_{\parallel}^{\alpha} \to \infty} Q^{\alpha}(r_{\parallel}^{\alpha}\hat{\boldsymbol{v}}^{\alpha}, 0) \equiv Q_{\infty}^{\alpha} \tag{5.81}$$

exists, we obtain for $q_{\parallel}^{\alpha} \to 0$

$$\delta n_{\boldsymbol{q}}^{\alpha} \sim \mathrm{sgn}(q_{\parallel}^{\alpha}) \frac{2}{\pi} \int_0^{\infty} \mathrm{d}x' \frac{\sin(x')}{x'} \mathrm{e}^{Q_{\infty}^{\alpha}} = \mathrm{sgn}(q_{\parallel}^{\alpha}) \mathrm{e}^{Q_{\infty}^{\alpha}} \quad . \tag{5.82}$$

But from Eqs.(5.31)–(5.33) we have $Q^{\alpha}(r_{\parallel}^{\alpha}\hat{\boldsymbol{v}}^{\alpha}, 0) = R^{\alpha} - S^{\alpha}(r_{\parallel}^{\alpha}\hat{\boldsymbol{v}}^{\alpha}, 0)$, where the constant term is simply given by $R^{\alpha} = S^{\alpha}(0,0)$. A sufficient condition for the existence of the limit Q_{∞}^{α} is the existence of R^{α}. Recall that according to Eq.(5.32) R^{α} is for $\beta, V \to \infty$ given by

$$R^{\alpha} = \int \frac{\mathrm{d}\boldsymbol{q}}{(2\pi)^d} \int_{-\infty}^{\infty} \frac{\mathrm{d}\omega}{2\pi} \frac{f_{\boldsymbol{q},\mathrm{i}\omega}^{\mathrm{RPA},\alpha}}{(\mathrm{i}\omega - \boldsymbol{v}^{\alpha} \cdot \boldsymbol{q})^2} \quad . \tag{5.83}$$

If this integral exists, then the Fourier integral theorem [5.5] implies that the integral

$$S^{\alpha}(r_{\parallel}^{\alpha}\hat{\boldsymbol{v}}^{\alpha}, 0) = \int \frac{\mathrm{d}\boldsymbol{q}}{(2\pi)^d} \int_{-\infty}^{\infty} \frac{\mathrm{d}\omega}{2\pi} \frac{f_{\boldsymbol{q},\mathrm{i}\omega}^{\mathrm{RPA},\alpha} \cos(r_{\parallel}^{\alpha}\hat{\boldsymbol{v}}^{\alpha} \cdot \boldsymbol{q})}{(\mathrm{i}\omega - \boldsymbol{v}^{\alpha} \cdot \boldsymbol{q})^2} \tag{5.84}$$

exists as well, *and vanishes*[4] *in the limit* $r_{\parallel}^{\alpha} \to \infty$. Hence, the finiteness of the integral in Eq.(5.83) implies that $Q_{\infty}^{\alpha} = R^{\alpha}$. In this case we obtain

$$\delta n_{\boldsymbol{q}}^{\alpha} \sim Z^{\alpha} \mathrm{sgn}(q_{\parallel}^{\alpha}) \quad , \quad q_{\parallel}^{\alpha} \to 0 \quad , \tag{5.85}$$

with the quasi-particle residue given by

$$Z^{\alpha} = \mathrm{e}^{R^{\alpha}} \quad . \tag{5.86}$$

Because we know that Z^{α} must be a real number between zero and unity, R^{α} should be real and negative. In Chap. 6.1 we shall show with the help of the dynamic structure factor that this is indeed the case.

[4] As shown in [5.5], a sufficient condition for the vanishing of the Fourier transform $G(\omega) = \int_{-\infty}^{\infty} \mathrm{d}\omega \mathrm{e}^{-\mathrm{i}\omega\tau} F(\tau)$ of a function $F(\tau)$ for $\omega \to \pm\infty$ is that $F(\tau)$ is (at least improperly) integrable on every finite interval, and that $\int_{-\infty}^{\infty} |F(\tau)| \mathrm{d}\tau < \infty$. In our case these conditions have to be satisfied by the function $F(q_{\parallel}^{\alpha}) = \int \mathrm{d}\boldsymbol{q}_{\perp}^{\alpha} \int_{-\infty}^{\infty} \mathrm{d}\omega f_{\boldsymbol{q},\mathrm{i}\omega}^{\mathrm{RPA},\alpha} (\mathrm{i}\omega - |\boldsymbol{v}^{\alpha}| q_{\parallel}^{\alpha})^{-2}$, where $q_{\parallel}^{\alpha}$ and $\boldsymbol{q}_{\perp}^{\alpha}$ are defined as in Eq. (5.46). Due to the rather singular structure of the integrand, it is by no means obvious that for arbitrary interactions the Fourier integral theorem is applicable. However, in all physical applications discussed in the second part of this book we find that $S^{\alpha}(r_{\parallel}^{\alpha}\hat{\boldsymbol{v}}^{\alpha}, 0)$ indeed vanishes as $r_{\parallel}^{\alpha} \to \pm\infty$.

The effective mass renormalization Z^α_{m}

Because the spatial dependence of the Debye-Waller factor $Q^\alpha(r^\alpha_\| \hat{\boldsymbol{v}}^\alpha, \tau)$ enters only via the projection $r^\alpha_\| = \hat{\boldsymbol{v}}^\alpha \cdot \boldsymbol{r}$, the direction of the renormalized Fermi velocity $\tilde{\boldsymbol{v}}^\alpha$ is *for linearized energy dispersion* always parallel to the direction of the bare Fermi velocity $\boldsymbol{v}^\alpha$. From Eq.(2.27) we see that in this case the effective mass renormalization factor associated with patch P^α_Λ is given by $Z^\alpha_{\mathrm{m}} = |\tilde{\boldsymbol{v}}^\alpha| / |\boldsymbol{v}^\alpha|$. The renormalized Fermi velocity can be directly obtained from the real space imaginary time sector Green's function $G^\alpha(\boldsymbol{r}, \tau)$ by taking *first* the limit $\tau \to \infty$ and *then* the limit $r^\alpha_\| \to \infty$. From Eq.(5.48) we obtain for the non-interacting sector Green's function in this limit

$$G^\alpha_0(\boldsymbol{r}, \tau) \sim -\delta^{(d-1)}(\boldsymbol{r}^\alpha_\perp) \frac{1}{2\pi |\boldsymbol{v}^\alpha| \tau} \quad , \quad \frac{\tau}{r^\alpha_\|} \to \infty \quad , \quad r^\alpha_\| \to \infty \quad . \tag{5.87}$$

Assuming the existence of the limit

$$S^\alpha_\infty = \lim_{r^\alpha_\| \to \infty} \left[\lim_{\tau \to \infty} S^\alpha(r^\alpha_\| \hat{\boldsymbol{v}}^\alpha, \tau) \right] \quad , \tag{5.88}$$

the relation analogous to Eq.(5.87) for the interacting sector Green's function given in Eq.(5.52) is

$$G^\alpha(\boldsymbol{r}, \tau) \sim -\delta^{(d-1)}(\boldsymbol{r}^\alpha_\perp) \frac{Z^\alpha \mathrm{e}^{-S^\alpha_\infty}}{2\pi |\boldsymbol{v}^\alpha| \tau} \quad , \quad \frac{\tau}{r^\alpha_\|} \to \infty \quad , \quad r^\alpha_\| \to \infty \quad . \tag{5.89}$$

But in an interacting Fermi liquid we should have

$$G^\alpha(\boldsymbol{r}, \tau) \sim -\delta^{(d-1)}(\boldsymbol{r}^\alpha_\perp) \frac{Z^\alpha}{2\pi |\tilde{\boldsymbol{v}}^\alpha| \tau} \quad , \quad \frac{\tau}{r^\alpha_\|} \to \infty \quad , \quad r^\alpha_\| \to \infty \quad , \tag{5.90}$$

where $\tilde{\boldsymbol{v}}^\alpha$ is the renormalized Fermi velocity. Comparing Eqs.(5.89) and (5.90) with (2.30), we conclude that the effective mass renormalization factor is given by

$$Z^\alpha_{\mathrm{m}} = \mathrm{e}^{S^\alpha_\infty} \quad . \tag{5.91}$$

The analytic evaluation of the limit S^α_∞ in Eq.(5.88) is rather difficult. We have not been able to obtain for general interactions a simple analytic expression for S^α_∞, which explicitly contains only the parameters $f_{\underline{q}}$ and $\xi_{\boldsymbol{k}}$ that appear in the definition of the original action[5]. Note that the naive application of the Fourier integral theorem [5.5] to $S^\alpha(0, \tau)$ implies that S^α_∞ should vanish, so that bosonization with linearized energy dispersion does not incorporate effective mass renormalizations. To examine this point more carefully, let us substitute the Dyson equation (5.65) into Eq.(5.67), and then solve for $\Sigma^\alpha(\tilde{q})$ as functional of $G^\alpha(\tilde{q})$. After some trivial algebra we obtain

$$\Sigma^\alpha(\tilde{q}) = [\mathrm{i}\tilde{\omega}_n - \boldsymbol{v}^\alpha \cdot \boldsymbol{q}] \frac{T^\alpha_{\tilde{q}}}{1 + T^\alpha_{\tilde{q}}} \quad , \tag{5.92}$$

---

[5] In the case of Z^α such an expression is given in Eqs.(5.86) and (5.83).

with

$$T^{\alpha}_{\tilde{q}} = \frac{1}{\beta V} \sum_{\tilde{q}'} X^{\alpha}_{\tilde{q}'-\tilde{q}} G^{\alpha}(\tilde{q}') \quad , \tag{5.93}$$

where we have used $X^{\alpha}_{-q} = -X^{\alpha}_{q}$, see Eqs.(5.56) and (5.57). At $q = 0$ Eq.(5.93) reduces to

$$T^{\alpha}_{0} = \frac{1}{\beta V} \sum_{q,m} \frac{f^{\mathrm{RPA},\alpha}_{q,\mathrm{i}\omega_m}}{(\mathrm{i}\omega_m - \boldsymbol{v}^{\alpha} \cdot \boldsymbol{q})} G^{\alpha}(\boldsymbol{q}, \mathrm{i}\tilde{\omega}_m) \quad , \tag{5.94}$$

which should be compared with

$$R^{\alpha} = \frac{1}{\beta V} \sum_{q,m} \frac{f^{\mathrm{RPA},\alpha}_{q,\mathrm{i}\omega_m}}{(\mathrm{i}\omega_m - \boldsymbol{v}^{\alpha} \cdot \boldsymbol{q})} \frac{1}{(\mathrm{i}\omega_m - \boldsymbol{v}^{\alpha} \cdot \boldsymbol{q})} \quad . \tag{5.95}$$

Obviously the only difference between T^{α}_{0} and R^{α} is that the full Green's function $G^{\alpha}(\boldsymbol{q}, \mathrm{i}\tilde{\omega}_m)$ on the right-hand side of Eq.(5.94) is replaced by a factor of $(\mathrm{i}\omega_m - \boldsymbol{v}^{\alpha} \cdot \boldsymbol{q})^{-1}$ in Eq.(5.95). Keeping in mind that in a Fermi liquid the integral in Eq.(5.95) remains finite in the limit $\beta, V \to \infty$ (recall Eq.(5.83)), it is tempting to speculate that the finiteness of R^{α} implies that also the expression for T^{α}_{0} in Eq.(5.94) must be finite. Defining the retarded function $T^{\alpha}(\boldsymbol{q}, \omega) = T^{\alpha}_{\boldsymbol{q},\mathrm{i}\tilde{\omega}_n}|_{\mathrm{i}\tilde{\omega}_n \to \omega + \mathrm{i}0^+}$, let us now *assume* that $T^{\alpha}(0,0)$ is finite, and that for small $\boldsymbol{q}$ and ω the corrections vanish with some positive power,

$$T^{\alpha}(\boldsymbol{q}, \omega) \sim T^{\alpha}(0,0) + O(|\boldsymbol{q}|^{\mu_1}, |\omega|^{\mu_2}) \quad , \quad \mu_1, \mu_2 > 0 \quad . \tag{5.96}$$

We would like to emphasize that at this point Eq.(5.96) should be considered as an assumption, which is motivated by the similarity between Eqs.(5.94) and (5.95), and by the fact that in a Fermi liquid R^{α} is finite. From Eq.(5.92) we see that the retarded self-energy can then be written as

$$\Sigma^{\alpha}(\boldsymbol{q}, \omega + \mathrm{i}0^+) = [\omega - \boldsymbol{v}^{\alpha} \cdot \boldsymbol{q}] \frac{T^{\alpha}(\boldsymbol{q}, \omega)}{1 + T^{\alpha}(\boldsymbol{q}, \omega)} \quad , \tag{5.97}$$

and satisfies

$$\left. \frac{\partial \Sigma^{\alpha}(0, \omega + \mathrm{i}0^+)}{\partial \omega} \right|_{\omega = 0} = \frac{T^{\alpha}(0,0)}{1 + T^{\alpha}(0,0)} \quad , \tag{5.98}$$

so that according to Eq.(2.21) the quasi-particle residue exists and is given by

$$Z^{\alpha} = \mathrm{e}^{R^{\alpha}} = 1 + T^{\alpha}(0,0) \quad . \tag{5.99}$$

From Eq.(5.99) we see that the replacement of the last factor $(\mathrm{i}\omega_m - \boldsymbol{v}^{\alpha} \cdot \boldsymbol{q})^{-1}$ in Eq.(5.95) by $G^{\alpha}(\boldsymbol{q}, \mathrm{i}\tilde{\omega}_m)$ in Eq.(5.94) amounts to an exponentiation. Substituting now Eqs.(5.99) and (5.93) into Eq.(2.23) we obtain for the renormalization of the Fermi velocity

$$\delta \boldsymbol{v}^{\alpha} = (Z^{\alpha} - 1)\boldsymbol{v}^{\alpha} + Z^{\alpha} \, \nabla_{\boldsymbol{q}} \Sigma^{\alpha}(\boldsymbol{q}, i0^{+})\big|_{\boldsymbol{q}=0}$$
$$= T^{\alpha}(0,0)\boldsymbol{v}^{\alpha} - (1 + T^{\alpha}(0,0))\boldsymbol{v}^{\alpha} \frac{T^{\alpha}(0,0)}{1 + T^{\alpha}(0,0)} = 0 \ . \tag{5.100}$$

Hence, under the assumption (5.96) the Fermi velocity is not renormalized, so that $Z^{\alpha}_{\mathrm{m}} = 1$. Although we have not proven Eq.(5.96), the similarity between Eqs.(5.94) and (5.95) strongly suggests that it is indeed correct. This is also in accordance with the Fourier integral theorem, which implies that S^{α}_{∞} should vanish if $S^{\alpha}(0,0) = R^{\alpha}$ exists. We thus conclude that higher-dimensional bosonization ***with linearized energy dispersion*** does not contain effective mass renormalizations. We shall come back to this point in Chap. 6.1.3, where we shall show that this is closely related to the fact that for linearized energy dispersion the Fermi surface is approximated by a *finite* number M of completely flat patches.

5.2 Beyond the Gaussian approximation

We now describe a general method for including the non-linear terms in the energy dispersion into our background field approach. This enables us to include the effects of the curvature of the Fermi surface into our non-perturbative expression for the single-particle Green's function. A brief description of our method has been published in the Letter [1.38]. Here we present for the first time the details.

One of the main approximations in Sect. 5.1 was the replacement of the Fermi surface by a collection of flat hyper-planes, which amounts to setting $1/m^{\alpha} = 0$ in the expansion (2.66) of the energy dispersion close to the Fermi surface. Although we have intuitively justified this approximation for sufficiently long-range interactions, we have not given a quantitative estimate of the corrections due to non-linear terms in the energy dispersion. Recall that for the density-density correlation function such a quantitative estimate has been given in Chap. 4.3; in this case the corrections due to the non-linear terms could be explicitly calculated, and in Eq.(4.115) we have identified the relevant small parameter.

In the context of conventional one-dimensional bosonization Haldane [1.15] has speculated that it should be possible to develop some kind of perturbation theory around the non-perturbative bosonization solution for linearized energy dispersion, using the inverse effective mass $1/m^{\alpha}$ as a small parameter. However, even in $d = 1$ a practically useful formulation of such a perturbation theory has not been developed. This seems to be due to the fact that the naive expansion of the conventional bosonization formula for the Green's function in powers of $1/m^{\alpha}$ becomes rather awkward in the absence of interactions [1.43], because in this case we can trivially write down the exact solution $G_0^\alpha = [\mathrm{i}\tilde{\omega}_n - \xi_{\boldsymbol{q}}^\alpha]^{-1}$. This expression contains infinite orders in $1/m^\alpha$, so that it can only be recovered by means of the $1/m^\alpha$-expansion suggested in [1.15] if all terms in the series are summed. This is of course an impossible task. In this chapter we shall develop a new method for including the non-linear terms in the energy dispersion into the bosonization procedure, which *in the non-interacting limit reproduces the exact free Green's function.* Thus, our method is *not* based on a direct expansion in powers of $1/m^\alpha$. It should be mentioned that in the special case of one dimension an alternative algebraic bosonization approach, which includes arbitrary non-linear terms in the dispersion relation, has recently been developed by Zemba and collaborators [5.6]. In higher dimensions the non-linear terms in the energy dispersion have also been discussed by Khveshchenko [1.48] within his "geometric" bosonization approach. However, his formalism is based on a rather complicated mathematical construction, and so far has not been of practical use for the explicit calculation of curvature effects on the bosonization result for the Green's function with linearized energy dispersion.

In dimensions $d > 1$ it is certainly more important to retain the non-linear terms in the energy dispersion than in $d = 1$, because only in higher dimensions the Fermi surface has a curvature. To see this more clearly, let us assume for the moment that locally the Fermi surface can be approximated by a quadratic form, and that in an appropriately oriented coordinate system the energy dispersion $\xi_{\boldsymbol{q}}^\alpha$ defined in Eq.(2.16) can be written as

$$\xi_{\boldsymbol{q}}^\alpha = \boldsymbol{v}^\alpha \cdot \boldsymbol{q} + \frac{(\hat{\boldsymbol{v}}^\alpha \cdot \boldsymbol{q})^2}{2m_{\parallel}^\alpha} + \frac{(\boldsymbol{q}_{\perp}^\alpha)^2}{2m_{\perp}^\alpha} \quad , \tag{5.101}$$

where $\boldsymbol{q}_\perp^\alpha = \boldsymbol{q} - (\boldsymbol{q} \cdot \hat{\boldsymbol{v}}^\alpha)\hat{\boldsymbol{v}}^\alpha$, and $m_\parallel^\alpha$ and $m_\perp^\alpha$ are the effective masses for the motion parallel and perpendicular to the local normal $\hat{\boldsymbol{v}}^\alpha$. The important point is now that only the last term in Eq.(5.101) describes the curvature of the patches. In other words, for $1/m_\perp^\alpha = 0$ but finite $1/m_\parallel^\alpha$ we still have completely flat patches. Obviously in $d > 1$ there exist hyper-planes in momentum space (defined by $\hat{\boldsymbol{v}}^\alpha \cdot \boldsymbol{q} = 0$) where the last term in Eq.(5.101) is the dominant contribution in the expansion of the energy dispersion. As already mentioned in the second footnote in Sect. 5.1.1, a priori it is not clear whether the contribution from these hyper-planes to some physical quantity of interest is negligible or not. From the previous section we expect that the curvature of the Fermi surface will certainly play an important role to obtain the correct effective mass renormalization in a Fermi liquid (recall the discussion in Sect. 5.1.5). As we shall see in Chap. 6.1.3, this problem is closely related to the existence of a *double pole* in the integrand of the linearized bosonization result for the Debye-Waller factor, see Eqs.(5.32) and (5.33).

Let us point out two more rather peculiar features of the higher-dimensional bosonization result for the Green's function with linearized energy dispersion. First of all, for any finite number M of patches the real space Green's function is of the form $G(\boldsymbol{r},\tau) = \sum_{\alpha=1}^{M} \mathrm{e}^{\mathrm{i}\boldsymbol{k}^\alpha \cdot \boldsymbol{r}} G^\alpha(\boldsymbol{r},\tau)$ where $G^\alpha(\boldsymbol{r},\tau)$ is proportional to a $d-1$-dimensional δ-function[6] $\delta^{(d-1)}(\boldsymbol{r}_{\perp}^{\alpha})$ of the components of $\boldsymbol{r}$ that are perpendicular to the local Fermi velocity $\boldsymbol{v}^{\alpha}$ (see Eqs.(5.43) and (5.53)). As a consequence, we may replace $\boldsymbol{r} \rightarrow (\boldsymbol{r} \cdot \hat{\boldsymbol{v}}^{\alpha})\hat{\boldsymbol{v}}^{\alpha}$ in the expression for the Debye-Waller factor $Q^{\alpha}(\boldsymbol{r}, \tau)$ (see Eq.(5.52)). If we naively take the limit of infinite patch number $M \rightarrow \infty$, then the patch summation is turned into a $d-1$-dimensional integral over the Fermi surface, so that in this limit the singular function $\delta_{\Lambda}^{(d-1)}(\boldsymbol{r}_{\perp}^{\alpha})$ appears under a $d-1$-dimensional integral, and the final result for the real space Green's function does not exhibit any singularities. However, because $M \rightarrow \infty$ implies a vanishing patch cutoff, $\Lambda \rightarrow 0$, and because the approximations made in deriving the above result can formally only be justified if Λ is held finite and large compared with the range q_c of the interaction in momentum space (see Fig. 2.5), one may wonder whether the above limiting procedure is justified. Of course, in momentum space this problem remains hidden, because the function $\delta_{\Lambda}^{(d-1)}(\boldsymbol{r}_{\perp}^{\alpha})$ is eliminated trivially via the Fourier transformation. Consequently the interacting Green's function for wave-vectors close to $\boldsymbol{k}^{\alpha}$ is simply $G(\boldsymbol{k}^{\alpha} + \boldsymbol{q}, \mathrm{i}\tilde{\omega}_n) = G^{\alpha}(\boldsymbol{q}, \mathrm{i}\tilde{\omega}_n)$ (see Eq.(5.40)), where $G^{\alpha}(\boldsymbol{q}, \mathrm{i}\tilde{\omega}_n)$ is the Fourier transform of $G^{\alpha}(\boldsymbol{r}, \tau)$. Nevertheless, it is legitimate to ask how the Green's function looks in real space, and the prediction of higher-dimensional bosonization with linearized energy dispersion is not quite satisfactory. Another shortcoming of the linearized theory will be discussed in detail in Chap. 7.2.4: the replacement of a curved Fermi surface by a finite number of flat patches can give rise to unphysical nesting singularities.

It is intuitively obvious that the problems mentioned above are related to the fact that we have ignored the curvature of the Fermi surface within a given patch. To cure these drawbacks of higher-dimensional bosonization, we shall now generalize our background field approach to the case of finite masses m_i^{α}.

5.2.1 The Green's function for fixed background field

We develop an imaginary time eikonal expansion for the single-particle Green's function $\mathcal{G}^{\alpha}(\boldsymbol{r}, \boldsymbol{r}', \tau, \tau')$ at fixed background field, which takes the non-linear terms in the energy dispersion non-perturbatively into account. In this way we obtain the generalization of the Schwinger ansatz given in Eq.(5.17) for non-linear energy dispersions.

Generalization of the Schwinger ansatz

We would like to invert the infinite matrix $\hat{G}^{-1}$ in Eq.(5.2) for general energy dispersion $\epsilon_{\boldsymbol{k}}$. As explained in Sect. 5.1.1, it is convenient to measure wave-vectors with respect to a coordinate system centered at $\boldsymbol{k}^{\alpha}$ and define

$$[\hat{G}]_{\boldsymbol{k}^{\alpha}+\boldsymbol{q},\mathrm{i}\tilde{\omega}_n;\boldsymbol{k}^{\alpha}+\boldsymbol{q}',\mathrm{i}\tilde{\omega}_{n'}} = [\hat{G}^{\alpha}]_{\boldsymbol{q},\mathrm{i}\tilde{\omega}_n;\boldsymbol{q}',\mathrm{i}\tilde{\omega}_{n'}} \equiv [\hat{G}^{\alpha}]_{\tilde{q}\tilde{q}'} \quad . \tag{5.102}$$

[6] As discussed in Sect. 5.1.3, at short distances the δ-function should actually be replaced by the cutoff-dependent function $\delta_{\Lambda}^{(d-1)}(\boldsymbol{r}_{\perp}^{\alpha})$ defined in Eq.(5.51).

Then the infinite matrix $[\hat{G}^{\alpha}]_{\tilde{q}\tilde{q}'}$ is determined by an equation of the form (5.11), with $[G_0^{\alpha}(\tilde{q})]^{-1}$ now given by $[G_0^{\alpha}(\tilde{q})]^{-1} = \mathrm{i}\tilde{\omega}_n - \epsilon_{\boldsymbol{k}^{\alpha}+\boldsymbol{q}} + \mu$. Note that the above transformations are valid for an arbitrary sectorization of momentum space (see Sect. 2.5), including the special case that we identify the entire momentum space with a single sector (then we just shift the coordinate origin in momentum space to $\boldsymbol{k}^{\alpha}$, as shown in Fig. 2.8.). Defining the Fourier transforms $\mathcal{G}^{\alpha}(\boldsymbol{r},\boldsymbol{r}',\tau,\tau')$ and $V^{\alpha}(\boldsymbol{r},\tau)$ of $[\hat{G}^{\alpha}]_{\tilde{q}\tilde{q}'}$ and V_q^{α} as in Eqs.(5.12) and (5.13), it is easy to see that $\mathcal{G}^{\alpha}(\boldsymbol{r},\boldsymbol{r}',\tau,\tau')$ satisfies the partial differential equation

$$[-\partial_{\tau} - \epsilon_{\boldsymbol{k}^{\alpha}+\boldsymbol{P}_r} + \mu - V^{\alpha}(\boldsymbol{r},\tau)]\,\mathcal{G}^{\alpha}(\boldsymbol{r},\boldsymbol{r}',\tau,\tau') = \delta(\boldsymbol{r}-\boldsymbol{r}')\delta^{*}(\tau-\tau')\ , \quad (5.103)$$

where $\boldsymbol{P}_r = -\mathrm{i}\nabla_r$ is the momentum operator. Eq.(5.103) is the generalization of Eq.(5.14) to arbitrary energy dispersions $\epsilon_{\boldsymbol{k}}$. This partial differential equation together with the Kubo-Martin-Schwinger boundary condition (5.16) uniquely determines the function $\mathcal{G}^{\alpha}(\boldsymbol{r},\boldsymbol{r}',\tau,\tau')$. We now truncate the expansion of $\epsilon_{\boldsymbol{k}^{\alpha}+\boldsymbol{q}} - \mu$ for small $\boldsymbol{q}$ at the second order, see Eqs.(2.65) and (2.66). Then Eq.(5.14) is of second order in the spatial derivatives. Note that for free fermions with energy dispersion $\epsilon_{\boldsymbol{k}} = \boldsymbol{k}^2/(2m)$ the truncation at the second order is exact, but for more complicated Fermi surfaces we are assuming that the sectors have been chosen sufficiently small such that the *local curvature* can be approximated by a constant. Linearization of the energy dispersion amounts to ignoring the quadratic terms in the expansion of $\epsilon_{\boldsymbol{k}^{\alpha}+\boldsymbol{q}}$ for small $\boldsymbol{q}$, in which case the Schwinger ansatz (5.17) solves Eq.(5.103). It is not difficult to see that for non-linear energy dispersion this ansatz does *not* lead to a consistent solution of Eq.(5.14). In order to develop a systematic method for treating the non-linear terms in the energy dispersion in a non-perturbative way, we need a generalization of the Schwinger ansatz (5.17) which in the limit $1/m_i^{\alpha} \to 0$ reduces to the solution of the linearized differential equation. The crucial observation is that the quantity $\mathcal{G}^{\alpha}(\boldsymbol{r},\boldsymbol{r},\tau,\tau)$ (which is obtained by setting $\boldsymbol{r} = \boldsymbol{r}'$ and $\tau = \tau'$ in the solution of Eq.(5.103)) represents physically a contribution to the ***density of the system***. Moreover, on physical grounds it is also clear that the external potential $V^{\alpha}(\boldsymbol{r},\tau)$ should lead to a deviation of the density from its equilibrium value. Evidently the Schwinger ansatz (5.17) predicts that the external potential does not lead to any modulation of the density, which is of course an unphysical artefact of the linearization. For non-linear energy dispersion, our generalized Schwinger ansatz should allow for density fluctuations. The simplest possible way to incorporate the physics of density fluctuations without changing the important exponential factor in the Schwinger ansatz is to set[7]

$$\mathcal{G}^{\alpha}(\boldsymbol{r},\boldsymbol{r}',\tau,\tau') = \mathcal{G}_1^{\alpha}(\boldsymbol{r},\boldsymbol{r}',\tau,\tau')\mathrm{e}^{\Phi^{\alpha}(\boldsymbol{r},\tau)-\Phi^{\alpha}(\boldsymbol{r}',\tau')} \ . \quad (5.104)$$

The KMS boundary conditions are satisfied by requiring that $\Phi^{\alpha}(\boldsymbol{r},\tau)$ should be periodic in τ, while $\mathcal{G}_1^{\alpha}(\boldsymbol{r},\boldsymbol{r}',\tau,\tau')$ should be antiperiodic in τ and τ'. Setting $\boldsymbol{r} = \boldsymbol{r}'$ and $\tau = \tau'$, we conclude that $\mathcal{G}_1^{\alpha}(\boldsymbol{r},\boldsymbol{r},\tau,\tau)$ is the contribution

[7] I would have never tried this ansatz without a hint from Lorenz Bartosch.

from states with momenta in sector $K^{\alpha}_{\Lambda,\lambda}$ to the density of the system. From the arguments given above it is therefore clear that for non-linear energy dispersion $\mathcal{G}^{\alpha}_{1}(\boldsymbol{r},\boldsymbol{r}',\tau,\tau')$ must be a non-trivial function of the external potential. Of course, in Eq.(5.104) we could always choose $\Phi^{\alpha}=0$ and $\mathcal{G}^{\alpha}=\mathcal{G}^{\alpha}_{1}$, so that nothing would be gained. The crucial point is, however, that there exists another non-trivial choice of Φ^{α} and $\mathcal{G}^{\alpha}_{1}$ which leads to the natural generalization of the Schwinger ansatz (5.17) to systems with energy dispersions of the type (2.65) and (2.66). To see this, we substitute Eq.(5.104) into Eq.(5.103) and obtain after a simple calculation

$$
\begin{aligned}
\left[-\partial_{\tau}-\epsilon_{\boldsymbol{k}^{\alpha}+\boldsymbol{P}_{r}}+\mu-\boldsymbol{u}^{\alpha}(\boldsymbol{r},\tau)\cdot\boldsymbol{P}_{r}\right]\mathcal{G}^{\alpha}_{1}(\boldsymbol{r},\boldsymbol{r}',\tau,\tau') &= \delta(\boldsymbol{r}-\boldsymbol{r}')\delta^{*}(\tau-\tau') \\
+\mathcal{G}^{\alpha}_{1}(\boldsymbol{r},\boldsymbol{r}',\tau,\tau')\left\{\left[-\partial_{\tau}-\xi^{\alpha}_{\boldsymbol{P}_{r}}\right]\Phi^{\alpha}(\boldsymbol{r},\tau)-V^{\alpha}(\boldsymbol{r},\tau)\right. \\
\left.-[\boldsymbol{P}_{r}\Phi^{\alpha}(\boldsymbol{r},\tau)](2\mathsf{M}^{\alpha})^{-1}[\boldsymbol{P}_{r}\Phi^{\alpha}(\boldsymbol{r},\tau)]\right\}\ . \qquad (5.105)
\end{aligned}
$$

where the $d\times d$-matrix M^{α} contains the effective masses,

$$
[\mathsf{M}^{\alpha}]_{ij}=\delta_{ij}m^{\alpha}_{i}\quad , \qquad (5.106)
$$

and the components $u^{\alpha}_{i}(\boldsymbol{r},\tau)$ of the velocity $\boldsymbol{u}^{\alpha}(\boldsymbol{r},\tau)$ are given by

$$
u^{\alpha}_{i}(\boldsymbol{r},\tau)\equiv\boldsymbol{e}_{i}\cdot\boldsymbol{u}^{\alpha}(\boldsymbol{r},\tau)=\frac{\boldsymbol{e}_{i}\cdot\boldsymbol{P}_{r}\Phi^{\alpha}(\boldsymbol{r},\tau)}{m^{\alpha}_{i}}\quad . \qquad (5.107)
$$

Here the unit vectors $\boldsymbol{e}_{1},\ldots,\boldsymbol{e}_{d}$ match the axes of the local coordinate system attached to $\boldsymbol{k}^{\alpha}$ in which the effective mass tensor M^{α} is diagonal. The crucial observation is now that, apart from the trivial solution $\Phi^{\alpha}=0$ and $\mathcal{G}^{\alpha}_{1}=\mathcal{G}^{\alpha}$, we obtain another *exact* solution of Eq.(5.105) by choosing $\Phi^{\alpha}(\boldsymbol{r},\tau)$ and $\mathcal{G}^{\alpha}_{1}(\boldsymbol{r},\boldsymbol{r},\tau,\tau')$ such that

$$
\left[-\partial_{\tau}-\xi^{\alpha}_{\boldsymbol{P}_{r}}\right]\Phi^{\alpha}(\boldsymbol{r},\tau)=V^{\alpha}(\boldsymbol{r},\tau)+[\boldsymbol{P}_{r}\Phi^{\alpha}(\boldsymbol{r},\tau)](2\mathsf{M}^{\alpha})^{-1}[\boldsymbol{P}_{r}\Phi^{\alpha}(\boldsymbol{r},\tau)]\ , \qquad (5.108)
$$

$$
\left[-\partial_{\tau}-\epsilon_{\boldsymbol{k}^{\alpha}+\boldsymbol{P}_{r}}+\mu-\boldsymbol{u}^{\alpha}(\boldsymbol{r},\tau)\cdot\boldsymbol{P}_{r}\right]\mathcal{G}^{\alpha}_{1}(\boldsymbol{r},\boldsymbol{r}',\tau,\tau')=\delta(\boldsymbol{r}-\boldsymbol{r}')\delta^{*}(\tau-\tau')\ . \qquad (5.109)
$$

Thus, $\mathcal{G}^{\alpha}_{1}(\boldsymbol{r},\boldsymbol{r}',\tau,\tau')$ is again a fermionic Green's function. Note that the differential equation (5.108) is non-linear, but contains only first-order derivatives. In contrast, the original Eq.(5.103) is linear but involves second-order derivatives. Differential equations of the type (5.108) are called *eikonal equations*, and appear in many fields of physics, such as classical mechanics[8], geometrical optics [5.8], quantum mechanical scattering theory [5.9], and relativistic quantum field theories [2.4, 5.10]. The functional $\Phi^{\alpha}(\boldsymbol{r},\tau)$ is called the *eikonal*. In the limit $1/m^{\alpha}_{i}\to 0$ the eikonal equation (5.108) reduces to the corresponding equation (5.21) of the linearized theory, which can be solved

[8] Recall the Hamilton-Jacobi equation [5.7] $-\partial S/\partial t=V(\boldsymbol{r},t)+\frac{(\nabla S)^{2}}{2m}$ for the action $S(\boldsymbol{r},t)$ of a particle with mass m that moves under the influence of an external potential $V(\boldsymbol{r},t)$.

exactly via Fourier transformation, see Eq.(5.23). Furthermore, in this case the velocity $\boldsymbol{u}^{\alpha}(\boldsymbol{r},\tau)$ vanishes, so that $\mathcal{G}_1^{\alpha}(\boldsymbol{r},\boldsymbol{r}',\tau,\tau') = G_0^{\alpha}(\boldsymbol{r}-\boldsymbol{r}',\tau-\tau')$. However, if one of the masses m_i^{α} is finite, the eikonal equation (5.108) is non-linear and cannot be solved exactly. We shall discuss a method to obtain an approximate solution shortly.

The differential equation (5.109) describes the motion of a fermion under the influence of a space- and time-dependent *random velocity* $\boldsymbol{u}^{\alpha}(\boldsymbol{r},\tau)$. At the first sight it seems that this problem is just as difficult to solve as the original Eq.(5.103). The crucial point is, however, that perturbation theory in terms of the *derivative potential* $\boldsymbol{u}^{\alpha}(\boldsymbol{r},\tau)\cdot\boldsymbol{P}_{\boldsymbol{r}}$ in Eq.(5.109) is *less infrared singular* than perturbation theory in terms of the original random potential $V^{\alpha}(\boldsymbol{r},\tau)$ in Eq.(5.103). Moreover, for large effective masses m_i^{α} the random velocity $\boldsymbol{u}^{\alpha}(\boldsymbol{r},\tau)$ is small, so that the perturbation theory in powers of the derivative potential is justified. Such a small parameter is absent in Eq.(5.103).

The eikonal equation

Although it is impossible to solve the non-linear partial differential equation (5.108) exactly, we can follow the pioneering work of E. S. Fradkin [5.10] to obtain the solution as series in powers of V^{α}. We would like to emphasize, however, that our imaginary time eikonal equation is *not identical* with the real time eikonal equation discussed by Fradkin [5.10]. The latter has recently been applied by Khveshchenko and Stamp [5.11] to the problem of fermions coupled to gauge fields, and involves an additional time-like auxiliary variable. For a d-dimensional quantum system one thus has to deal with a $d+2$-dimensional partial differential equation, which leads to rather complicated expressions for the higher-order terms in the eikonal expansion [5.10]. In contrast, our imaginary time eikonal equation (5.108) is $d+1$-dimensional and does not depend on additional auxiliary variables. This facilitates the calculation of corrections to the leading term. See the work [5.12] for a detailed discussion of the real time eikonal method and a comparison with our functional bosonization approach.

Following Fradkin [5.10], we obtain the solution of Eq.(5.108) by making the ansatz

$$\Phi^{\alpha}(\boldsymbol{r},\tau) = \sum_{n=1}^{\infty} \Phi_n^{\alpha}(\boldsymbol{r},\tau) \quad , \tag{5.110}$$

where $\Phi_n^{\alpha}(\boldsymbol{r},\tau)$ involves by assumption n powers of V^{α}. Substituting Eq.(5.110) into Eq.(5.108) and comparing powers of V^{α}, it is easy to see that the nth order term $\Phi_n^{\alpha}(\boldsymbol{r},\tau)$ is determined by the inhomogeneous linear differential equation

$$\left[-\partial_{\tau} - \xi^{\alpha}_{\boldsymbol{P}_{\boldsymbol{r}}}\right] \Phi_n^{\alpha}(\boldsymbol{r},\tau) = V_n^{\alpha}(\boldsymbol{r},\tau) \quad , \quad n = 1,2,\ldots \quad , \tag{5.111}$$

where the first order potential is simply

$$V_1^{\alpha}(\boldsymbol{r},\tau) = V^{\alpha}(\boldsymbol{r},\tau) \quad , \tag{5.112}$$

and the higher orders are

$$V_n^\alpha(\boldsymbol{r},\tau) = \sum_{n'=1}^{n-1} [\boldsymbol{P}_{\boldsymbol{r}}\Phi_{n'}^\alpha(\boldsymbol{r},\tau)](2\mathrm{M}^\alpha)^{-1}[\boldsymbol{P}_{\boldsymbol{r}}\Phi_{n-n'}^\alpha(\boldsymbol{r},\tau)] \ , \ n = 2,3,\ldots \ . \tag{5.113}$$

Note that the inhomogeneity $V_n^\alpha(\boldsymbol{r},\tau)$ in the differential equation (5.111) for $\Phi_n^\alpha(\boldsymbol{r},\tau)$ depends only on solutions $\Phi_{n'}^\alpha(\boldsymbol{r},\tau)$ with $n' < n$, so that we can calculate the functionals $\Phi_n^\alpha(\boldsymbol{r},\tau)$ iteratively. Because Eq.(5.111) is linear, its solution is easily obtained by means of the Green's function of the differential operator on the left-hand side,

$$\Phi_n^\alpha(\boldsymbol{r},\tau) = \int \mathrm{d}\boldsymbol{r}' \int_0^\beta \mathrm{d}\tau' G_{\mathrm{b}}^\alpha(\boldsymbol{r}-\boldsymbol{r}',\tau-\tau')V_n^\alpha(\boldsymbol{r}',\tau') \ , \tag{5.114}$$

where

$$G_{\mathrm{b}}^\alpha(\boldsymbol{r},\tau) = \frac{1}{\beta V}\sum_q \mathrm{e}^{\mathrm{i}(\boldsymbol{q}\cdot\boldsymbol{r}-\omega_m\tau)} G_{\mathrm{b}}^\alpha(q) \ , \quad G_{\mathrm{b}}^\alpha(q) = \frac{1}{\mathrm{i}\omega_m - \xi_{\boldsymbol{q}}^\alpha} \ . \tag{5.115}$$

This Green's function should not be confused with the corresponding free fermionic Green's function

$$G_0^\alpha(\boldsymbol{r},\tau) = \frac{1}{\beta V}\sum_{\tilde{q}} \mathrm{e}^{\mathrm{i}(\boldsymbol{q}\cdot\boldsymbol{r}-\tilde{\omega}_n\tau)} G_0^\alpha(\tilde{q}) \ , \quad G_0^\alpha(\tilde{q}) = \frac{1}{\mathrm{i}\tilde{\omega}_n - \epsilon_{\boldsymbol{k}^\alpha+\boldsymbol{q}} + \mu} \ , \tag{5.116}$$

which for linearized energy dispersion and $\epsilon_{\boldsymbol{k}^\alpha} = \mu$ reduces to Eq.(5.22). Note that the Fourier transform $G_{\mathrm{b}}^\alpha(q)$ of $G_{\mathrm{b}}^\alpha(\boldsymbol{r},\tau)$ involves *bosonic* Matsubara frequencies and depends on the *excitation energy* $\xi_{\boldsymbol{q}}^\alpha = \epsilon_{\boldsymbol{k}^\alpha+\boldsymbol{q}}^\alpha - \epsilon_{\boldsymbol{k}^\alpha}^\alpha$. The bosonic frequencies insure that the functional $\Phi^\alpha(\boldsymbol{r},\tau)$ is periodic in τ, so that our ansatz (5.104) satisfies the KMS boundary conditions (see also the discussion in Sect. 5.1.1). In contrast, the Fourier transform $G_0^\alpha(\tilde{q})$ of $G_0^\alpha(\boldsymbol{r},\tau)$ depends on fermionic frequencies and involves the usual combination $\epsilon_{\boldsymbol{k}^\alpha+\boldsymbol{q}}^\alpha - \mu$. Recall that in general we may choose $\epsilon_{\boldsymbol{k}^\alpha} \neq \mu$.

To carry out the above iterative procedure in practice, we find it more convenient to work in Fourier space. Defining the Fourier transforms V_q^α and Φ_q^α as in Eq.(5.13), it is easy to show that Eq.(5.108) implies for the Fourier components

$$\left[\mathrm{i}\omega_m - \xi_{\boldsymbol{q}}^\alpha\right] \Phi_q^\alpha = V_q^\alpha + \sum_{q'} (\boldsymbol{q}-\boldsymbol{q}')(2\mathrm{M}^\alpha)^{-1}\boldsymbol{q}'\Phi_{q-q'}^\alpha \Phi_{q'}^\alpha \ . \tag{5.117}$$

Keeping in mind that $V_q^\alpha = \frac{\mathrm{i}}{\beta}\phi_q^\alpha$ (see Eq.(3.31)), it is convenient to define

$$\Psi_q^\alpha = \frac{\beta}{\mathrm{i}}[\mathrm{i}\omega_m - \xi_{\boldsymbol{q}}^\alpha]\Phi_q^\alpha \ , \tag{5.118}$$

so that Eq.(5.117) can also be written as

$$\Psi_q^\alpha = \phi_q^\alpha + \sum_{q'q''} \delta_{q,q'+q''} \gamma_{q',q''}^\alpha \Psi_{q'}^\alpha \Psi_{q''}^\alpha \ , \tag{5.119}$$

where the dimensionless kernel is

$$\gamma^{\alpha}_{q',q''} = \frac{\mathrm{i}}{\beta} \boldsymbol{q}'(2\mathsf{M}^{\alpha})^{-1}\boldsymbol{q}'' G^{\alpha}_{\mathrm{b}}(q') G^{\alpha}_{\mathrm{b}}(q'') \quad . \tag{5.120}$$

Note that this kernel is symmetric under the exchange $q' \leftrightarrow q''$. The $q \equiv [\boldsymbol{q}, \mathrm{i}\omega_m] = 0$-term requires a special treatment. Setting $q = 0$ on both sides of Eq.(5.117), we obtain

$$0 = V^{\alpha}_0 - \sum_{q'} \boldsymbol{q}'(2\mathsf{M}^{\alpha})^{-1}\boldsymbol{q}' \Phi^{\alpha}_{-q'} \Phi^{\alpha}_{q'} \quad . \tag{5.121}$$

Subtracting this from Eq.(5.117), we see that $\Psi^{\alpha}_0 = 0$. With the above definitions, the eikonal can be written as

$$\Phi^{\alpha}(\boldsymbol{r},\tau) - \Phi^{\alpha}(\boldsymbol{r}',\tau') = \sum_{q} \mathcal{J}^{\alpha}_{-q}(\boldsymbol{r},\boldsymbol{r}',\tau,\tau') \Psi^{\alpha}_{q} \quad , \tag{5.122}$$

where

$$\mathcal{J}^{\alpha}_{-q}(\boldsymbol{r},\boldsymbol{r}',\tau,\tau') = \frac{\mathrm{i}}{\beta} G^{\alpha}_{\mathrm{b}}(q) \left[\mathrm{e}^{\mathrm{i}(\boldsymbol{q}\cdot\boldsymbol{r} - \omega_m \tau)} - \mathrm{e}^{\mathrm{i}(\boldsymbol{q}\cdot\boldsymbol{r}' - \omega_m \tau')} \right] \quad . \tag{5.123}$$

Note that for linearized energy dispersion $\mathcal{J}^{\alpha}_{q}(\boldsymbol{r},0,\tau,0)$ reduces precisely to the function $\mathcal{J}^{\alpha}_{q}(\boldsymbol{r},\tau)$ defined in Eq.(5.27). By iteration of the non-linear integral equation (5.119) it is easy to obtain an expansion of the functional Ψ^{α}_{q} in powers of the Hubbard-Stratonovich field $\phi^{\alpha}_{q} = \frac{\beta}{\mathrm{i}} V^{\alpha}_{q}$,

$$\Psi^{\alpha}_{q} = \sum_{n=1}^{\infty} \Psi^{\alpha}_{n,q} \quad , \tag{5.124}$$

where for $q \neq 0$ the functional $\Psi^{\alpha}_{n,q}$ is of the form

$$\Psi^{\alpha}_{n,q} = \sum_{q_1 \cdots q_n} \delta_{q,q_1+\ldots+q_n} \tilde{U}^{\alpha}_{n}(q_1 \ldots q_n) \phi^{\alpha}_{q_1} \cdots \phi^{\alpha}_{q_n} \quad . \tag{5.125}$$

For practical calculations beyond the Gaussian approximation it is useful to have a diagrammatic representation of Eq.(5.122), which is defined in Fig. 5.4. The dimensionless vertices $\tilde{U}^{\alpha}_{n}$ are proportional to $(1/m^{\alpha})^{n-1}$. The first three vertices are

$$\tilde{U}^{\alpha}_{1}(q_1) = 1 \quad , \quad \tilde{U}^{\alpha}_{2}(q_1 q_2) = \gamma^{\alpha}_{q_1,q_2} \quad , \tag{5.126}$$

$$\tilde{U}^{\alpha}_{3}(q_1 q_2 q_3) = \frac{2}{3} \left[\gamma^{\alpha}_{q_1,q_2} \gamma^{\alpha}_{q_1+q_2,q_3} + \gamma^{\alpha}_{q_2,q_3} \gamma^{\alpha}_{q_2+q_3,q_1} + \gamma^{\alpha}_{q_3,q_1} \gamma^{\alpha}_{q_3+q_1,q_2} \right] . \tag{5.127}$$

We have used the invariance of Eq.(5.125) under relabeling of the fields to symmetrize the vertices with respect to the interchange of any two labels. Substituting Eqs.(5.124) and (5.125) into Eq.(5.122), we obtain the desired expansion of the eikonal in powers of the Hubbard-Stratonovich field ϕ^{α}. Note that each iteration involves an additional power of ϕ^{α}/m^{α}. Because the Gaussian propagator of ϕ^{α}-field is proportional to the RPA interaction (see

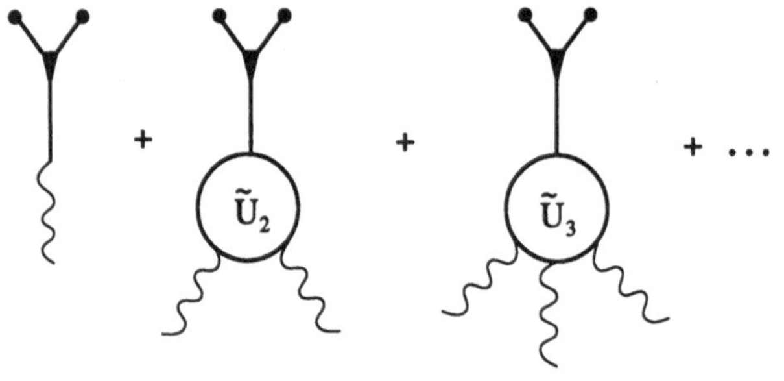

Fig. 5.4. Graphical representation of the functional $\Phi^\alpha(\boldsymbol{r},\tau) - \Phi^\alpha(\boldsymbol{r}',\tau')$ defined in Eq.(5.122). As in Fig. 4.1, the ϕ^α-fields are represented by wavy lines. The Y-shaped symbol represents the function $\mathcal{J}^\alpha_{-q}(\boldsymbol{r},\boldsymbol{r}',\tau,\tau')$ defined in Eq.(5.123). The solid dots represent the external points $\boldsymbol{r},\tau$ and $\boldsymbol{r}',\tau'$.

Eq.(4.32)), the small parameter controlling this expansion is proportional to $f_q^{\mathrm{RPA},\alpha}/m^\alpha$. This will become more evident in Sect. 5.2.2, where we explicitly calculate the leading corrections to the Gaussian approximation for the average eikonal.

The Dyson equation for the prefactor Green's function

Having solved Eq.(5.119) to a certain order in ϕ^α, we know also the random velocity $\boldsymbol{u}^\alpha(\boldsymbol{r},\tau)$ in Eq.(5.107) (and hence the derivative potential $\boldsymbol{u}^\alpha(\boldsymbol{r},\tau)\cdot\boldsymbol{P}_{\boldsymbol{r}}$ in Eq.(5.109)) to the same order in ϕ^α. For practical calculations we find it again more convenient to work in Fourier space. Defining the Fourier transform $[\hat{G}_1^\alpha]_{\tilde{q}\tilde{q}'}$ of $\mathcal{G}_1^\alpha(\boldsymbol{r},\boldsymbol{r}',\tau,\tau')$ as in Eq.(5.12), it is easy to see that in Fourier space Eq.(5.109) is equivalent with the Dyson equation

$$[\hat{G}_1^\alpha]_{\tilde{q}\tilde{q}'} = \delta_{\tilde{q}\tilde{q}'} G_0^\alpha(\tilde{q}) + G_0^\alpha(\tilde{q}) \sum_{\tilde{q}''} [\hat{D}^\alpha]_{\tilde{q}\tilde{q}''} [\hat{G}_1^\alpha]_{\tilde{q}''\tilde{q}'} \quad , \tag{5.128}$$

where the matrix elements of the derivative potential are

$$[\hat{D}^\alpha]_{\tilde{q}\tilde{q}'} = \Psi^\alpha_{\tilde{q}-\tilde{q}'} \lambda^\alpha_{\tilde{q},\tilde{q}'} \quad . \tag{5.129}$$

Here Ψ_q^α is defined as functional of the ϕ^α-field via the non-linear integral equation (5.119), and the vertex $\lambda^\alpha_{\tilde{q},\tilde{q}'}$ is given by

$$\lambda^\alpha_{\tilde{q},\tilde{q}'} = \frac{\mathrm{i}}{\beta}(\boldsymbol{q}-\boldsymbol{q}')(\mathsf{M}^\alpha)^{-1}\boldsymbol{q}' G_\mathrm{b}^\alpha(\tilde{q}-\tilde{q}') \quad . \tag{5.130}$$

Iteration of Eq.(5.128) generates an expansion of $\hat{G}_1^\alpha$ in powers of the derivative potential. A graphical representation of Eq.(5.128) is shown in Fig. 5.5.

5.2.2 Non-Gaussian averaging

We now average the Green's function $\mathcal{G}^\alpha(\boldsymbol{r},\boldsymbol{r}',\tau,\tau')$ with respect to the probability distribution of the background field.

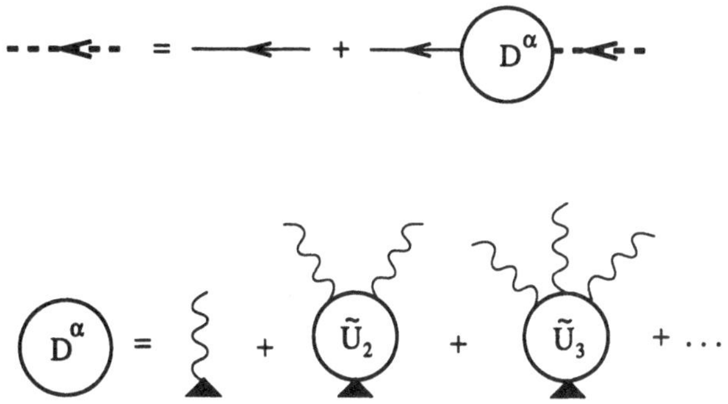

Fig. 5.5. Diagrammatic representation of the Dyson equation for $\hat{G}_1^\alpha$, which is represented by a thick dashed line. The solid triangle denotes the vertex $\lambda^\alpha_{\bar{q},\bar{q}'}$ defined in Eq.(5.130).

To obtain the Green's function of the many-body system, we need to average the Green's function $\mathcal{G}^\alpha(\boldsymbol{r},\boldsymbol{r}',\tau,\tau')$ given in Eq.(5.104) with respect to the probability distribution $\mathcal{P}\{\phi^\alpha\}$ defined in Eq.(3.35). Because for finite masses m_i^α the effective action $S_{\text{eff}}\{\phi^\alpha\}$ in Eqs.(3.36) and (3.37) is not Gaussian, we have to use perturbation theory to perform the averaging procedure. Recall that the leading non-Gaussian corrections to $S_{\text{eff}}\{\phi^\alpha\}$ have been explicitly calculated in Chap. 4.3.2, see Eq.(4.77). Because averaging restores translational invariance, we may set $\boldsymbol{r}' = \tau' = 0$ and calculate

$$G^\alpha(\boldsymbol{r},\tau) = \langle \mathcal{G}^\alpha(\boldsymbol{r},0,\tau,0)\rangle_{S_{\text{eff}}} \quad . \tag{5.131}$$

Let us parameterize the average Green's function as

$$G^\alpha(\boldsymbol{r},\tau) = [G_1^\alpha(\boldsymbol{r},\tau) + G_2^\alpha(\boldsymbol{r},\tau)]\mathrm{e}^{Q^\alpha(\boldsymbol{r},\tau)} \quad , \tag{5.132}$$

where

$$Q^\alpha(\boldsymbol{r},\tau) = \ln\left\langle \mathrm{e}^{\Phi^\alpha(\boldsymbol{r},\tau)-\Phi^\alpha(0,0)}\right\rangle_{S_{\text{eff}}} \quad , \tag{5.133}$$

$$G_1^\alpha(\boldsymbol{r},\tau) = \langle \mathcal{G}_1^\alpha(\boldsymbol{r},0,\tau,0)\rangle_{S_{\text{eff}}} \quad , \tag{5.134}$$

and the function $G_2^\alpha(\boldsymbol{r},\tau)$ contains all correlations between the two factors in Eq.(5.104),

$$G_2^\alpha(\boldsymbol{r},\tau) = \frac{\left\langle \delta\mathcal{G}_1^\alpha(\boldsymbol{r},0,\tau,0)\delta\mathrm{e}^{\Phi^\alpha(\boldsymbol{r},\tau)-\Phi^\alpha(0,0)}\right\rangle_{S_{\text{eff}}}}{\left\langle \mathrm{e}^{\Phi^\alpha(\boldsymbol{r},\tau)-\Phi^\alpha(0,0)}\right\rangle_{S_{\text{eff}}}} \quad . \tag{5.135}$$

Here $\delta X = X - < X >_{S_{\text{eff}}}$. We emphasize that Eqs.(5.132)–(5.135) are an exact decomposition of the different contributions to Eq.(5.131), the usefulness of which will become evident shortly.

Let us now consider in some detail the calculation of the function $Q^\alpha(\boldsymbol{r},\tau)$ defined in Eq.(5.133). By definition we have[9]

[9] Note that the label α on the right-hand side of Eq.(5.136) is an external label which is not summed over. The summation labels are denoted by α'.

$$Q^{\alpha}(\boldsymbol{r},\tau) = \ln\left(\frac{\int \mathcal{D}\{\phi^{\alpha'}\}\exp\left[-S_{\mathrm{eff}}\{\phi^{\alpha'}\} + \sum_q \mathcal{J}^{\alpha}_{-q}(\boldsymbol{r},\tau)\phi^{\alpha}_q + \mathcal{F}^{\alpha}\{\mathcal{J}^{\alpha},\phi^{\alpha}\}\right]}{\int \mathcal{D}\{\phi^{\alpha'}\}\exp\left[-S_{\mathrm{eff}}\{\phi^{\alpha'}\}\right]}\right) , \tag{5.136}$$

where $\mathcal{J}^{\alpha}_{-q}(\boldsymbol{r},\tau) \equiv \mathcal{J}^{\alpha}_{-q}(\boldsymbol{r},0,\tau,0)$, and the functional $\mathcal{F}^{\alpha}\{\mathcal{J}^{\alpha},\phi^{\alpha}\}$ is defined as the sum of all terms on the right-hand side of Eq.(5.122) involving more than one power of the ϕ^{α}-field. Explicitly, the first two terms are

$$\mathcal{F}^{\alpha}\{\mathcal{J}^{\alpha},\phi^{\alpha}\} = \sum_{q,q_1,q_2} \mathcal{J}^{\alpha}_{-q}(\boldsymbol{r},\tau)\delta_{q,q_1+q_2}\tilde{U}^{\alpha}_2(q_1q_2)\phi^{\alpha}_{q_1}\phi^{\alpha}_{q_2}$$
$$+ \sum_{q,q_1,q_2,q_3} \mathcal{J}^{\alpha}_{-q}(\boldsymbol{r},\tau)\delta_{q,q_1+q_2+q_3}\tilde{U}^{\alpha}_3(q_1q_2q_3)\phi^{\alpha}_{q_1}\phi^{\alpha}_{q_2}\phi^{\alpha}_{q_3} + \ldots . \tag{5.137}$$

These terms correspond precisely to the diagrams with two and three wavy lines in Fig. 5.4. Following the procedure outlined in Chap. 4.3, we write

$$S_{\mathrm{eff}}\{\phi^{\alpha'}\} = \mathrm{i}\sum_{\alpha'} \phi^{\alpha'}_0 N^{\alpha'}_0 + S_{\mathrm{eff},2}\{\phi^{\alpha'}\} + S'_{\mathrm{kin}}\{\phi^{\alpha'}\} , \tag{5.138}$$

where the Gaussian part $S_{\mathrm{eff},2}\{\phi^{\alpha'}\}$ of the effective action is given in Eq.(4.30), and the non-Gaussian part $S'_{\mathrm{kin}}\{\phi^{\alpha'}\}$ is defined in Eq.(4.58). The leading contributions to $S'_{\mathrm{kin}}\{\phi^{\alpha'}\}$ are explicitly given in Eq.(4.77). After eliminating the second term in the numerator of Eq.(5.136) by means of the shift-transformation

$$\phi^{\alpha'}_q \to \phi^{\alpha'}_q + [\underline{\tilde{f}}^{\mathrm{RPA}}_q]^{\alpha'\alpha}\mathcal{J}^{\alpha}_q(\boldsymbol{r},\tau) , \tag{5.139}$$

where $\underline{\tilde{f}}^{\mathrm{RPA}}_q \equiv \frac{\beta}{V}\underline{f}^{\mathrm{RPA}}_q$ is the rescaled RPA interaction matrix (see Eq.(4.33)), we obtain

$$Q^{\alpha}(\boldsymbol{r},\tau) = \frac{1}{2}\sum_q \tilde{f}^{\mathrm{RPA},\alpha}_q \mathcal{J}^{\alpha}_{-q}(\boldsymbol{r},\tau)\mathcal{J}^{\alpha}_q(\boldsymbol{r},\tau)$$
$$+\ln\left\langle \exp\left[-S'_{\mathrm{kin}}\{\phi^{\alpha'}_q + [\underline{\tilde{f}}^{\mathrm{RPA}}_q]^{\alpha'\alpha}\mathcal{J}^{\alpha}_q\} + \mathcal{F}^{\alpha}\{\mathcal{J}^{\alpha}_q, \phi^{\alpha}_q + \tilde{f}^{\mathrm{RPA},\alpha}_q\mathcal{J}^{\alpha}_q\}\right]\right\rangle_{S_{\mathrm{eff},2}}$$
$$-\ln\left\langle \exp\left[-S'_{\mathrm{kin}}\{\phi^{\alpha'}_q\}\right]\right\rangle_{S_{\mathrm{eff},2}} . \tag{5.140}$$

We have used the notation $\tilde{f}^{\mathrm{RPA},\alpha}_q = [\underline{\tilde{f}}^{\mathrm{RPA}}_q]^{\alpha\alpha} = \frac{\beta}{V}[\underline{f}^{\mathrm{RPA}}_q]^{\alpha\alpha}$, see also Eq.(5.34). Finally, we use the linked cluster theorem [2.10] and obtain, in complete analogy with Eq.(4.71),

$$Q^{\alpha}(\boldsymbol{r},\tau) = \frac{1}{2}\sum_q \tilde{f}^{\mathrm{RPA},\alpha}_q \mathcal{J}^{\alpha}_{-q}(\boldsymbol{r},\tau)\mathcal{J}^{\alpha}_q(\boldsymbol{r},\tau) + \sum_{n=1}^{\infty}\frac{(-1)^n}{n}$$
$$\times\left\{\left\langle\left[S'_{\mathrm{kin}}\{\phi^{\alpha'}_q + [\underline{\tilde{f}}^{\mathrm{RPA}}_q]^{\alpha'\alpha}\mathcal{J}^{\alpha}_q\} - \mathcal{F}^{\alpha}\{\mathcal{J}^{\alpha}_q, \phi^{\alpha}_q + \tilde{f}^{\mathrm{RPA},\alpha}_q\mathcal{J}^{\alpha}_q\}\right]^n\right\rangle^{\mathrm{con}}_{S_{\mathrm{eff},2}}\right.$$

$$- \left\langle \left[S'_{\mathrm{kin}} \{ \phi_q^{\alpha'} \} \right]^n \right\rangle_{S_{\mathrm{eff},2}}^{\mathrm{con}} \Bigg\} \quad . \tag{5.141}$$

From this expression it is obvious that in general the function $Q^\alpha(\boldsymbol{r}, \tau)$ can be written as

$$Q^\alpha(\boldsymbol{r}, \tau) = \sum_{n=1}^{\infty} Q_n^\alpha(\boldsymbol{r}, \tau) \quad , \tag{5.142}$$

where $Q_n^\alpha(\boldsymbol{r}, \tau)$ involves $n+1$ powers of the function $\mathcal{J}_q^\alpha(\boldsymbol{r}, \tau)$,

$$Q_n^\alpha(\boldsymbol{r}, \tau) = \sum_{q q_1 \dots q_n} \delta_{q, q_1 + \dots + q_n} W_n^\alpha(q_1 \dots q_n) \\ \times \mathcal{J}_{-q}^\alpha(\boldsymbol{r}, \tau) \mathcal{J}_{q_1}^\alpha(\boldsymbol{r}, \tau) \cdots \mathcal{J}_{q_n}^\alpha(\boldsymbol{r}, \tau) \quad . \tag{5.143}$$

The vertices $W_n^\alpha(q_1 \dots q_n)$ can be calculated perturbatively in powers of the RPA interaction. Evidently the first term in Eq.(5.141) corresponds to the following contribution to W_1^α,

$$W_{1,1}^\alpha(q_1) = \frac{1}{2} \tilde{f}_{q_1}^{\mathrm{RPA},\alpha} \quad . \tag{5.144}$$

The crucial point is now that all other terms contain at least two powers of $f^{\mathrm{RPA},\alpha}$, so that, to first order in the RPA interaction, higher order terms can be neglected. In general, each vertex can be expanded as

$$W_n^\alpha(q_1 \dots q_n) = \sum_{m=n}^{\infty} W_{n,m}^\alpha(q_1 \dots q_n) \quad , \tag{5.145}$$

where the second subscript gives the power of the RPA interaction. Because the nth-order vertex W_n^α involves at least n powers of the RPA interaction, the m-sum in Eq.(5.145) starts at $m = n$. This is due to the fact that each of the higher-order diagrams in Fig. 5.4 contains a single function $\mathcal{J}^\alpha$, but additional powers of the ϕ^α-field. Actually, from Chap. 4.3.4 we expect that the true small parameter which controls the corrections to the Gaussian approximation should also involve the local curvature of the Fermi surface (i.e. the inverse effective masses m_i^α) and the range of the interaction in momentum space. To investigate this point, it is convenient to visualize the structure of the perturbation expansion for the higher-order terms with the help of the graphical elements introduced in Fig. 5.4. To order $(f^{\mathrm{RPA},\alpha})^2$, we should retain

$$W_1^\alpha(q_1) \approx W_{1,1}^\alpha(q_1) + W_{1,2}^\alpha(q_1) \quad , \tag{5.146}$$

$$W_2^\alpha(q_1, q_2) \approx W_{2,2}^\alpha(q_1, q_2) \quad , \tag{5.147}$$

and neglect all W_n^α with $n \geq 3$. The diagrams contributing to $W_{1,2}^\alpha(q_1)$ and $W_{2,2}^\alpha(q_1, q_2)$ are shown in Fig. 5.6. The explicit expressions are

$$W_{1,2}^\alpha(q_1) = 3 \tilde{f}_{q_1}^{\mathrm{RPA},\alpha} \sum_{q_2} \tilde{U}_3^\alpha(q_1, q_2, -q_2) \tilde{f}_{q_2}^{\mathrm{RPA},\alpha}$$

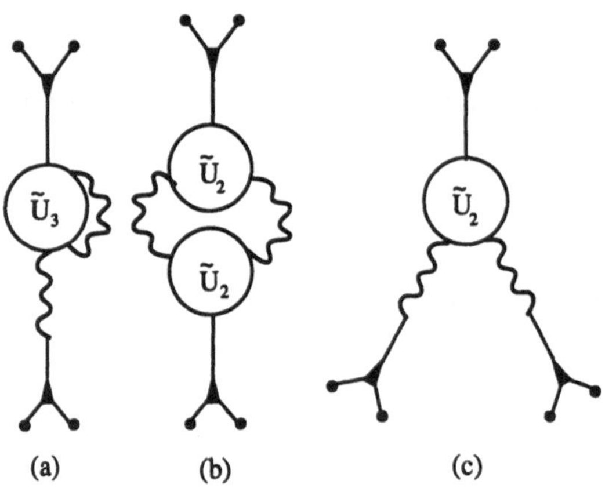

Fig. 5.6. The leading non-Gaussian corrections to the average eikonal. The thick wavy line is the Gaussian propagator of the ϕ^α-field, i.e. the RPA screened interaction. The other symbols are defined in Fig. 5.4. The diagrams (a) and (b) contain two factors of $\mathcal{J}^\alpha$ and hence contribute to $W^\alpha_{1,2}$. Diagrams (a) and (b) represent the first and second term in Eq.(5.148). Diagram (c) contains three factors of $\mathcal{J}^\alpha$ and is the only contribution to $W^\alpha_{2,2}$, see Eq.(5.149).

$$+\sum_{q_2} \tilde{U}^\alpha_2(q_2, q_1 - q_2)\tilde{U}^\alpha_2(-q_2, q_2 - q_1) f^{\mathrm{RPA},\alpha}_{q_1-q_2} f^{\mathrm{RPA},\alpha}_{q_2} \quad , \tag{5.148}$$

$$W^\alpha_{2,2}(q_1, q_2) = \tilde{U}^\alpha_2(q_1, q_2) f^{\mathrm{RPA},\alpha}_{q_1} f^{\mathrm{RPA},\alpha}_{q_2} \quad . \tag{5.149}$$

Because by construction $\tilde{U}^\alpha_n$ is proportional to $(1/m^\alpha)^{n-1}$, it is clear that $W_{1,2} \propto (f^{\mathrm{RPA},\alpha}/m^\alpha)^2$, while $W_{2,2} \propto (f^{\mathrm{RPA},\alpha})^2/m^\alpha$. Thus, the corrections to the first order term in the average eikonal are not only controlled by higher powers of the RPA interaction, but also by higher powers of the inverse effective mass $1/m^\alpha$. Note that $1/m^\alpha$ is a measure[10] for the local curvature of the Fermi surface close to $\boldsymbol{k}^\alpha$. Moreover, for interactions that are sufficiently well behaved for $\boldsymbol{q} \to 0$ and have a natural cutoff $q_c \ll k_F$ in momentum space, each additional loop integration in Eq.(5.143) gives rise to a factor of $(q_c/k_F)^d$. We therefore conclude that the leading correction to the Gaussian approximation for the average eikonal is controlled by the same dimensionless small parameter that appears in the calculation of the non-Gaussian correction to the density-density correlation function, see Eq.(4.115).

The analysis of next-to-leading terms is rather complicated. Clearly, all contributions to the functions $W^\alpha_{n,n}$ that involve only the vertices $\tilde{U}^\alpha_n$ are at tree-level proportional to $(f^{\mathrm{RPA},\alpha})^n/(m^\alpha)^{n-1}$. This is due to the fact that by construction the vertex $\tilde{U}^\alpha_n$ is proportional to $1/(m^\alpha)^{n-1}$. However, at order $(f^{\mathrm{RPA},\alpha})^3$ corrections due to the non-Gaussian part $S'_{\mathrm{kin}}\{\phi^\alpha\}$ of the effective action for the ϕ^α-field must also be taken into account. These involve the vertices U_n defined in Eq.(4.5). The leading contributions of this type are shown in Fig. 5.7. Certainly, a subset of these diagrams leads to the replacement of the Gaussian propagator $< \phi^\alpha_q \phi^\alpha_{-q} >_{S_{\mathrm{eff},2}} = f^{\mathrm{RPA},\alpha}_q$ by the exact effective propagator $< \phi^\alpha_q \phi^\alpha_{-q} >_{S_{\mathrm{eff}}}$, which depends on the exact dielectric function

[10] The relevant *dimensionless* parameter C^α which measures the local curvature of the Fermi surface has been identified in Chap. 4.3.4 (see Eqs.(4.113) and (4.114)).

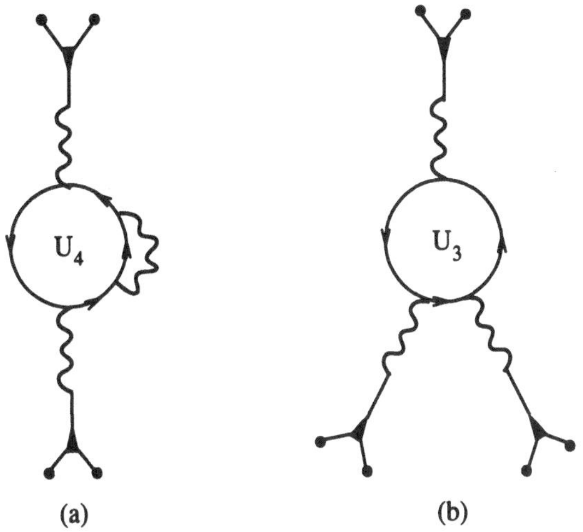

Fig. 5.7.
Lowest order corrections to the average eikonal due to the non-Gaussian terms of the probability distribution. Both diagrams represent corrections of order $(f^{\mathrm{RPA},\alpha})^3$. The diagram (a) involves the vertex U_4 (see Eq.(4.77)) and renormalizes the RPA interaction in Eq.(5.144). The diagram (b) involves U_3 and leads to a renormalization of the vertex $\tilde{U}_2^{\alpha}$ in Eq.(5.149).

of the many-body system. However, the non-Gaussian part $S'_{\mathrm{kin}}\{\phi^{\alpha}\}$ of our effective action will also give rise to more complicated vertex corrections.

Although the vertices U_n do not explicitly contain the curvature parameter $1/m^{\alpha}$, the closed loop theorem discussed in Chap. 4.1 implies that in the infrared limit the closed fermion loops in Fig. 5.7 lead to large-scale cancellations, so that we expect that also these higher order terms are proportional to powers of the inverse effective masses. Fortunately, the diagrams in Fig. 5.7 are of order $(f^{\mathrm{RPA},\alpha})^3$ and therefore do not contribute to the leading correction to the Gaussian approximation.

Finally we would like to point out that for models with spin the non-Gaussian corrections to the average eikonal lead to a mixing of the density fields with spin fields. This implies that in the one-dimensional Tomonaga-Luttinger model the non-linear terms in the energy dispersion destroy the spin-charge separation [5.13].

5.3 The Gaussian approximation with non-linear energy dispersion

We now perform all averaging operations to first order in the RPA interaction. We emphasize again that we do not expand in powers of $1/m^{\alpha}$, so that curvature effects are taken into account non-perturbatively.

5.3.1 The average eikonal

From now on we shall restrict ourselves to the first order in the RPA interaction. Then it is sufficient to retain only the term $Q_1^{\alpha}(\boldsymbol{r},\tau)$ in Eq.(5.142), and approximate the vertex $W_1^\alpha(q_1)$ by Eq.(5.144). Using the definition of $\mathcal{J}_{-q}^\alpha(\boldsymbol{r},\tau)$ in Eq.(5.123), we have

$$\mathcal{J}_{-q}^\alpha(\boldsymbol{r},\tau)\mathcal{J}_q^\alpha(\boldsymbol{r},\tau) = -\frac{2}{\beta^2}G_{\mathrm{b}}^\alpha(-q)G_{\mathrm{b}}^\alpha(q)\left[1-\cos\left(\boldsymbol{q}\cdot\boldsymbol{r}-\omega_m\tau\right)\right]$$
$$= \frac{2}{\beta^2}\frac{1-\cos\left(\boldsymbol{q}\cdot\boldsymbol{r}-\omega_m\tau\right)}{[\mathrm{i}\omega_m-\xi_{\boldsymbol{q}}^\alpha][\mathrm{i}\omega_m+\xi_{-\boldsymbol{q}}^\alpha]} \quad , \tag{5.150}$$

so that we obtain, in complete analogy with Eqs.(5.31)–(5.33),

$$Q_1^\alpha(\boldsymbol{r},\tau) = R_1^\alpha - S_1^\alpha(\boldsymbol{r},\tau) \quad , \tag{5.151}$$

with

$$R_1^\alpha = \frac{1}{\beta V}\sum_q \frac{f_{\boldsymbol{q}}^{\mathrm{RPA},\alpha}}{[\mathrm{i}\omega_m-\xi_{\boldsymbol{q}}^\alpha][\mathrm{i}\omega_m+\xi_{-\boldsymbol{q}}^\alpha]} = S_1^\alpha(0,0) \quad , \tag{5.152}$$

$$S_1^\alpha(\boldsymbol{r},\tau) = \frac{1}{\beta V}\sum_q \frac{f_{\boldsymbol{q}}^{\mathrm{RPA},\alpha}\cos(\boldsymbol{q}\cdot\boldsymbol{r}-\omega_m\tau)}{[\mathrm{i}\omega_m-\xi_{\boldsymbol{q}}^\alpha][\mathrm{i}\omega_m+\xi_{-\boldsymbol{q}}^\alpha]} \quad . \tag{5.153}$$

Note that these expressions contain the non-linear terms in the energy dispersion via $\xi_{\boldsymbol{q}}^\alpha$ *and* $f_{\boldsymbol{q}}^{\mathrm{RPA},\alpha}$ in a non-perturbative way. Moreover, all problems due to the double pole in the corresponding expressions for linearized energy dispersion (see the discussion in the introduction to Sect. 5.2) have disappeared in Eqs.(5.151)–(5.153) in an almost trivial way, because

$$\xi_{-\boldsymbol{q}}^\alpha = -\xi_{\boldsymbol{q}}^\alpha + \boldsymbol{q}(\mathsf{M}^\alpha)^{-1}\boldsymbol{q} \quad , \tag{5.154}$$

so that

$$\frac{1}{[\mathrm{i}\omega_m-\xi_{\boldsymbol{q}}^\alpha][\mathrm{i}\omega_m+\xi_{-\boldsymbol{q}}^\alpha]} = \frac{1}{\boldsymbol{q}(\mathsf{M}^\alpha)^{-1}\boldsymbol{q}}\left[\frac{1}{\mathrm{i}\omega_m-\xi_{\boldsymbol{q}}^\alpha} - \frac{1}{\mathrm{i}\omega_m+\xi_{-\boldsymbol{q}}^\alpha}\right] \quad . \tag{5.155}$$

Hence, as long as at least one of the inverse effective masses $1/m_i^\alpha$ is finite, the denominator in Eqs.(5.152) and (5.153) gives only rise to *simple poles* in the complex frequency plane. In fact, as will be discussed in more detail in Chaps. 6.1.3 and 9.4, the double pole that appears in the Debye-Waller factor for linearized energy dispersion gives rise to some rather peculiar and probably unphysical features in the analytic structure of the Green's function in Fourier space.

5.3.2 The prefactor Green's functions

We use the impurity diagram technique to calculate the leading non-trivial contributions to the Green's functions G_1^α and G_2^α defined in Eqs.(5.134) and (5.135).

Calculation of G_1^α

Let us first consider $G_1^\alpha(\boldsymbol{r},\tau) = \langle \mathcal{G}_1^\alpha(\boldsymbol{r},0,\tau,0)\rangle_{S_{\text{eff}}}$. Naively one might try a direct expansion of $G_1^\alpha(\boldsymbol{r},\tau)$ in powers of $f^{\text{RPA},\alpha}$. The terms in this expansion are easily generated by iterating the Dyson equation (5.128) and then averaging. Because for $q \neq 0$ the Gaussian average $< \phi_q^\alpha >_{S_{\text{eff},2}}$ vanishes, the leading term (of order $f^{\text{RPA},\alpha}$) arises from the second iteration of Eq.(5.128). However, as already mentioned in Chap. 1.1, the direct expansion of a single-particle Green's function in powers of the interaction is usually ill-defined, because a truncation at any finite order generates unphysical multiple poles in Fourier space. Within a perturbative approach, this problem is avoided by calculating the irreducible self-energy to some finite order in the interaction, and extrapolating the perturbation series by solving the Dyson equation. Thus, introducing the Fourier transform of $G_1^\alpha(\boldsymbol{r},\tau)$ as usual (see Eq.(5.116)),

$$G_1^\alpha(\boldsymbol{r},\tau) = \frac{1}{\beta V}\sum_{\tilde{q}} e^{i(\boldsymbol{q}\cdot\boldsymbol{r}-\tilde{\omega}_n\tau)} G_1^\alpha(\tilde{q}) \quad , \tag{5.156}$$

we define the irreducible self-energy $\Sigma_1^\alpha(\tilde{q})$ via the Dyson equation for $G_1^\alpha(\tilde{q})$,

$$[G_1^\alpha(\tilde{q})]^{-1} = [G_0^\alpha(\tilde{q})]^{-1} - \Sigma_1^\alpha(\tilde{q}) \quad . \tag{5.157}$$

We now use the self-consistent Born approximation to calculate the self-energy $\Sigma_1^\alpha(\tilde{q})$. This is a standard approximation in the theory of disordered systems [1.3], which is expected to be accurate if interference terms are negligible. The corresponding Feynman diagram is shown in Fig. 5.8 (a), and yields

$$\Sigma_1^\alpha(\tilde{q}) = -\frac{1}{\beta V}\sum_{\tilde{q}'} \lambda^\alpha_{\tilde{q},\tilde{q}'}\lambda^\alpha_{\tilde{q}',\tilde{q}} f^{\text{RPA},\alpha}_{\tilde{q}-\tilde{q}'} G_1^\alpha(\tilde{q}') \quad , \tag{5.158}$$

where the dimensionless vertex $\lambda^\alpha_{\tilde{q},\tilde{q}'}$ is defined in Eq.(5.130). At the first sight

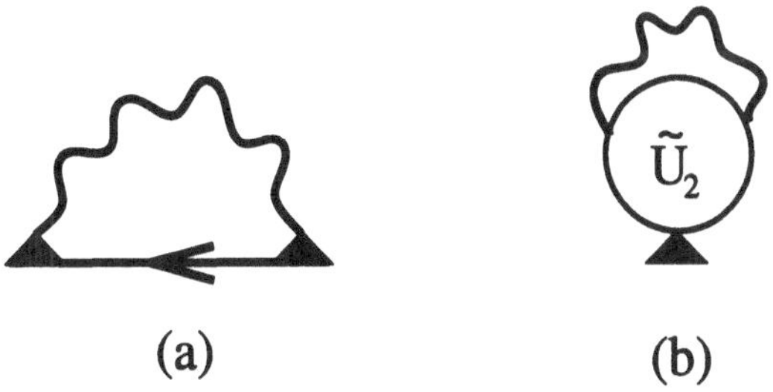

Fig. 5.8. (a) Self-consistent Born approximation for the self-energy $\Sigma_1^\alpha(\tilde{q})$. The thick solid arrow denotes the self-consistent Green's function $G_1^\alpha(\tilde{q})$. (b) This contribution to $\Sigma_1^\alpha(\tilde{q})$ vanishes.

it seems that the averaging procedure gives also rise to another contribution of order $f^{\text{RPA},\alpha}$ to Σ_1^α, which is shown in Fig. 5.8(b). This contribution is generated by averaging the $\tilde{U}_2^\alpha$-term in Fig. 5.5, and physically describes a renormalization of the chemical potential. However, according to Eq.(5.130) the vertex $\lambda^{\alpha}_{\tilde{q},\tilde{q}'}$ vanishes for $\boldsymbol{q} = \boldsymbol{q}'$, implying that the contribution from diagram (b) in Fig. 5.8 vanishes as well. In the language of many-body theory, the diagram (a) in Fig. 5.8 is the self-consistent GW diagram for the self-energy associated with $G^{\alpha}_{1}(\tilde{q})$. Comparing Eq.(5.158) with the expression (5.74) for the usual GW self-energy associated with the full Green's function $G^{\alpha}(\tilde{q})$, we see that the GW approximation for $\Sigma^{\alpha}_{1}(\tilde{q})$ involves two additional powers of the vertex $\lambda^{\alpha}_{\tilde{q},\tilde{q}'}$ defined in Eq.(5.130). The crucial point is now that *this additional vertex makes the GW self-energy associated with G^{α}_{1} less infrared singular than the corresponding GW self-energy of the full Green's function G^{α}*. To see this more clearly, we substitute Eq.(5.130) into Eq.(5.158) and shift the summation variable according to $\tilde{q} - \tilde{q}' = -q'$. Then we obtain

$$\Sigma^{\alpha}_{1}(\tilde{q}) = \frac{1}{\beta V} \sum_{q'} \left[\boldsymbol{q}'(\mathsf{M}^{\alpha})^{-1}(\boldsymbol{q}+\boldsymbol{q}')\right] \left[\boldsymbol{q}'(\mathsf{M}^{\alpha})^{-1}\boldsymbol{q}\right]$$
$$\times\, G^{\alpha}_{\mathrm{b}}(q') G^{\alpha}_{\mathrm{b}}(-q') f^{\mathrm{RPA},\alpha}_{q'} G^{\alpha}_{1}(\tilde{q}+q') \quad . \tag{5.159}$$

Using the symmetries of the integrand under renaming $q' \to -q'$, we find that Eq.(5.159) can also be written as

$$\Sigma^{\alpha}_{1}(\tilde{q}) = -\frac{1}{\beta V} \sum_{q'} \frac{f^{\mathrm{RPA},\alpha}_{q'}}{[\mathrm{i}\omega_{m'} - \xi^{\alpha}_{\boldsymbol{q}'}][\mathrm{i}\omega_{m'} + \xi^{\alpha}_{-\boldsymbol{q}'}]}$$
$$\times \frac{1}{2}\Big\{ \left[\boldsymbol{q}(\mathsf{M}^{\alpha})^{-1}\boldsymbol{q}'\right]^2 \left[G^{\alpha}_{1}(\tilde{q}+q') + G^{\alpha}_{1}(\tilde{q}-q')\right]$$
$$+ \left[\boldsymbol{q}(\mathsf{M}^{\alpha})^{-1}\boldsymbol{q}'\right] \left[\boldsymbol{q}'(\mathsf{M}^{\alpha})^{-1}\boldsymbol{q}'\right] \left[G^{\alpha}_{1}(\tilde{q}+q') - G^{\alpha}_{1}(\tilde{q}-q')\right]\Big\} \quad . \tag{5.160}$$

To see that the infrared behavior of $\Sigma^{\alpha}_{1}(\tilde{q})$ is less singular than that of the self-energy $\Sigma^{\alpha}(\tilde{q})$ associated the full Green's function, note that the first line in Eq.(5.158) is (up to a sign) identical with the factor R^{α}_{1} given in Eq.(5.152). But we know from Sect. 5.1.5 that the finiteness of this factor implies a finite quasi-particle residue. Conversely, non-Fermi liquid behavior should manifest itself via infrared divergencies in R^{α}_{1}. The crucial point is now that the second and third lines in Eq.(5.160) contain additional powers of $\boldsymbol{q}'$, so that, at least for not too singular interactions, $\Sigma^{\alpha}_{1}(\tilde{q})$ is finite, even though the integral defining R^{α}_{1} does not exist. In particular, for one-dimensional systems with regular interactions, where R^{α}_{1} is only logarithmically divergent, $\Sigma^{\alpha}_{1}(\tilde{q})$ does not exhibit any divergencies.

Calculation of G^{α}_{2}

Next, let us calculate the interference contribution G^{α}_{2} defined in Eq.(5.135). Diagrammatically this function is the sum of all Feynman diagrams which combine the graphical elements defined in Figs. 5.4 and 5.5. To first order in the RPA interaction only the diagram shown in Fig. 5.9 contributes. Evaluation of this diagram yields

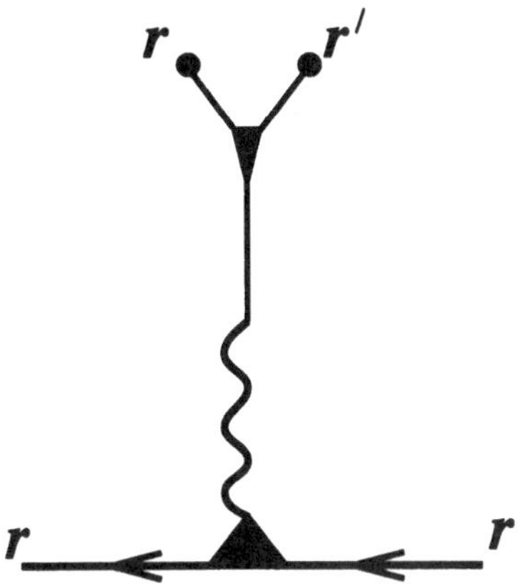

Fig. 5.9. Leading contribution to G_2^α. Because the function $\mathcal{J}_q^\alpha(\boldsymbol{r},\tau)$ depends on $\boldsymbol{r}$ and τ, the diagram has to be understood as a real space, imaginary time diagram. The spatial labels are written on the corresponding end-points. The thick solid arrows represent the self-consistent average Green's function G_1^α, as defined in Eqs.(5.156)–(5.158). To lowest order in $f^{\mathrm{RPA},\alpha}$ the thick arrows should be replaced by thin arrows (representing the non-interacting Green's function G_0^α). However, in the spirit of the self-consistent Born approximation, we have included disorder corrections to the Green's functions attached to the eikonal contribution.

$$G_2^\alpha(\boldsymbol{r},\tau) = \frac{1}{\beta V}\sum_{\tilde{q}} \mathrm{e}^{\mathrm{i}(\boldsymbol{q}\cdot\boldsymbol{r}-\tilde{\omega}_n\tau)} G_2^\alpha(\tilde{q}) \quad , \quad G_2^\alpha(\tilde{q}) = G_1^\alpha(\tilde{q}) Y^\alpha(\tilde{q}) \quad , \tag{5.161}$$

where, after symmetrization, the dimensionless function $Y^\alpha(\tilde{q})$ can be written as

$$\begin{aligned} Y^\alpha(\tilde{q}) = & \frac{1}{\beta V}\sum_{q'} \frac{f_{q'}^{\mathrm{RPA},\alpha}}{[\mathrm{i}\omega_{m'} - \xi_{q'}^\alpha][\mathrm{i}\omega_{m'} + \xi_{-q'}^\alpha]} \\ & \times \left\{ \frac{\boldsymbol{q}'(\mathsf{M}^\alpha)^{-1}\boldsymbol{q}'}{2} \left[G_1^\alpha(\tilde{q}+q') + G_1^\alpha(\tilde{q}-q')\right] \right. \\ & \left. + \boldsymbol{q}(\mathsf{M}^\alpha)^{-1}\boldsymbol{q}' \left[G_1^\alpha(\tilde{q}+q') - G_1^\alpha(\tilde{q}-q')\right] \right\} \quad . \end{aligned} \tag{5.162}$$

Again we see that the integrand of $Y^\alpha(\tilde{q})$ is less infrared singular than that of R_1^α in Eq.(5.152).

In summary, the total prefactor Green's function in Eq.(5.132) can be written as

$$G_1^\alpha(\boldsymbol{r},\tau) + G_2^\alpha(\boldsymbol{r},\tau) = \frac{1}{\beta V}\sum_{\tilde{q}} \mathrm{e}^{\mathrm{i}(\boldsymbol{q}\cdot\boldsymbol{r}-\tilde{\omega}_n\tau)} \frac{1+Y^\alpha(\tilde{q})}{\mathrm{i}\tilde{\omega}_n - \epsilon_{\boldsymbol{k}^\alpha+\boldsymbol{q}} + \mu - \Sigma_1^\alpha(\tilde{q})} \quad , \tag{5.163}$$

where $\Sigma_1^\alpha(\tilde{q})$ and $Y^\alpha(\tilde{q})$ can be calculated perturbatively in powers of the RPA interaction. The leading contributions are given in Eqs.(5.160) and (5.162). In the limit of infinite effective masses (corresponding to linearized energy dispersion) the functions Σ_1^α and Y^α are identically zero. Then the right-hand side of Eq.(5.163) reduces to the non-interacting Green's function, and we recover the result for linearized energy dispersion. Furthermore, in the absence of interactions Q_1^α, Σ_1^α and Y^α vanish identically, so that we recover the exact non-interacting Green's function, which contains of course infinite orders in $1/m^\alpha$. This shows that we have not performed a naive expansion in powers of $1/m^\alpha$, as originally suggested in [1.15]. We would also like to emphasize that corrections to the above expressions involve an additional power of the RPA-interaction, so that *in the weak-coupling regime* we may truncate our expansion at the leading order, *even if the interaction is not dominated by small momentum transfers.* In other words, as long as the RPA interaction is finite and small, the above expressions remain valid to first order in the interaction even in the presence of Umklapp and back-scattering processes!

Our result (5.151) for the leading term $Q_1^\alpha(\boldsymbol{r}, \tau)$ in the expansion of the average eikonal (which can be viewed as the natural generalization of the Debye-Waller (5.31)–(5.33) to the case of non-linear energy dispersion) and the corrections (5.160) and (5.162) to the prefactor Green's function cure all pathologies that are generated by the linearization of the energy dispersion and the concomitant replacement of a curved Fermi surface by a collection of flat hyper-planes. First of all, the singular function $\delta^{(d-1)}(\boldsymbol{r}_\perp^\alpha)$ in Eq.(5.53) has disappeared, because now $G_0^\alpha(\boldsymbol{r}, \tau)$ is replaced by $G_1^\alpha(\boldsymbol{r}, \tau) + G_2^\alpha(\boldsymbol{r}, \tau)$. Due to the finite curvature term, this prefactor is a non-singular function of all of its arguments. Of course, now the Fourier transformation involves a full $d+1$-dimensional integration, so that from a numerical point of view the problem in $d > 1$ is more difficult than in the case of linearized energy dispersion. Furthermore, possible problems associated with the double pole in the expression for the Debye-Waller factor of the linearized theory are solved trivially, because the non-linear terms in the energy dispersion split the double pole into two isolated poles that are separated by a distance $\boldsymbol{q}(\mathsf{M}^\alpha)^{-1}\boldsymbol{q}$ on the real frequency axis (see Eq.(5.154)).

5.3.3 Connection with lowest order perturbation theory

We show that the expansion of our result for $G^\alpha(\boldsymbol{r}, \tau)$ to first order in the RPA interaction exactly reproduces perturbation theory.

By construction all corrections to our result for the average eikonal in Eqs.(5.151)–(5.153) and the expressions (5.160) and (5.162) for the functions $\Sigma_1^\alpha(\tilde{q})$ and $Y^\alpha(\tilde{q})$ involve at least two powers of $f^{\mathrm{RPA},\alpha}$. Therefore a direct expansion of these expressions to first order in $f^{\mathrm{RPA},\alpha}$ should *exactly* reproduce the usual perturbative result, i.e. the GW self-energy with full non-linear energy dispersion. In this section we show by explicit calculation that this is indeed the case. For simplicity we shall assume here that the matrix M^α is proportional to the unit matrix, so that $\boldsymbol{q}(\mathsf{M}^\alpha)^{-1}\boldsymbol{q}' = \boldsymbol{q} \cdot \boldsymbol{q}'/m^\alpha$.

Expanding Eq.(5.132) to first order in the RPA interaction, we have

$$\begin{aligned} G^{\alpha}(\boldsymbol{r},\tau) &\equiv [G_1^{\alpha}(\boldsymbol{r},\tau) + G_2^{\alpha}(\boldsymbol{r},\tau)] e^{Q^{\alpha}(\boldsymbol{r},\tau)} \\ &\approx G_0^{\alpha}(\boldsymbol{r},\tau) + G_0^{\alpha}(\boldsymbol{r},\tau) Q_1^{\alpha}(\boldsymbol{r},\tau) \\ &+ \frac{1}{\beta V}\sum_{\tilde{q}} e^{i(\boldsymbol{q}\cdot\boldsymbol{r} - \tilde{\omega}_n \tau)} G_0^{\alpha}(\tilde{q}) \Sigma_1^{\alpha}(\tilde{q}) G_0^{\alpha}(\tilde{q}) \\ &+ \frac{1}{\beta V}\sum_{\tilde{q}} e^{i(\boldsymbol{q}\cdot\boldsymbol{r} - \tilde{\omega}_n \tau)} G_0^{\alpha}(\tilde{q}) Y^{\alpha}(\tilde{q}) + \ldots \quad . \end{aligned} \tag{5.164}$$

Note that to first order in $f^{\mathrm{RPA},\alpha}$ we may replace $G_1^{\alpha} \to G_0^{\alpha}$ on the right-hand sides of Eqs.(5.160) and (5.162). On the other hand, to leading order in the interaction we have for the Fourier transform of the full Green's function

$$\begin{aligned} G^{\alpha}(\tilde{q}) &\equiv \int d\boldsymbol{r} \int_0^{\beta} d\tau e^{-i(\boldsymbol{q}\cdot\boldsymbol{r} - \tilde{\omega}_n \tau)} G^{\alpha}(\boldsymbol{r},\tau) \\ &= G_0^{\alpha}(\tilde{q}) + G_0^{\alpha}(\tilde{q}) \Sigma^{\alpha}(\tilde{q}) G_0^{\alpha}(\tilde{q}) + \ldots \quad . \end{aligned} \tag{5.165}$$

Substituting our first-order result (5.151) for the Debye-Waller factor into Eq.(5.164), Fourier transforming, and comparing with Eq.(5.165), it is not difficult to show that within our bosonization approach the first-order self-energy is approximated by

$$\Sigma^{\alpha}(\tilde{q}) = \Sigma_{\mathrm{Q}}^{\alpha}(\tilde{q}) + \Sigma_1^{\alpha}(\tilde{q}) + \Sigma_{\mathrm{Y}}^{\alpha}(\tilde{q}) \quad , \tag{5.166}$$

where the self-energy $\Sigma_1^{\alpha}(\tilde{q})$ is given in Eq.(5.158), and the contribution $\Sigma_{\mathrm{Q}}^{\alpha}(\tilde{q})$ due to the Debye-Waller factor on the right-hand side of Eq.(5.164) is

$$\begin{aligned} \Sigma_{\mathrm{Q}}^{\alpha}(\tilde{q}) &= [i\tilde{\omega}_n - \epsilon_{\boldsymbol{k}^{\alpha}+\boldsymbol{q}} + \mu]^2 \frac{1}{\beta V} \sum_{q'} \frac{f_{q'}^{\mathrm{RPA},\alpha}}{[i\omega_{m'} - \xi_{\boldsymbol{q}'}^{\alpha}][i\omega_{m'} + \xi_{-\boldsymbol{q}'}^{\alpha}]} \\ &\times \left\{ \frac{1}{i\tilde{\omega}_n - \epsilon_{\boldsymbol{k}^{\alpha}+\boldsymbol{q}} + \mu} \right. \\ &\left. -\frac{1}{2}\left[\frac{1}{i\tilde{\omega}_{n+m'} - \epsilon_{\boldsymbol{k}^{\alpha}+\boldsymbol{q}+\boldsymbol{q}'} + \mu} + \frac{1}{i\tilde{\omega}_{n-m'} - \epsilon_{\boldsymbol{k}^{\alpha}+\boldsymbol{q}-\boldsymbol{q}'} + \mu} \right] \right\} . \end{aligned} \tag{5.167}$$

Similarly, we obtain from (5.162) for the last term in Eq.(5.166)

$$\begin{aligned} \Sigma_{\mathrm{Y}}^{\alpha}(\tilde{q}) &= [i\tilde{\omega}_n - \epsilon_{\boldsymbol{k}^{\alpha}+\boldsymbol{q}} + \mu] \frac{1}{\beta V} \sum_{q'} \frac{f_{q'}^{\mathrm{RPA},\alpha}}{[i\omega_{m'} - \xi_{\boldsymbol{q}'}^{\alpha}][i\omega_{m'} + \xi_{-\boldsymbol{q}'}^{\alpha}]} \\ &\times \left\{ \frac{\boldsymbol{q}'^2}{2m^{\alpha}} [G_0^{\alpha}(\tilde{q}+q') + G_0^{\alpha}(\tilde{q}-q')] + \frac{\boldsymbol{q}\cdot\boldsymbol{q}'}{m^{\alpha}} [G_0^{\alpha}(\tilde{q}+q') - G_0^{\alpha}(\tilde{q}-q')] \right\} . \end{aligned} \tag{5.168}$$

We now show that the self-energy $\Sigma^{\alpha}(\tilde{q})$ given in Eq.(5.166) is identical with the usual GW self-energy. At the first sight this is not at all obvious, because the three terms on the right-hand side of Eq.(5.166) have no resemblance to the usual perturbative result for the GW self-energy, which can be written as (see Eq.(5.74))

$$\Sigma^{\alpha}_{\mathrm{GW}}(\tilde{q}) = -\frac{1}{\beta V}\sum_{q'} f^{\mathrm{RPA},\alpha}_{q'} \frac{1}{2}\left[\frac{1}{\mathrm{i}\tilde{\omega}_{n+m'} - \epsilon_{k^{\alpha}+q+q'} + \mu} + \frac{1}{\mathrm{i}\tilde{\omega}_{n-m'} - \epsilon_{k^{\alpha}+q-q'} + \mu}\right] \ . \tag{5.169}$$

We have used the invariance of $f^{\mathrm{RPA},\alpha}_{q'}$ with respect to relabeling $q' \rightarrow -q'$ to symmetrize the rest of the integrand.

Let us begin by manipulating $\Sigma^{\alpha}_{\mathrm{Q}}(\tilde{q})$ in precisely the same way as one would proceed in the case of linearized energy dispersion. Then one would partial fraction the differences of two non-interacting Green's functions in the second line of Eq.(5.167). For linearized energy dispersion the result can be expressed again in terms of non-interacting Green's function[11]. For energy dispersions with a quadratic term, the generalization of Eq.(4.13) is

$$\frac{1}{\mathrm{i}\tilde{\omega}_n - \epsilon^{\alpha}_{q}} - \frac{1}{\mathrm{i}\tilde{\omega}_{n+m'} - \epsilon^{\alpha}_{q+q'}} = \frac{\mathrm{i}\omega_{m'} - \xi^{\alpha}_{q'} - \frac{q\cdot q'}{m^{\alpha}}}{[\mathrm{i}\tilde{\omega}_n - \epsilon^{\alpha}_{q}][\mathrm{i}\tilde{\omega}_{n+m'} - \epsilon^{\alpha}_{q+q'}]} \ , \tag{5.170}$$

and similarly

$$\frac{1}{\mathrm{i}\tilde{\omega}_n - \epsilon^{\alpha}_{q}} - \frac{1}{\mathrm{i}\tilde{\omega}_{n-m'} - \epsilon^{\alpha}_{q-q'}} = \frac{-\mathrm{i}\omega_{m'} - \xi^{\alpha}_{-q'} + \frac{q\cdot q'}{m^{\alpha}}}{[\mathrm{i}\tilde{\omega}_n - \epsilon^{\alpha}_{q}][\mathrm{i}\tilde{\omega}_{n-m'} - \epsilon^{\alpha}_{q-q'}]} \ , \tag{5.171}$$

where for simplicity we have introduced the notation $\epsilon^{\alpha}_{q} = \epsilon_{k^{\alpha}+q} - \mu$. With the help of these identities it is easy to show that Eq.(5.167) can also be written as

$$\Sigma^{\alpha}_{\mathrm{Q}}(\tilde{q}) = \frac{1}{\beta V}\sum_{q'} f^{\mathrm{RPA},\alpha}_{q'} \frac{1}{2}\left[\frac{\mathrm{i}\tilde{\omega}_n - \epsilon^{\alpha}_{q}}{[\mathrm{i}\omega_{m'} + \xi^{\alpha}_{-q'}][\mathrm{i}\tilde{\omega}_{n+m'} - \epsilon^{\alpha}_{q+q'}]} - \frac{\mathrm{i}\tilde{\omega}_n - \epsilon^{\alpha}_{q}}{[\mathrm{i}\omega_{m'} - \xi^{\alpha}_{q'}][\mathrm{i}\tilde{\omega}_{n-m'} - \epsilon^{\alpha}_{q-q'}]}\right]$$
$$- \left[\mathrm{i}\tilde{\omega}_n - \epsilon^{\alpha}_{q}\right] \frac{1}{\beta V}\sum_{q'} \frac{f^{\mathrm{RPA},\alpha}_{q'}}{[\mathrm{i}\omega_{m'} - \xi^{\alpha}_{q'}][\mathrm{i}\omega_{m'} + \xi^{\alpha}_{-q'}]} \times \frac{q\cdot q'}{2m^{\alpha}}\left[\frac{1}{\mathrm{i}\tilde{\omega}_{n+m'} - \epsilon^{\alpha}_{q+q'}} - \frac{1}{\mathrm{i}\tilde{\omega}_{n-m'} - \epsilon^{\alpha}_{q-q'}}\right] \ . \tag{5.172}$$

Next, we use the following two exact identities,

$$\frac{\mathrm{i}\tilde{\omega}_n - \epsilon^{\alpha}_{q}}{[\mathrm{i}\omega_{m'} + \xi^{\alpha}_{-q'}][\mathrm{i}\tilde{\omega}_{n+m'} - \epsilon^{\alpha}_{q+q'}]} = \frac{1}{\mathrm{i}\omega_{m'} + \xi^{\alpha}_{-q'}} - \frac{1}{\mathrm{i}\tilde{\omega}_{n+m'} - \epsilon^{\alpha}_{q+q'}} + \frac{q'\cdot(q+q')}{m^{\alpha}} \frac{1}{[\mathrm{i}\omega_{m'} + \xi^{\alpha}_{-q'}][\mathrm{i}\tilde{\omega}_{n+m'} - \epsilon^{\alpha}_{q+q'}]} \ , \tag{5.173}$$

[11] Recall the partial fraction decomposition (4.13), which was crucial in the proof of the closed loop theorem.

$$\frac{i\tilde{\omega}_n - \epsilon^\alpha_q}{[i\omega_{m'} - \xi^\alpha_{q'}][i\tilde{\omega}_{n-m'} - \epsilon^\alpha_{q-q'}]} = \frac{1}{i\omega_{m'} - \xi^\alpha_{q'}} - \frac{1}{i\tilde{\omega}_{n-m'} - \epsilon^\alpha_{q-q'}}$$
$$- \frac{q' \cdot (q - q')}{m^\alpha} \frac{1}{[i\omega_{m'} - \xi^\alpha_{q'}][i\tilde{\omega}_{n-m'} - \epsilon^\alpha_{q-q'}]} \quad , \tag{5.174}$$

and obtain from Eq.(5.172)

$$\Sigma^\alpha_{\mathrm{Q}}(\tilde{q}) = \Sigma^\alpha_{\mathrm{GW}}(\tilde{q}) - \frac{1}{\beta V}\sum_{q'} f^{\mathrm{RPA},\alpha}_{q'} \frac{1}{2}\left[\frac{1}{i\omega_{m'} - \xi^\alpha_{q'}} - \frac{1}{i\omega_{m'} + \xi^\alpha_{-q'}}\right]$$
$$+ \frac{1}{\beta V}\sum_{q'} \frac{f^{\mathrm{RPA},\alpha}_{q'}}{[i\omega_{m'} - \xi^\alpha_{q'}][i\omega_{m'} + \xi^\alpha_{-q'}]}\left[\frac{q' \cdot (q + q')}{m^\alpha} \frac{i\omega_{m'} - \xi^\alpha_{q'}}{i\tilde{\omega}_{n+m'} - \epsilon^\alpha_{q+q'}}\right.$$
$$\left. + \frac{q' \cdot (q - q')}{m^\alpha} \frac{i\omega_{m'} + \xi^\alpha_{-q'}}{i\tilde{\omega}_{n-m'} - \epsilon^\alpha_{q-q'}}\right]$$
$$+ \left[i\tilde{\omega}_n - \epsilon^\alpha_q\right] \frac{1}{\beta V}\sum_{q'} \frac{f^{\mathrm{RPA},\alpha}_{q'}}{[i\omega_{m'} - \xi^\alpha_{q'}][i\omega_{m'} + \xi^\alpha_{-q'}]}$$
$$\times \frac{q \cdot q'}{2m^\alpha}\left[\frac{1}{i\tilde{\omega}_{n+m'} - \epsilon^\alpha_{q+q'}} - \frac{1}{i\tilde{\omega}_{n-m'} - \epsilon^\alpha_{q-q'}}\right] \quad . \tag{5.175}$$

Here the function $\Sigma^\alpha_{\mathrm{GW}}(\tilde{q})$ is given in Eq.(5.169). Finally we write in the third term on the right-hand side of Eq.(5.175)

$$i\omega_{m'} - \xi^\alpha_{q'} = -\left[i\tilde{\omega}_n - \epsilon^\alpha_q\right] + \left[i\tilde{\omega}_{n+m'} - \epsilon^\alpha_{q+q'}\right] + \frac{q \cdot q'}{m^\alpha} \quad , \tag{5.176}$$

$$i\omega_{m'} + \xi^\alpha_{-q'} = \left[i\tilde{\omega}_n - \epsilon^\alpha_q\right] - \left[i\tilde{\omega}_{n-m'} - \epsilon^\alpha_{q-q'}\right] + \frac{q \cdot q'}{m^\alpha} \quad , \tag{5.177}$$

and arrive at

$$\Sigma^\alpha_{\mathrm{Q}}(\tilde{q}) = \Sigma^\alpha_{\mathrm{GW}}(\tilde{q}) - \Sigma^\alpha_{\mathrm{GW}}(\tilde{q} = 0) - \Sigma^\alpha_{\mathrm{I}}(\tilde{q}) - \Sigma^\alpha_{\mathrm{Y}}(\tilde{q}) \quad , \tag{5.178}$$

where we have used the fact that

$$\Sigma^\alpha_{\mathrm{GW}}(\tilde{q} = 0) = -\frac{1}{\beta V}\sum_{q'} f^{\mathrm{RPA},\alpha}_{q'} \frac{1}{2}\left[\frac{1}{i\omega_{m'} - \xi^\alpha_{q'}} - \frac{1}{i\omega_{m'} + \xi^\alpha_{-q'}}\right]$$
$$= -\frac{1}{\beta V}\sum_{q'} f^{\mathrm{RPA},\alpha}_{q'} \frac{\frac{q^2}{2m^\alpha}}{[i\omega_{m'} - \xi^\alpha_{q'}][i\omega_{m'} + \xi^\alpha_{-q'}]} \quad . \tag{5.179}$$

The last two terms in Eq.(5.178), which arise due to the above partial fraction decompositions of $\Sigma^\alpha_{\mathrm{Q}}$, are exactly cancelled by the last two terms in Eq.(5.166), so that the final result for the first order self-energy calculated within our bosonization approach is

$$\Sigma^\alpha(\tilde{q}) = \Sigma^\alpha_{\mathrm{GW}}(\tilde{q}) - \Sigma^\alpha_{\mathrm{GW}}(\tilde{q} = 0) \quad . \tag{5.180}$$

The term $\Sigma^{\alpha}_{\mathrm{GW}}(\tilde{q} = 0)$ subtracts the renormalization of the chemical potential contained in the first term. This is in agreement with the fact that by definition we start from the exact chemical potential of the many-body system, so that μ should not be renormalized. Note, however, that in general the ***shape*** of the Fermi surface will be renormalized by the interaction. This effect is lost if one linearizes the energy dispersion. The crucial role of the terms Σ^{α}_{1} and Y^{α} is now evident. If we had ignored these corrections, we would have obtained a discrepancy with lowest order perturbation theory, because for finite m^{α} the exponentiation $\mathrm{e}^{Q^{\alpha}}$ of the perturbation series is not quite correct. In a sense, we have exponentiated "too much", so that it is necessary to introduce correction terms in the prefactor.

5.4 Summary and outlook

In this chapter we have developed a new method for calculating the single-particle Green's function of an interacting Fermi system. Our result within the Gaussian approximation can be considered as the natural generalization of the non-perturbative bosonization solution of the Tomonaga-Luttinger model [1.17–1.19] to arbitrary dimensions. Because in $d > 1$ the curvature of the Fermi surface leads to qualitatively new effects which do not exist in $d = 1$, we have developed a systematic method for including the non-linear terms in the energy dispersion into the bosonization procedure in arbitrary dimensions.

Let us summarize our main result for the Green's function for the special case of a spherical Fermi surface of radius $k_{\mathrm{F}} = v_{\mathrm{F}}/m$ and a patch-independent bare interaction $f_{\boldsymbol{q}}$. As discussed in Chap. 2.5, in this case it is not necessary to subdivide the Fermi surface into several sectors – instead, if we are interested in the Matsubara Green's function $G(\boldsymbol{k}, \mathrm{i}\tilde{\omega}_n)$ for a given $\boldsymbol{k}$, we choose a special coordinate system centered at $\boldsymbol{k}^{\alpha}$ on the Fermi surface shown in Fig. 2.8. As discussed at the end of Chap. 2.5, due to the spherical symmetry, $G(\boldsymbol{k}, \mathrm{i}\tilde{\omega}_n)$ depends on $\boldsymbol{k}$ exclusively via the combination $|\boldsymbol{k}| - k_{\mathrm{F}}$. Then we may write

$$G(\boldsymbol{k}, \mathrm{i}\tilde{\omega}_n) = G^{\alpha}(|\boldsymbol{k}|\hat{\boldsymbol{k}}^{\alpha} - \boldsymbol{k}^{\alpha}, \mathrm{i}\tilde{\omega}_n) \quad , \tag{5.181}$$

with

$$G^{\alpha}(\boldsymbol{q}, \mathrm{i}\tilde{\omega}_n) = \int \mathrm{d}\boldsymbol{r} \int_0^{\beta} \mathrm{d}\tau \mathrm{e}^{-\mathrm{i}(\boldsymbol{q}\cdot\boldsymbol{r} - \tilde{\omega}_n \tau)} \tilde{G}^{\alpha}(\boldsymbol{r}, \tau) \mathrm{e}^{Q^{\alpha}_1(\boldsymbol{r}, \tau)} \quad , \tag{5.182}$$

where the Debye-Waller factor is

$$Q^{\alpha}_1(\boldsymbol{r}, \tau) = \frac{1}{\beta V} \sum_{q} f^{\mathrm{RPA}}_{q} \frac{1 - \cos(\boldsymbol{q}\cdot\boldsymbol{r} - \omega_m \tau)}{[\mathrm{i}\omega_m - \xi^{\alpha}_{\boldsymbol{q}}][\mathrm{i}\omega_m + \xi^{\alpha}_{-\boldsymbol{q}}]} \quad , \tag{5.183}$$

and the prefactor Green's function $\tilde{G}^{\alpha}(\boldsymbol{r}, \tau)$ has the Fourier expansion

$$\tilde{G}^{\alpha}(\boldsymbol{r}, \tau) = \frac{1}{\beta V} \sum_{\tilde{q}} \mathrm{e}^{\mathrm{i}(\boldsymbol{q}\cdot\boldsymbol{r} - \tilde{\omega}_n \tau)} \tilde{G}^{\alpha}(\tilde{q}) \quad , \tag{5.184}$$

$$\tilde{G}^{\alpha}(\tilde{q}) = \frac{1 + Y^{\alpha}(\tilde{q})}{\mathrm{i}\tilde{\omega}_n - \epsilon_{\boldsymbol{k}^{\alpha}+\boldsymbol{q}} + \mu - \Sigma_1^{\alpha}(\tilde{q})} \quad , \tag{5.185}$$

with the prefactor self-energy

$$\Sigma_1^{\alpha}(\tilde{q}) = -\frac{1}{\beta V} \sum_{q'} f_{q'}^{\mathrm{RPA}} G_1^{\alpha}(\tilde{q} + q') \times \frac{(\boldsymbol{q} \cdot \boldsymbol{q}')\boldsymbol{q}'^2 + (\boldsymbol{q} \cdot \boldsymbol{q}')^2}{m^2[\mathrm{i}\omega_{m'} - \xi_{\boldsymbol{q}'}^{\alpha}][\mathrm{i}\omega_{m'} + \xi_{-\boldsymbol{q}'}^{\alpha}]} \quad , \tag{5.186}$$

and the vertex function

$$Y^{\alpha}(\tilde{q}) = \frac{1}{\beta V} \sum_{q'} f_{q'}^{\mathrm{RPA}} G_1^{\alpha}(\tilde{q} + q') \times \frac{\boldsymbol{q}'^2 + 2\boldsymbol{q} \cdot \boldsymbol{q}'}{m[\mathrm{i}\omega_{m'} - \xi_{\boldsymbol{q}'}^{\alpha}][\mathrm{i}\omega_{m'} + \xi_{-\boldsymbol{q}'}^{\alpha}]} \quad . \tag{5.187}$$

Note that Q_1^{α}, Σ_1^{α}, Y^{α} are of the first order in the RPA interaction and involve a single fermionic loop summation (apart from the infinite series of bubble diagrams contained in f^{RPA}). The above expressions can be considered as a new extrapolation of the perturbation series, which involves a partial exponentiation in real space, a partial geometric resummation in Fourier space, and an intricate mixed Fourier representation. Our extrapolation scheme is quite different from the usual geometric extrapolation of the perturbation series for the Green's function in momentum space, which is implicitly performed if one first calculates the irreducible self-energy $\Sigma(\boldsymbol{k}, \mathrm{i}\tilde{\omega}_n)$ to some finite order in f^{RPA} and then solves the Dyson equation. As shown in Sect. 5.3.3, our resummation scheme has the important property that the expansion of our result for the Green's function to first order in f^{RPA} *exactly* reproduces the leading term in naive a perturbative expansion. Moreover, in Chap. 6.3 we shall show that in one dimension and for linearized energy dispersion Eqs.(5.181)–(5.187) correctly reproduce the exact solution of the Tomonaga-Luttinger model [1.17–1.19].

In the second part of this book we shall partially evaluate the above expressions in some simple limiting cases where we can make progress without resorting to numerical methods. However, our analysis will not be complete, because in general the integrations in Eqs.(5.181)–(5.187) are very difficult to perform. It particular, the calculation of the full momentum- and frequency-dependent **spectral function**

$$A(\boldsymbol{k}, \omega) = -\frac{1}{\pi} \mathrm{Im} G(\boldsymbol{k}, \omega + \mathrm{i}0^+) \tag{5.188}$$

from Eqs.(5.181)–(5.187) in a non-trivial interacting Fermi system in $d > 1$ is an interesting open problem, which seems to require extensive numerical work. We would like to emphasize that such a calculation would yield a highly non-perturbative result for the spectral function. In particular, Eqs.(5.181)–(5.187) can be used to determine by direct calculation whether an interacting

Fermi system is a Fermi liquid or not. In both cases these equations are well-defined (at least for not too singular interactions, see Chap. 6.2.5), and provide an explicit expression for the single-particle Green's function which can serve as a basis for quantitative calculations.

Before embarking on applications of our formalism to problems of physical interest, let us briefly mention two more open problems, which will not be further discussed in this book[12]. First of all, the problem of **back-scattering**: Because in Eqs.(5.181)–(5.187) we have not made use of the patching construction and have identified the entire momentum space with a single sector, the restriction that the maximum momentum transfer q_c of the interaction must be smaller than the size of the sectors (see Fig.2.5) does not exclude processes with large momentum transfer any more. Therefore Eqs.(5.181)–(5.187) are also valid for short-range interactions, *provided the dimensionless parameter* A_0^α *given in Eq.(4.115) is small.* Of course, in this case we loose the small factor $(q_c/k_F)^d$ in Eq.(4.115), so that our non-perturbative expression for the Green's function can only be accurate for sufficiently small interactions. However, in the weak-coupling regime Eqs.(5.181)–(5.187) can be considered as the leading term in a non-perturbative expansion in powers of the RPA interaction, which includes the effect of scattering processes involving large momentum transfers, such as back-scattering or Umklapp-scattering.

The second interesting direction for further research is the generalization of our formalism to include **broken symmetries**. Note that throughout this work we are assuming that the electrons remain normal, i.e. that they do not undergo a phase transition to a state with spontaneously broken symmetry. In particular, we have ignored the tendencies towards superconductivity and antiferromagnetism, which are known to exist in many strongly correlated Fermi systems at sufficiently low temperatures. It seems, however, that it is not difficult to include these effects into our formalism, at least at the level of the Gaussian approximation. In fact, functional integration and Hubbard-Stratonovich transformation are the ideal formal starting point to study spontaneous symmetry breaking in Fermi systems [2.6, 5.14]. Therefore we expect that it is straightforward to generalize the non-perturbative methods developed in this book to incorporate superconductivity and various types of itinerant magnetism. In particular, our methods might provide a non-perturbative microscopic approach to nearly antiferromagnetic Fermi liquids [5.15].

In this context we would also like to point out that for systems with special spin symmetries or other internal symmetries it might be necessary to decouple the relevant operators by means of matrix-field Hubbard-Stratonovich transformations which preserve the symmetries. This could lead to higher-dimensional generalizations **of non-abelian bosonization** [5.16, 5.17]. An

[12] I would like to encourage all readers to contribute to the solution of these problems.

attempt to develop such an approach has recently been made by Schmeltzer [5.18].

Part II

Applications to physical systems

6. Singular interactions ($f_q \sim |q|^{-\eta}$)

We analyze singular density-density interactions that diverge in d dimensions as $|q|^{-\eta}$ for $q \to 0$. For linearized energy dispersion we explicitly calculate the asymptotic long-distance behavior of $Q^{\alpha}(r,0)$. For regular interactions ($\eta = 0$) in one dimension it is possible to calculate the full Debye-Waller factor $Q^{\alpha}(r,\tau)$ if a certain cutoff procedure is adopted. Then we reproduce the well-known bosonization result for the Tomonaga-Luttinger model.

In this chapter we shall study in some detail the Debye-Waller factor $Q^{\alpha}(r,\tau)$ derived in Chap. 5 for singular density-density interactions of the form

$$f_q = \frac{g_c^2}{|q|^{\eta}} e^{-|q|/q_c} \quad , \quad \eta > 0 \quad , \quad q_c \ll k_F \quad , \tag{6.1}$$

where g_c is some coupling constant with the correct units. The long-range part of the physical Coulomb interaction in d dimensions corresponds to $g_c = -e$ (the charge of the electron), $\eta = d-1$, and $q_c = \infty$, see Appendix A.3.1. As recently noticed by Bares and Wen [6.1], in the more singular case $\eta = 2(d-1)$ one obtains an instability of the Fermi liquid state. Although for general η interactions of the above type are unphysical, it is instructive study them as model systems which exhibit non-Fermi liquid behavior.

From Eq.(4.35) we know that for patch-independent bare interaction the screened interaction $f_q^{\mathrm{RPA},\alpha}$ in Eqs.(5.31)–(5.33) and (5.151)–(5.153) can be identified with the usual RPA interaction $f_q^{\mathrm{RPA}} = f_q[1 + f_q \Pi_0(q)]^{-1}$. For practical calculations it is convenient to express f_q^{RPA} in terms of the dynamic structure factor $S_{\mathrm{RPA}}(q,\omega)$, which is the spectral function of the RPA polarization[1]

$$\Pi_{\mathrm{RPA}}(q) = \frac{\Pi_0(q)}{1 + f_q \Pi_0(q)} = \int_0^{\infty} d\omega S_{\mathrm{RPA}}(q,\omega) \frac{2\omega}{\omega^2 + \omega_m^2} \quad , \tag{6.2}$$

[1] We would like to point out that the relation (2.42) between the imaginary part of the polarization and the dynamic structure factor is only valid if the shape of the Fermi surface is invariant with respect to inversion $k \to -k$. If we approximate the Fermi surface by a finite number of flat patches, then Eqs.(6.2) and (6.3) are only valid if for each patch P_{Λ}^{α} with Fermi velocity v^{α} there exists an opposite patch $P_{\Lambda}^{\bar{\alpha}}$ with $v^{\bar{\alpha}} = -v^{\alpha}$, see Appendix A.4.

see Eqs.(2.42)–(2.46) with $\beta \to \infty$. Hence

$$
\begin{aligned}
f_q^{\mathrm{RPA}} &= \frac{f_q}{1+f_q\Pi_0(q)} = f_q - f_q^2 \frac{\Pi_0(q)}{1+f_q\Pi_0(q)} \\
&= f_q - f_q^2 \int_0^\infty \mathrm{d}\omega S_{\mathrm{RPA}}(q,\omega)\frac{2\omega}{\omega^2+\omega_m^2} \quad . \qquad (6.3)
\end{aligned}
$$

The advantage of introducing the dynamic structure factor is that it is by construction a real non-negative function, see Eq.(2.43). Furthermore, the qualitative behavior of the dynamic structure factor can be understood from simple intuitive arguments [1.7], which is very helpful for the evaluation of complicated integrals.

6.1 Manipulations with the help of the dynamic structure factor

By introducing the spectral function of the RPA polarization (i.e. the dynamic structure factor), we can perform the Matsubara sum at the very beginning of the calculation, and then make some general statements which are valid irrespective of the precise form of the interaction.

6.1.1 Non-linear energy dispersion

Although in the rest of this chapter we shall for simplicity work with linearized energy dispersion, it is convenient to consider first the Debye-Waller factor $Q^\alpha(\boldsymbol{r},\tau)$ for quadratic energy dispersion. Substituting the spectral representation (6.3) into Eqs.(5.151)–(5.153), the Matsubara sum over ω_m can be performed trivially, and we obtain[2] for $\beta \to \infty$

$$
\begin{aligned}
R_1^\alpha &= \frac{1}{V}\sum_{\boldsymbol{q}} f_{\boldsymbol{q}} \frac{\mathrm{sgn}(\xi_{\boldsymbol{q}}^\alpha)}{(-\frac{\boldsymbol{q}^2}{m^\alpha})} \\
&- \frac{1}{V}\sum_{\boldsymbol{q}} f_{\boldsymbol{q}}^2 \int_0^\infty \mathrm{d}\omega S_{\mathrm{RPA}}(\boldsymbol{q},\omega) \frac{2\mathrm{sgn}(\xi_{\boldsymbol{q}}^\alpha)}{(-\frac{\boldsymbol{q}^2}{m^\alpha})(\omega+|\xi_{\boldsymbol{q}}^\alpha|)} \quad . \qquad (6.4)
\end{aligned}
$$

$$
\begin{aligned}
\mathrm{Re}S_1^\alpha(\boldsymbol{r},\tau) &= \frac{1}{V}\sum_{\boldsymbol{q}} \cos(\boldsymbol{q}\cdot\boldsymbol{r}) f_{\boldsymbol{q}} \frac{\mathrm{sgn}(\xi_{\boldsymbol{q}}^\alpha)}{(-\frac{\boldsymbol{q}^2}{m^\alpha})} \mathrm{e}^{-|\xi_{\boldsymbol{q}}^\alpha||\tau|} \\
&- \frac{1}{V}\sum_{\boldsymbol{q}} \cos(\boldsymbol{q}\cdot\boldsymbol{r}) f_{\boldsymbol{q}}^2 \int_0^\infty \mathrm{d}\omega S_{\mathrm{RPA}}(\boldsymbol{q},\omega) \frac{2\mathrm{sgn}(\xi_{\boldsymbol{q}}^\alpha)}{(-\frac{\boldsymbol{q}^2}{m^\alpha})(\omega+|\xi_{\boldsymbol{q}}^\alpha|)} \\
&\times \left[\frac{\omega \mathrm{e}^{-|\xi_{\boldsymbol{q}}^\alpha||\tau|} - |\xi_{\boldsymbol{q}}^\alpha| \mathrm{e}^{-\omega|\tau|}}{\omega - |\xi_{\boldsymbol{q}}^\alpha|}\right] \quad , \qquad (6.5)
\end{aligned}
$$

[2] For simplicity we assume hat the effective mass tensor M^α is proportional to the unit matrix. For general anisotropic effective mass tensor one should simply make the replacement $\frac{\boldsymbol{q}^2}{m^\alpha} \to \xi_{\boldsymbol{q}}^\alpha + \xi_{-\boldsymbol{q}}^\alpha = \boldsymbol{q}(\mathsf{M}^\alpha)^{-1}\boldsymbol{q}$ in Eqs.(6.4)–(6.6).

$$\begin{aligned}\mathrm{Im}S_1^\alpha(\boldsymbol{r},\tau) &= \frac{\mathrm{sgn}(\tau)}{V}\sum_{\boldsymbol{q}}\sin(\boldsymbol{q}\cdot\boldsymbol{r})f_{\boldsymbol{q}}\frac{\mathrm{e}^{-|\xi_{\boldsymbol{q}}^\alpha||\tau|}}{(-\frac{\boldsymbol{q}^2}{m^\alpha})} \\ &- \frac{\mathrm{sgn}(\tau)}{V}\sum_{\boldsymbol{q}}\sin(\boldsymbol{q}\cdot\boldsymbol{r})f_{\boldsymbol{q}}^2\int_0^\infty \mathrm{d}\omega S_{\mathrm{RPA}}(\boldsymbol{q},\omega)\frac{2\omega}{(-\frac{\boldsymbol{q}^2}{m^\alpha})(\omega+|\xi_{\boldsymbol{q}}^\alpha|)} \\ &\times\left[\frac{\mathrm{e}^{-|\xi_{\boldsymbol{q}}^\alpha||\tau|}-\mathrm{e}^{-\omega|\tau|}}{\omega-|\xi_{\boldsymbol{q}}^\alpha|}\right] \quad . \end{aligned} \tag{6.6}$$

6.1.2 The limit of linear energy dispersion

We now carefully take the limit $1/m^\alpha \to 0$ in Eqs.(6.4)–(6.6). In this way we obtain the spectral representation of the Debye-Waller factor for linearized energy dispersion.

At the first sight it seems that Eqs.(6.4)–(6.6) diverge for $1/m^\alpha \to 0$, because the integrand is proportional to m^α. However, this factor is cancelled when we perform the integration, because the contribution from the regimes $\boldsymbol{v}^\alpha\cdot\boldsymbol{q} \geq 0$ and $\boldsymbol{v}^\alpha\cdot\boldsymbol{q} \leq 0$ to Eqs.(6.4)–(6.6) almost perfectly cancel in such a way that the integral is finite. To obtain the constant part R^α of the Debye-Waller factor for linearized energy dispersion ($\xi_{\boldsymbol{q}}^\alpha \approx \boldsymbol{v}^\alpha\cdot\boldsymbol{q}$), we expand the second term in Eq.(6.4) to first order in $1/m^\alpha$,

$$\begin{aligned}\frac{1}{\omega+|\xi_{\boldsymbol{q}}^\alpha|} &= \frac{1}{\omega+|\boldsymbol{v}^\alpha\cdot\boldsymbol{q}+\frac{\boldsymbol{q}^2}{2m^\alpha}|} \\ &= \frac{1}{\omega+|\boldsymbol{v}^\alpha\cdot\boldsymbol{q}|} - \frac{\boldsymbol{q}^2}{2m^\alpha}\frac{\mathrm{sgn}(\boldsymbol{v}^\alpha\cdot\boldsymbol{q})}{(\omega+|\boldsymbol{v}^\alpha\cdot\boldsymbol{q}|)^2} + O\left(1/(m^\alpha)^2\right) \quad . \end{aligned} \tag{6.7}$$

By symmetry the first term yields a vanishing contribution to Eq.(6.4), but the contribution from the second term in Eq.(6.7) is finite and independent of m^α. The expansion of the term $\mathrm{sgn}(\xi_{\boldsymbol{q}}^\alpha)$ in Eq.(6.4) to first order in $1/m^\alpha$ does *not* contribute to the Debye-Waller factor in the limit $|m^\alpha| \to \infty$. This is perhaps not so obvious, because the expansion of $\mathrm{sgn}(\xi_{\boldsymbol{q}}^\alpha)$ in powers of $1/m^\alpha$ produces also a term of order $1/m^\alpha$,

$$\begin{aligned}\mathrm{sgn}\left(\boldsymbol{v}^\alpha\cdot\boldsymbol{q}+\frac{\boldsymbol{q}^2}{2m^\alpha}\right) &= \Theta\left(\boldsymbol{v}^\alpha\cdot\boldsymbol{q}+\frac{\boldsymbol{q}^2}{2m^\alpha}\right) - \Theta\left(-\boldsymbol{v}^\alpha\cdot\boldsymbol{q}-\frac{\boldsymbol{q}^2}{2m^\alpha}\right) \\ &\approx \Theta(\boldsymbol{v}^\alpha\cdot\boldsymbol{q}) - \Theta(-\boldsymbol{v}^\alpha\cdot\boldsymbol{q}) + \delta(\boldsymbol{v}^\alpha\cdot\boldsymbol{q})\frac{\boldsymbol{q}^2}{m^\alpha} \quad . \end{aligned} \tag{6.8}$$

In the limit $m^\alpha \to \infty$ the last term in Eq.(6.8) gives rise to the following contribution to R^α,

$$\delta R^\alpha = -\frac{1}{V}\sum_{\boldsymbol{q}}\delta(\boldsymbol{v}^\alpha\cdot\boldsymbol{q})f_{\boldsymbol{q}}\left[1-f_{\boldsymbol{q}}\int_0^\infty \mathrm{d}\omega\frac{2S_{\mathrm{RPA}}(\boldsymbol{q},\omega)}{\omega}\right] \quad . \tag{6.9}$$

The two terms in the square braces are due to the first and second term in Eq.(6.4). From Eq.(6.2) we have

$$\int_0^\infty \mathrm{d}\omega \frac{2S_{\mathrm{RPA}}(\boldsymbol{q},\omega)}{\omega} = \Pi_{\mathrm{RPA}}(\boldsymbol{q},0) \quad , \tag{6.10}$$

so that δR^α can also be written as

$$\delta R^\alpha = -\frac{1}{V}\sum_{\boldsymbol{q}} \delta(\boldsymbol{v}^\alpha \cdot \boldsymbol{q}) f^{\mathrm{RPA}}_{\boldsymbol{q},0} \quad , \tag{6.11}$$

where $f^{\mathrm{RPA}}_{\boldsymbol{q},0} = f^{\mathrm{RPA}}_{\boldsymbol{q},\mathrm{i}\omega_m=0}$ is the static RPA interaction. Although the contribution (6.11) is non-zero, it is exactly cancelled by a corresponding contribution δS^α that is generated by expanding $\mathrm{sgn}(\xi^\alpha_{\boldsymbol{q}})$ in Eq.(6.5),

$$\delta S^\alpha = -\frac{1}{V}\sum_{\boldsymbol{q}} \delta(\boldsymbol{v}^\alpha \cdot \boldsymbol{q}) f^{\mathrm{RPA}}_{\boldsymbol{q},0} \cos(\boldsymbol{q}\cdot\boldsymbol{r}) \quad . \tag{6.12}$$

Noting that for linearized energy dispersion we may replace $\boldsymbol{r} \to r^\alpha_\parallel \hat{\boldsymbol{v}}^\alpha$ in the Debye-Waller factor (see Eqs.(5.48) and (5.52)), and using

$$\delta(\boldsymbol{v}^\alpha \cdot \boldsymbol{q}) \cos(\hat{\boldsymbol{v}}^\alpha \cdot \boldsymbol{q} r^\alpha_\parallel) = \delta(\boldsymbol{v}^\alpha \cdot \boldsymbol{q}) \quad , \tag{6.13}$$

it is obvious that $\delta R^\alpha - \delta S^\alpha = 0$. We conclude that in the limit $1/m^\alpha \to 0$ the constant part of the Debye-Waller factor is given by

$$R^\alpha = -\frac{1}{V}\sum_{\boldsymbol{q}} f^2_{\boldsymbol{q}} \int_0^\infty \mathrm{d}\omega \frac{S_{\mathrm{RPA}}(\boldsymbol{q},\omega)}{(\omega + |\boldsymbol{v}^\alpha \cdot \boldsymbol{q}|)^2} \quad . \tag{6.14}$$

Recall that the dynamic structure factor is real and positive by construction (see Eq.(2.43)), so that it is clear that R^α is a real negative number. Because for linearized energy dispersion the quasi-particle residue is given by $Z^\alpha = \mathrm{e}^{R^\alpha}$ (see Eq.(5.86)), the bosonization result for the Green's function is for arbitrary interactions in accordance with the requirement

$$0 \leq Z^\alpha \leq 1 \quad . \tag{6.15}$$

Note also that in a weak-coupling expansion the leading term in Eq.(6.14) is of the second order in the bare interaction, so that the leading interaction contribution to the quasi-particle residue $Z^\alpha \approx 1 + R^\alpha$ is of order $f^2_{\boldsymbol{q}}$. This is in agreement with perturbation theory. For non-linear energy dispersion the term R^α_1 has a non-vanishing contribution that is first order in $f_{\boldsymbol{q}}$. This is not in contradiction with perturbation theory, because the quantity $\mathrm{e}^{R^\alpha_1}$ cannot be identified with the quasi-particle residue Z^α any more; the function $Y^\alpha(\tilde{q})$ gives rise to an additional contribution[3] to Z^α. From Chap. 5.3.3 we know that by construction our method produces the correct perturbative result, so that the leading corrections to Z^α are of the second order in the bare interaction.

[3] From Eq.(5.160) we see that, at least for not too singular interactions, $\Sigma^\alpha_1(\boldsymbol{q} = 0, \mathrm{i}\tilde{\omega}_n) = 0$, so that G^α_1 does not renormalize the quasi-particle residue.

Similarly we obtain from Eqs.(6.5) and (6.6) after a tedious but straightforward calculation in the limit $1/m^\alpha \to 0$

$$\mathrm{Re}S^\alpha(r_\parallel^\alpha \hat{\boldsymbol{v}}^\alpha, \tau) = \frac{1}{V}\sum_{\boldsymbol{q}} \cos(\hat{\boldsymbol{v}}^\alpha \cdot \boldsymbol{q} r_\parallel^\alpha)$$
$$\times \left\{ L_{\boldsymbol{q}}^\alpha(\tau) - f_{\boldsymbol{q}}^2 \int_0^\infty \mathrm{d}\omega \frac{S_{\mathrm{RPA}}(\boldsymbol{q},\omega)}{(\omega + |\boldsymbol{v}^\alpha \cdot \boldsymbol{q}|)^2} \right.$$
$$\left. \times \frac{[(\boldsymbol{v}^\alpha \cdot \boldsymbol{q})^2 + \omega^2]\mathrm{e}^{-\omega|\tau|} - 2|\boldsymbol{v}^\alpha \cdot \boldsymbol{q}|\omega \mathrm{e}^{-|\boldsymbol{v}^\alpha \cdot \boldsymbol{q}||\tau|}}{(\omega - |\boldsymbol{v}^\alpha \cdot \boldsymbol{q}|)^2} \right\} \quad , \tag{6.16}$$

$$\mathrm{Im}S^\alpha(r_\parallel^\alpha \hat{\boldsymbol{v}}^\alpha, \tau) = \frac{\mathrm{sgn}(\tau)}{V}\sum_{\boldsymbol{q}} \sin(|\hat{\boldsymbol{v}}^\alpha \cdot \boldsymbol{q}| r_\parallel^\alpha)$$
$$\times \left\{ L_{\boldsymbol{q}}^\alpha(\tau) - f_{\boldsymbol{q}}^2 \int_0^\infty \mathrm{d}\omega \frac{S_{\mathrm{RPA}}(\boldsymbol{q},\omega)}{(\omega + |\boldsymbol{v}^\alpha \cdot \boldsymbol{q}|)^2} \right.$$
$$\left. \times 2|\boldsymbol{v}^\alpha \cdot \boldsymbol{q}|\omega \frac{\mathrm{e}^{-\omega|\tau|} - \mathrm{e}^{-|\boldsymbol{v}^\alpha \cdot \boldsymbol{q}||\tau|}}{(\omega - |\boldsymbol{v}^\alpha \cdot \boldsymbol{q}|)^2} \right\} \quad , \tag{6.17}$$

where we have defined

$$L_{\boldsymbol{q}}^\alpha(\tau) = \frac{|\tau|}{2} f_{\boldsymbol{q}} \mathrm{e}^{-|\boldsymbol{v}^\alpha \cdot \boldsymbol{q}||\tau|} \left[1 - f_{\boldsymbol{q}} \int_0^\infty \mathrm{d}\omega S_{\mathrm{RPA}}(\boldsymbol{q},\omega) \frac{2\omega}{\omega^2 - (\boldsymbol{v}^\alpha \cdot \boldsymbol{q})^2} \right] \quad . \tag{6.18}$$

We emphasize again that after the linearization we may replace $\boldsymbol{r} \to r_\parallel^\alpha \hat{\boldsymbol{v}}^\alpha$ in the argument of the Debye-Waller factor, because in this case the prefactor Green's function $G_0^\alpha(\boldsymbol{r},\tau)$ is proportional to $\delta^{(d-1)}(\boldsymbol{r}_\perp^\alpha)$ (see Eqs.(5.48) and (5.52)). In contrast, Eqs.(6.4)–(6.6) should be considered for all $\boldsymbol{r}$.

6.1.3 Finite versus infinite patch number

Now comes a really subtle point related to the fact that for linearized energy dispersion we cover the Fermi surface with a finite number of patches.

The term $L_{\boldsymbol{q}}^\alpha(\tau)$ in Eqs.(6.16) and (6.17) is mathematically closely related to the existence of a *double pole* in the integrand defining the Debye-Waller factor for linear energy dispersion. When the ω_m-integral in Eqs.(5.32) and (5.33) is done by means of contour integration, the double pole at $\mathrm{i}\omega_m = \boldsymbol{v}^\alpha \cdot \boldsymbol{q}$ gives rise to a contribution proportional to the derivative of the rest of the integrand with respect the to frequency; the resulting term is therefore proportional to τ, and can be identified with $L_{\boldsymbol{q}}^\alpha(\tau)$. However, as long as *the Fermi surface is covered by a finite number M of patches* we have exactly

$$L_{\boldsymbol{q}}^\alpha(\tau) = 0 \quad . \tag{6.19}$$

To prove this, we use Eq.(2.46) to rewrite Eq.(6.18) as

$$
\begin{aligned}
L^{\alpha}_{q}(\tau) &= \frac{|\tau|}{2} f_q \left[1 - f_q \Pi_{\mathrm{RPA}}(q, v^{\alpha} \cdot q)\right] \mathrm{e}^{-|v^{\alpha} \cdot q||\tau|} \\
&= \frac{|\tau|}{2} \frac{f_q}{\epsilon_{\mathrm{RPA}}(q, v^{\alpha} \cdot q)} \mathrm{e}^{-|v^{\alpha} \cdot q||\tau|} \quad , \qquad (6.20)
\end{aligned}
$$

where the RPA dielectric function at frequency $\omega = v^{\alpha} \cdot q$ is (see Eqs.(2.52), (3.13) and (4.24))

$$
\epsilon_{\mathrm{RPA}}(q, v^{\alpha} \cdot q) = 1 + f_q \Pi_0(q, v^{\alpha} \cdot q) \quad , \qquad (6.21)
$$

with

$$
\Pi_0(q, v^{\alpha} \cdot q) = \sum_{\alpha'=1}^{M} \nu^{\alpha'} \frac{v^{\alpha'} \cdot q}{(v^{\alpha'} - v^{\alpha}) \cdot q} \quad . \qquad (6.22)
$$

Evidently the term $\alpha' = \alpha$ in Eq.(6.22) is divergent, so that $\Pi_0(q, v^{\alpha} \cdot q)$ and hence also the dielectric function at frequency $\omega = v^{\alpha} \cdot q$ are infinite. It follows that

$$
\frac{f_q}{\epsilon^{RPA}(q, v^{\alpha} \cdot q)} = 0 \quad , \qquad (6.23)
$$

so that from Eq.(6.20) we can conclude that $L^{\alpha}_{q}(\tau) = 0$.

This proof does not go through any more if we take the limit of an infinite number of patches, because then the α'-summation in Eq.(6.22) is for $d > 1$ replaced by an angular integration, and the singularity in the integrand must be regularized via the usual pole prescription $v^{\alpha} \cdot q \to v^{\alpha} \cdot q + \mathrm{i}0^{+}$. Then in $d > 1$ the function $\Pi_0(q, v^{\alpha} \cdot q + \mathrm{i}0^{+})$ is finite. For example, for a spherical Fermi surface $\Pi_0(q, v^{\alpha} \cdot q + \mathrm{i}0^{+}) = \nu g_d(\hat{v}^{\alpha} \cdot \hat{q} + \mathrm{i}0^{+})$, where the function $g_d(x + \mathrm{i}0^{+})$ is given in Eq.(A.3). In other words, in the limit $M \to \infty$ the singularity in $\Pi_0(q, v^{\alpha} \cdot q)$ is regularized by the finite imaginary part of the function $g_d(x + i0^{+})$ for $x < 1$, see Eq.(A.18).

The above difference between the cases $M < \infty$ and $M = \infty$ is due to qualitatively different behavior of the dynamic structure factor in both cases. As discussed in detail in Appendix A.4, for $M < \infty$ the dynamic structure factor $S_{\mathrm{RPA}}(q, \omega)$ exhibits M delta-function peaks. For $M \to \infty$ only two of these peaks survive and can be identified with the undamped plasmon mode at frequencies $\pm\omega_q$, while the other peaks merge into the particle-hole continuum. From the formal point of view the procedure of substituting the infinite-patch limit for the dynamic structure factor into the Debye-Waller factor for linearized energy dispersion (see Eqs.(6.16) and (6.17)) is certainly not satisfactory, because the approximations used to derive these equations are only valid as long as the sector cutoffs Λ and λ are kept finite and large compared with the range q_c of the interaction in momentum space, see Fig.2.5. But $M \to \infty$ implies that we are taking the limit $\Lambda \to 0$, so that for fixed q_c the condition $q_c \ll \Lambda$ (see Eq.(2.63)) cannot be satisfied.

Obviously the problem associated with the limit of infinite patch number does not arise in our more general results (6.4)–(6.6) for non-linear energy dispersion, because in this case the dynamic structure factor exhibits the particle-hole continuum even if we work with a finite number of patches, and a term similar to $L^{\alpha}_{\boldsymbol{q}}(\tau)$ that is linear in τ simply does not appear, because there is no double pole in the Debye-Waller factor. The disadvantage of Eqs.(6.4)–(6.6) is that these expressions are more difficult to evaluate than the corresponding expressions for linearized energy dispersion. Fortunately, at $\tau = 0$ we have $L^{\alpha}_{\boldsymbol{q}}(0) = 0$, so that possible ambiguities related to the limit of infinite patch number in the linearized theory do not appear in all quantities involving the *static* Debye-Waller factor $Q^{\alpha}(r^{\alpha}_{\parallel}\hat{\boldsymbol{v}}^{\alpha}, 0)$. In this case the use of the $M = \infty$ limit for the dynamic structure factor in the Debye-Waller factor for linearized energy dispersion seems to be justified[4], at least as long the patch cutoffs are small compared with k_{F}. In the rest of this chapter we shall therefore focus on the static Debye-Waller factor $Q^{\alpha}(r^{\alpha}_{\parallel}\hat{\boldsymbol{v}}^{\alpha}, 0)$ for linearized energy dispersion, and use the $M \to \infty$ limit for the dynamic structure factor.

6.2 The static Debye-Waller factor for linearized energy dispersion

We now explicitly evaluate $Q^{\alpha}(r^{\alpha}_{\parallel}\hat{\boldsymbol{v}}^{\alpha}, 0)$ for singular interactions of the form (6.1) for a spherically symmetric d-dimensional system. We show that the Fermi liquid state is only stable for $\eta < 2(d-1)$, but that in the interval $2(d-2) < \eta < 2(d-1)$ the sub-leading corrections are anomalously large. We then consider the regime $\eta \geq 2(d-1)$, and show that for $\eta \geq 2(d+1)$ the bosonization result for the equal-time Debye-Waller factor $Q^{\alpha}(r^{\alpha}_{\parallel}\hat{\boldsymbol{v}}^{\alpha}, 0)$ is mathematically not well-defined.

6.2.1 Consequences of spherical symmetry

For a spherically symmetric Fermi surface we have $|\boldsymbol{v}^{\alpha}| = v_{\mathrm{F}}$ for all α, so that the non-interacting sector Green's function given in Eq.(5.48) can be written as

$$G^{\alpha}_{0}(\boldsymbol{r}, \tau) = \delta^{(d-1)}(\boldsymbol{r}^{\alpha}_{\perp}) G_{0}(r^{\alpha}_{\parallel}, \tau) \quad , \quad r^{\alpha}_{\parallel} = \hat{\boldsymbol{v}}^{\alpha} \cdot \boldsymbol{r} \quad , \tag{6.24}$$

where

$$G_{0}(x, \tau) = \left(\frac{-\mathrm{i}}{2\pi} \right) \frac{1}{x + \mathrm{i} v_{\mathrm{F}} \tau} \tag{6.25}$$

[4] By taking the limit $M \to \infty$ in the Debye-Waller factor, we also eliminate artificial nesting singularities, which are generated if the covering of the Fermi surface contains at least two parallel patches, see Chap. 7.2.4. In this sense the limit $M \to \infty$ is really the physical limit of interest, although for linearized energy dispersion it is not possible to give a formally convincing justification for this limiting procedure. Of course, in case of ambiguities we can always go back to our more general results (6.4)–(6.6).

is the usual one-dimensional non-interacting Green's function. Note that for a spherical Fermi surface the polarization $\Pi_0(q)$ depends at long wavelengths only on the combination $\mathrm{i}\omega_m/(v_{\mathrm{F}}|\boldsymbol{q}|)$, see Eq.(A.1). It follows that the Debye-Waller factor (5.31) is actually of the form $Q^{\alpha}(r_{\|}^{\alpha}\hat{\boldsymbol{v}}^{\alpha}, \tau) = Q(r_{\|}^{\alpha}, \tau)$, where $Q(x, \tau)$ is the following function of two variables x and τ,

$$Q(x, \tau) = R - S(x, \tau) = \frac{1}{\beta V} \sum_{q} \frac{f_{\boldsymbol{q}}[1 - \cos(\hat{\boldsymbol{v}}^{\alpha} \cdot \boldsymbol{q}x - \omega_m \tau)]}{[1 + f_{\boldsymbol{q}} \Pi_0(q)](\mathrm{i}\omega_m - \boldsymbol{v}^{\alpha} \cdot \boldsymbol{q})^2} \quad . \tag{6.26}$$

Due to rotational invariance, the value of the integral is independent of the direction of the unit vector $\hat{\boldsymbol{v}}^{\alpha}$, as can be easily seen by introducing d-dimensional spherical coordinates, see Eq.(A.7). From Eqs.(5.37) and (5.52) we conclude that for rotationally invariant systems the interacting patch Green's function can be written as

$$G^{\alpha}(\boldsymbol{r}, \tau) = \delta^{(d-1)}(\boldsymbol{r}_{\perp}^{\alpha}) G_0(r_{\|}^{\alpha}, \tau) \mathrm{e}^{Q(r_{\|}^{\alpha}, \tau)} \quad , \quad r_{\|}^{\alpha} = \hat{\boldsymbol{v}}^{\alpha} \cdot \boldsymbol{r} \quad . \tag{6.27}$$

Let us study the constant part R of the Debye-Waller factor in more detail. The form of the RPA dynamic structure factor for spherical Fermi surfaces is discussed in detail in Appendix A.2. Using Eqs.(A.26), (A.27) and (A.32), and taking the limit $V \to \infty$ in Eq.(6.14), we obtain

$$R = -\int \frac{\mathrm{d}\boldsymbol{q}}{(2\pi)^d} f_{\boldsymbol{q}}^2 \left[\frac{Z_{\boldsymbol{q}}}{(\omega_{\boldsymbol{q}} + |\boldsymbol{v}^{\alpha} \cdot \hat{\boldsymbol{q}}|)^2} + \frac{\nu}{\pi} \int_0^{v_{\mathrm{F}}|\boldsymbol{q}|} \mathrm{d}\omega \frac{1}{(\omega + |\boldsymbol{v}^{\alpha} \cdot \hat{\boldsymbol{q}}|)^2} \mathrm{Im} \left\{ \frac{g_d(\frac{\omega}{v_{\mathrm{F}}|\boldsymbol{q}|} + \mathrm{i}0^+)}{1 + F_{\boldsymbol{q}} g_d(\frac{\omega}{v_{\mathrm{F}}|\boldsymbol{q}|} + \mathrm{i}0^+)} \right\} \right] , \tag{6.28}$$

where the energy $\omega_{\boldsymbol{q}}$ and the residue $Z_{\boldsymbol{q}}$ of the collective plasmon mode are given in Eqs.(A.29) and (A.33). Using Eq.(A.5) and the fact that according to Eqs.(A.34) and (A.35) the residue of the plasmon mode is of the form $Z_{\boldsymbol{q}} = \nu v_{\mathrm{F}} |\boldsymbol{q}| Z_d(F_{\boldsymbol{q}})$, we obtain

$$R = -\frac{1}{k_{\mathrm{F}}^{d-1} \Omega_d} \int \frac{\mathrm{d}\boldsymbol{q}}{|\boldsymbol{q}|} F_{\boldsymbol{q}}^2 \left[\frac{Z_d(F_{\boldsymbol{q}})}{\left(\frac{\omega_{\boldsymbol{q}}}{v_{\mathrm{F}}|\boldsymbol{q}|} + |\hat{\boldsymbol{v}}^{\alpha} \cdot \hat{\boldsymbol{q}}| \right)^2} + \int_0^1 \frac{\mathrm{d}x}{\pi} \frac{1}{(x + |\hat{\boldsymbol{v}}^{\alpha} \cdot \hat{\boldsymbol{q}}|)^2} \mathrm{Im} \left\{ \frac{g_d(x + \mathrm{i}0^+)}{1 + F_{\boldsymbol{q}} g_d(x + \mathrm{i}0^+)} \right\} \right] , \tag{6.29}$$

where we have introduced the usual dimensionless interaction $F_{\boldsymbol{q}} = \nu f_{\boldsymbol{q}}$.

Because by assumption $F_{\boldsymbol{q}}$ depends only on $|\boldsymbol{q}|$, the angular integration can be expressed in terms of the function

$$h_d(x) = \left\langle \frac{1}{(x + |\hat{\boldsymbol{v}}^{\alpha} \cdot \hat{\boldsymbol{q}}|)^2} \right\rangle_{\hat{\boldsymbol{q}}} , \tag{6.30}$$

where the angular average is defined as in Eqs.(A.4), (A.7) and (A.8). In $d = 1$ we obtain

$$h_1(x) = \frac{1}{(x+1)^2} \quad , \tag{6.31}$$

and in $d > 1$

$$h_d(x) = \gamma_d \int_0^\pi d\vartheta \frac{(\sin\vartheta)^{d-2}}{(x + |\cos\vartheta|)^2} \quad , \tag{6.32}$$

with γ_d given in Eq.(A.10). In particular, in $d = 2$ we have

$$h_2(x) = \frac{2}{\pi} \times \begin{cases} \frac{1}{1-x^2}\left[\frac{1}{x} - \frac{x}{\sqrt{1-x^2}} \ln\left(\frac{1+\sqrt{1-x^2}}{x}\right)\right] & \text{for } x < 1 \\ \frac{2}{3} & \text{for } x = 1 \\ \frac{1}{x^2-1}\left[-\frac{1}{x} + \frac{x}{\sqrt{x^2-1}} \arccos(\frac{1}{x})\right] & \text{for } x > 1 \end{cases} \quad , \tag{6.33}$$

while in $d = 3$ the result is simply

$$h_3(x) = \frac{1}{x(x+1)} \quad . \tag{6.34}$$

For large and small x we have

$$h_d(x) \sim \frac{1}{x^2} \quad , \quad x \to \infty \quad , \tag{6.35}$$

$$h_d(x) \sim \frac{2\gamma_d}{x} \quad , \quad x \to 0 \quad , \quad d > 1 \quad . \tag{6.36}$$

We are now ready to rewrite Eq.(6.29) in terms of rescaled variables. Using Eq.(A.35) and the fact that $\omega_{\boldsymbol{q}}/(v_{\mathrm{F}}|\boldsymbol{q}|)$ is according to Eq.(A.29) a function of $F_{\boldsymbol{q}}$, we obtain

$$R = -\frac{1}{k_{\mathrm{F}}^{d-1}} \int_0^\infty dq q^{d-2} \left[C_d(F_q) + L_d(F_q)\right] \quad , \tag{6.37}$$

where the dimensionless functions $C_d(F)$ and $L_d(F)$ are given by

$$C_d(F) = F^2 Z_d(F) h_d\left(g_d^{-1}(-\frac{1}{F})\right) = \frac{h_d\left(g_d^{-1}(-\frac{1}{F})\right)}{g_d'\left(g_d^{-1}(-\frac{1}{F})\right)} \quad , \tag{6.38}$$

$$L_d(F) = F^2 \int_0^1 \frac{dx}{\pi} h_d(x) \mathrm{Im}\left\{\frac{g_d(x+\mathrm{i}0^+)}{1 + F g_d(x + \mathrm{i}0^+)}\right\} \quad , \tag{6.39}$$

with $g_d(z)$ defined in Eq.(A.3). Note that $C_d(F)$ represents the *collective mode* contribution to the RPA dynamic structure factor (see Eq.(A.32)), while $L_d(F)$ represents the single-pair contribution due to *Landau damping* (see Eq.(A.27)). The asymptotic behavior of the functions $C_d(F)$ and $L_d(F)$ determines the parameter regime where the system is a Fermi liquid. For $F \to \infty$ we have to leading order (see Eqs.(A.39), (A.41) and (6.35))

$$C_d(F) \sim \frac{\sqrt{d}}{2}\sqrt{F} \quad , \quad F \to \infty \quad , \tag{6.40}$$

while the Landau damping contribution reduces to a finite constant,

$$L_d(F) \sim L_d^{\infty} \equiv - \int_0^1 \frac{\mathrm{d}x}{\pi} h_d(x) \mathrm{Im} \left\{ \frac{1}{g_d(x + \mathrm{i}0^+)} \right\} \quad , \quad F \to \infty \quad . \tag{6.41}$$

To see more clearly that L_d^{∞} is for all d a finite *positive* constant, note that from Eqs.(A.7) and (A.18)

$$\mathrm{Im} \left\{ \frac{1}{g_d(x + \mathrm{i}0^+)} \right\} = - \frac{\pi x \left\langle \delta(\hat{\boldsymbol{q}} \cdot \hat{\boldsymbol{k}} - x) \right\rangle_{\hat{\boldsymbol{k}}}}{|g_d(x + \mathrm{i}0^+)|^2}$$
$$= - \frac{\pi x \gamma_d \int_0^{\pi} \mathrm{d}\vartheta (\sin \vartheta)^{d-2} \delta(\cos \vartheta - x)}{|g_d(x + \mathrm{i}0^+)|^2} \quad , \tag{6.42}$$

so that from Eq.(6.41)

$$L_d^{\infty} = \gamma_d \int_0^{\pi/2} \mathrm{d}\vartheta (\sin \vartheta)^{d-2} \frac{\cos \vartheta h_d(\cos \vartheta)}{|g_d(\cos \vartheta + \mathrm{i}0^+)|^2} \quad . \tag{6.43}$$

The integrand in Eq.(6.43) is non-singular and positive for all ϑ, so that $0 < L_d^{\infty} < \infty$. The weak coupling behavior of $C_d(F)$ is easily obtained from Eq.(A.45),

$$C_d(F) \sim \begin{cases} 0 & \text{for } d > 3 \text{ and } F < |g_d(1)|^{-1} \\ \mathrm{e}^{-2/F} & \text{for } d = 3 \\ \frac{2h_d(1)}{(3-d)c_d} (c_d F)^{\frac{5-d}{3-d}} & \text{for } d < 3 \end{cases} \quad , \tag{6.44}$$

where the numerical constant c_d is defined via Eq.(A.22). The Landau damping part is at weak coupling proportional to F^2,

$$L_d(F) \sim L'_d F^2 \quad , \quad F \to 0 \quad , \tag{6.45}$$

where the numerical constant L'_d is given by

$$L'_d = \int_0^{\infty} \frac{\mathrm{d}x}{\pi} h_d(x) \mathrm{Im} g_d(x + \mathrm{i}0^+) \quad . \tag{6.46}$$

Note that at strong coupling

$$\frac{C_d(F)}{L_d(F)} \sim \frac{\sqrt{d}}{2L_d^{\infty}} \sqrt{F} \quad , \quad F \to \infty \quad , \tag{6.47}$$

so that the relative weight of the collective mode is always larger than that of the Landau damping part. In the other hand, at weak coupling it is easy to see from Eqs.(6.44) and (6.45) that the Landau damping part is dominant. In particular, for $1 < d < 3$ we have

$$\frac{C_d(F)}{L_d(F)} \sim \frac{2h_d(1) c_d^{\frac{2}{3-d}}}{(3-d) L'_d} F^{\frac{d-1}{3-d}} \quad , \quad F \to 0 \quad . \tag{6.48}$$

The important point is that for $1 < d < 3$ the exponent of F is always positive, so that for small F the right-hand side of Eq.(6.48) is indeed small. Hence, the collective mode contribution is negligible at weak coupling.

6.2.2 The existence of the quasi-particle residue

For singular interactions of the form (6.1) we have $F_{\boldsymbol{q}} = (\kappa/|\boldsymbol{q}|)^{\eta} e^{-|\boldsymbol{q}|/q_c}$, see Eq.(A.62). Having determined the weak and strong coupling behavior of the functions $C_d(F)$ and $L_d(F)$ in Eq.(6.37), it is now easy to calculate the quasi-particle residue for this type of interaction. Introducing in Eq.(6.37) the dimensionless integration variable $p = q/\kappa$ and setting $p_c = q_c/\kappa$ we obtain

$$R = -\left(\frac{\kappa}{k_F}\right)^{d-1} \tilde{R}(d, \eta, p_c) \quad , \tag{6.49}$$

where the dimensionless function $\tilde{R}(d, \eta, p_c)$ is given by

$$\tilde{R}(d, \eta, p_c) = \int_0^{\infty} \mathrm{d}p p^{d-2} \left[C_d(p^{-\eta} e^{-p/p_c}) + L_d(p^{-\eta} e^{-p/p_c}) \right] \quad . \tag{6.50}$$

Because the functions $C_d(F)$ and $L_d(F)$ do not have any singularities at finite values of F, the integral in Eq.(6.50) can only diverge due to possible infrared singularities at small p, or ultraviolet singularities at large p. Let us first consider the infrared limit. Because the exponent η is positive, this limit is determined by the strong coupling behavior of $C_d(F)$ and $L_d(F)$. From Eq.(6.47) we know that in this limit the collective mode is dominant, so that the most singular contribution arises from the first term in Eq.(6.50). Using Eq.(6.40), it is easy to see that this term yields

$$\begin{aligned} \tilde{R}(d, \eta, p_c) &\sim \frac{\sqrt{d}}{2} \int_0^{p_c} \mathrm{d}p p^{d-2-\frac{\eta}{2}} \\ &= \frac{\sqrt{d}}{2} \frac{p_c^{d-1-\frac{\eta}{2}}}{d-1-\frac{\eta}{2}} \quad , \quad \text{for } \eta < 2(d-1) \quad . \end{aligned} \tag{6.51}$$

Evidently $\tilde{R}(d, \eta, p_c) = \infty$ for $\eta \geq 2(d-1)$, so that in this case $R = -\infty$. We conclude that

$$Z = 0 \quad , \quad \text{for } \eta \geq \eta_{\mathrm{ir}} \equiv 2(d-1) \quad . \tag{6.52}$$

Therefore, the Fermi liquid is only stable for $\eta < 2(d-1)$. This result has first been derived by Bares and Wen [6.1].

Another special value for the exponent η is determined by the requirement that the integral in Eq.(6.50) is convergent even without ultraviolet cutoff p_c. Assuming that we have eliminated the high-energy degrees of freedom outside a thin shell of thickness λ around the Fermi surface, we should choose $q_c \approx \lambda$ and hence $p_c = \lambda/\kappa$. Because in practice we cannot explicitly perform the integration over the high-energy degrees of freedom, it is important that at the end of the calculation physical quantities do not depend on λ. This requirement is automatically satisfied if it is possible to take the limit $\lambda/\kappa \to \infty$, so that the final expression for the Green's function looses its dependence on the unphysical cutoff λ. We now determine the range of η where the integrand in Eq.(6.50) vanishes at large p sufficiently fast to insure convergence of the integral even without the cutoff p_c. Because for large p the arguments of the functions $C_d(F)$ and $L_d(F)$ in Eq.(6.50) are small, we need to know the behavior of these functions at weak coupling. From Eq.(6.48) it is clear that in this regime the Landau damping contribution $L_d(F)$ is dominant. Using Eq.(6.45), we find that the ultraviolet behavior of Eq.(6.50) is determined by

$$\tilde{R}(d,\eta,p_c) \sim L'_d \int_1^\infty \mathrm{d}p p^{d-2-2\eta} e^{-2p/p_c} \quad . \tag{6.53}$$

Setting $p_c = \infty$, we see that the integral exists only for

$$\eta > \eta_{\mathrm{uv}} \equiv \frac{d-1}{2} \quad . \tag{6.54}$$

If this condition is satisfied, the integrand falls off sufficiently fast to insure convergence of the integral. Note that $\eta_{\mathrm{uv}} < \eta_{\mathrm{ir}}$, so that there exists a finite interval for η where the quasi-particle residue is finite and the ultraviolet cutoff can be removed. Because we have rescaled $p = |q|/\kappa$, the convergence of the integral implies that the numerical value of the quasi-particle residue is determined by the regime $|q| \lesssim \kappa$. In this case κ (and not q_c) acts as the relevant screening wave-vector in the problem. In this sense an interaction of the form (A.62) with $\eta > \frac{d-1}{2}$ and $\kappa \ll q_c$ effectively replaces any unphysical ultraviolet cutoff q_c (which might have been generated by integrating out high energy modes) by the physical cutoff κ in the bosonization result for the quasi-particle residue. In summary, for singular density-density interactions of the form (6.1) the function $\tilde{R}(d,\eta,\infty)$ exists for

$$\frac{d-1}{2} < \eta < 2(d-1) \quad . \tag{6.55}$$

In this interval the interaction falls off sufficiently fast at large $|q|$ to insure convergence at short wavelengths, but diverges weak enough to lead to a stable Fermi liquid.

6.2.3 Why the Coulomb interaction is so nice

As discussed in Appendix A.3.1, the Coulomb interaction in $1 < d \leq 3$ corresponds to $\eta = d-1$ and $q_c = \infty$. Furthermore, κ can now be identified with the usual Thomas-Fermi screening wave-vector given in Eq.(A.50). Note that $\eta = d-1$ satisfies for all d the condition (6.55). Setting $\eta = d-1$ and $q_c = \infty$ in Eq.(6.49), and changing variables to $F = p^{-(d-1)}$ in Eq.(6.50), we obtain

$$R = -\left(\frac{\kappa}{k_F}\right)^{d-1} \frac{\tilde{r}_d}{d-1} \quad , \tag{6.56}$$

with

$$\tilde{r}_d \equiv (d-1)\tilde{R}(d, d-1, \infty) = \int_0^\infty \frac{\mathrm{d}F}{F^2} \left[C_d(F) + L_d(F)\right] \quad . \tag{6.57}$$

From the previous section we know that the integral in Eq.(6.57) exists for all $d > 1$. Note also that according to Eq.(A.54) the prefactor $(\kappa/k_{\mathrm{F}})^{d-1}$ is proportional to the Wigner-Seitz radius r_s, which is the relevant small parameter in the usual high-density expansion for the homogeneous electron gas [6.2]. We conclude that higher-dimensional bosonization predicts for the Coulomb interaction in dimensions $1 < d \leq 3$ a finite result for the quasi-particle residue, which in the limit of high densities (i.e. for $\kappa \ll k_{\mathrm{F}}$) is close to unity and independent of the unphysical sector cutoffs.

By isolating a factor of $\frac{1}{d-1}$ in Eq.(6.56) we have anticipated that $\tilde{r}_d$ has a finite limit for $d \to 1$. If we are only interested in the leading behavior of R for $d \to 1$, it is sufficient to calculate $\tilde{r}_1$. In this case $L_1(F) = 0$, and the functional form of $C_1(F)$ is simply obtained by replacing $F_0 \to F$ in the expression for the anomalous dimension of the Tomonaga-Luttinger model [1.13–1.15] (see Eq.(6.88) below),

$$C_1(F) = \frac{F^2}{2\sqrt{1+F}\left[\sqrt{1+F}+1\right]^2} \quad . \tag{6.58}$$

Substituting this into Eq.(6.57), we obtain

$$\tilde{r}_1 = \frac{1}{2}\int_0^\infty \mathrm{d}F \frac{1}{\sqrt{1+F}\left[\sqrt{1+F}+1\right]^2} = \frac{1}{2} \quad . \tag{6.59}$$

We conclude that for $d \to 1$

$$\begin{aligned} R &= -\left(\frac{\kappa}{k_{\mathrm{F}}}\right)^{d-1} \frac{1}{2(d-1)} + O(1) \\ &= -\frac{1}{2(d-1)} + \frac{1}{2}\ln\left(\frac{k_{\mathrm{F}}}{\kappa}\right) + O(1) \quad . \end{aligned} \tag{6.60}$$

Exponentiating Eq.(6.60), we see that quasi-particle residue vanishes as

$$Z \propto \left(\frac{k_{\mathrm{F}}}{\kappa}\right)^{\frac{1}{2}} \mathrm{e}^{-\frac{1}{2(d-1)}} \quad , \quad d \to 1 \quad . \tag{6.61}$$

A similar result has also been obtained by Castellani, Di Castro and Metzner [1.54].

6.2.4 The sub-leading corrections for $0 < \eta < 2(d-1)$

So far we have shown that for singular interactions of the type (6.1) the integral defining R does not exist if $\eta \geq 2(d-1)$. The divergence is due to the infrared regime of the collective mode contribution to the dynamic structure factor. On the other hand, for $\eta < 2(d-1)$ the quasi-particle residue is finite. In this case we know from Chap. 5.1.5 that $S(\boldsymbol{x},0)$ vanishes at large distances, so that in general we expect (ignoring possible logarithmic corrections)

$$S(\boldsymbol{x},0) \sim -\left(\frac{\kappa}{k_{\mathrm{F}}}\right)^{d-1} \frac{\tilde{S}(d,\eta,\frac{q_c}{\kappa})}{|\kappa \boldsymbol{x}|^{\zeta}} \quad , \quad \boldsymbol{x} \to \infty \quad , \quad \zeta > 0 \quad , \tag{6.62}$$

with some dimensionless function $\tilde{S}(d, \eta, p_c)$. In a Landau Fermi liquid we expect $\zeta = 1$, because otherwise the self-energy $\Sigma(\boldsymbol{k}^\alpha + \boldsymbol{q}, \omega)$ cannot have a power series expansion for small $\boldsymbol{q}$, see Eq.(2.18). However, if η is smaller than (but sufficiently close to) $2(d-1)$, we expect an exponent ζ smaller than unity. It turns out that there exists a critical value η_c such that $0 < \zeta < 1$ for $\eta_c < \eta < 2(d-1)$. In this regime the system is a Fermi liquid with anomalously large sub-leading corrections. We now determine the critical η_c for singular interactions in $d > 1$. Proceeding precisely as above, we obtain (see Eqs.(6.29) and (6.37))

$$S(x, 0) = -\frac{1}{k_F^{d-1} \Omega_d} \int \frac{d\boldsymbol{q}}{|\boldsymbol{q}|} \cos(\hat{\boldsymbol{v}}^\alpha \cdot \boldsymbol{q} x) \left[C_d(F_{\boldsymbol{q}}) + L_d(F_{\boldsymbol{q}}) \right] \quad . \tag{6.63}$$

From Sect. 6.2.2 we know that for singular interactions the integral in Eq.(6.63) is dominated by the strong-coupling limit of the function $C_d(F)$, which is given in Eq.(6.40). Introducing d-dimensional spherical coordinates (see Eqs.(A.7) and (A.9)), we obtain for the dominant part of Eq.(6.63) after a simple rescaling

$$S(x, 0) \sim -\left(\frac{\kappa}{k_F} \right)^{d-1} \frac{\sqrt{d}}{2} \frac{\gamma_d}{|\kappa x|^{d-1-\frac{\eta}{2}}} \int_0^{q_c |x|} dp p^{d-2-\frac{\eta}{2}}$$
$$\times \int_0^\pi d\vartheta (\sin \vartheta)^{d-2} \cos(p \cos \vartheta) \quad . \tag{6.64}$$

For $d - 2 - \frac{\eta}{2} < 0$ the integrand vanishes for large p sufficiently fast, so that the integral is convergent even if the cutoff q_c is removed. In this case we obtain for $\kappa x \to \infty$ and $q_c \gg \kappa$

$$S(x, 0) \sim -\left(\frac{\kappa}{k_F} \right)^{d-1} \frac{\tilde{S}(d, \eta, \infty)}{|\kappa x|^{d-1-\frac{\eta}{2}}} \quad , \quad x \to \infty \quad , \quad 0 < d-1-\frac{\eta}{2} < 1 \quad , \tag{6.65}$$

with

$$\tilde{S}(d, \eta, \infty) = \frac{\sqrt{d}}{2} \gamma_d \int_0^\infty dp p^{d-2-\frac{\eta}{2}} \int_0^\pi d\vartheta (\sin \vartheta)^{d-2} \cos(p \cos \vartheta) \quad . \tag{6.66}$$

This is precisely the asymptotic behavior given in Eq.(6.62), with exponent $\zeta = d-1-\frac{\eta}{2} < 1$. The integral in Eq.(6.66) can be done analytically [6.3,6.4], and we obtain after some rearrangements

$$\tilde{S}(d, \eta, \infty) = -\frac{\sqrt{\pi d}}{4} \frac{\Gamma(\frac{d}{2})}{\Gamma(\frac{1+\frac{\eta}{2}}{2}) \cos[\frac{\pi}{2}(d - \frac{\eta}{2})]} \quad . \tag{6.67}$$

On the other hand, if the exponent $d - 2 - \frac{\eta}{2}$ in Eq.(6.64) is positive, then the integral in Eq.(6.64) depends on the cutoff q_c. In this case we obtain for large x the asymptotic behavior predicted in Eq.(6.62) with $\zeta = 1$ and

$$\tilde{S}(d, \eta, \frac{q_c}{\kappa}) \propto \frac{1}{d - 2 - \frac{\eta}{2}} \left(\frac{q_c}{\kappa} \right)^{d-2-\frac{\eta}{2}} \quad , \quad d - 2 - \frac{\eta}{2} > 0 \quad . \tag{6.68}$$

We conclude that in the regime

$$\eta < 2(d-2) \equiv \eta_c \tag{6.69}$$

the correction to the leading constant term of the static Debye-Waller factor vanishes as x^{-1} at large distances, so that in real space we have analyticity around $x = \infty$. In Fourier space this implies analyticity around the origin, as postulated for the self-energy in a Landau Fermi liquid (see Eq.(2.18)). On the other hand, if η lies in the regime

$$2(d-2) < \eta < 2(d-1) \quad , \tag{6.70}$$

the system is not a conventional Landau Fermi liquid, because the corrections to the leading constant term R are not analytic. If η approaches the value $\eta_{\mathrm{ir}} = 2(d-1)$ from below, the constant term R diverges logarithmically, but the divergence is cancelled by $S(x, \tau)$, so that the total Debye-Waller factor $Q(x, \tau) = R - S(x, \tau)$ remains finite. Similarly, we expect logarithmic corrections to the leading x^{-1} decay of $S(x, 0)$ at the lower limit $\eta = \eta_c = 2(d-2)$ of the interval in Eq.(6.70). Interestingly, the Coulomb interaction, which in d dimensions corresponds to $\eta = d - 1$, satisfies the condition (6.70) for $d < 3$. In particular, in $d = 2$ the Coulomb interaction leads to a Fermi liquid with anomalously large sub-leading corrections.

6.2.5 The regime $\eta \geq 2(d-1)$

Finally, let us consider the regime $\eta \geq 2(d-1)$, where the integral (6.49) defining R is divergent. Clearly, if the exponent η is chosen sufficiently large, the divergence will be so strong that it cannot be regularized by means of the subtraction $Q(x, \tau) = R - S(x, \tau)$ in the Debye-Waller factor. Hence, there exists a critical value of η where the bosonization result in d dimensions is divergent. To investigate this point, we now calculate the long-distance behavior of $Q(x, 0)$ for $\eta \geq 2(d-1)$. Repeating the manipulations leading to Eq.(6.64), we obtain for $\frac{\eta}{2} - d + 1 \geq 0$

$$Q(x,0) \sim -\left(\frac{\kappa}{k_{\mathrm{F}}}\right)^{d-1} \frac{\sqrt{d}\gamma_d}{2} |\kappa x|^{\frac{\eta}{2}-d+1} \int_0^{q_c|x|} \mathrm{d}p p^{-(\frac{\eta}{2}-d+2)}$$
$$\times \int_0^{\pi} \mathrm{d}\vartheta (\sin\vartheta)^{d-2} [1 - \cos(p\cos\vartheta)] \quad , \tag{6.71}$$

From this expression it is easy to show that precisely at $\eta = 2(d-1)$ the Debye-Waller factor increases logarithmically at large distances,

$$Q(x,0) \sim -\gamma_{\mathrm{LL}} \ln(q_c|x|) \quad , \quad \eta = 2(d-1) \quad , \tag{6.72}$$

with the anomalous dimension given by

$$\gamma_{\mathrm{LL}} = \frac{\sqrt{d}}{2} \left(\frac{\kappa}{k_{\mathrm{F}}}\right)^{d-1} \quad . \tag{6.73}$$

The logarithmic divergence of the static Debye-Waller factor is familiar from the one-dimensional Tomonaga-Luttinger model (see Sect. 6.3). As a consequence, the momentum distribution n_k exhibits an algebraic singularity at the Fermi surface. The location of this singularity can be used to define the Fermi surface of the interacting system in a mathematically precise way.

For $\eta > 2(d-1)$ we find a *stretched exponential* divergence of the static Debye-Waller factor,

$$Q(x,0) \sim -\left(\frac{\kappa}{k_{\mathrm{F}}}\right)^{d-1} \tilde{Q}(d,\eta)|\kappa x|^{\frac{\eta}{2}-d+1} \quad , \quad \frac{\eta}{2}-d+1>0 \quad , \tag{6.74}$$

with

$$\tilde{Q}(d,\eta) = \frac{\sqrt{d}\Gamma(\frac{d}{2})}{\sqrt{\pi}[\eta-2(d-1)]\Gamma\left(\frac{1+\frac{\eta}{2}}{2}\right)} \cos\left(\frac{\pi}{2}(\frac{\eta}{2}-d+1)\right) \times \Gamma\left(1+(\frac{\eta}{2}-d+1)\right)\Gamma\left(1-(\frac{\eta}{2}-d+1)\right) \quad . \tag{6.75}$$

The important point is now that for $\frac{\eta}{2}-d+1 = 2$ the function $\tilde{Q}(d,\eta)$ diverges, because the argument of the second Γ-function in Eq.(6.75) becomes -1. Hence, for

$$\eta \geq 2(d+1) \tag{6.76}$$

bosonization cannot cure the divergence due to the singular interactions. The physical behavior of the system in this parameter regime cannot be discussed within the framework of our bosonization approach. Note that for $\eta = 2(d+1)$ the equal-time Debye-Waller factor in Eq.(6.74) would be quadratically divergent, so that the equal-time Green's function would vanish like a *Gaussian* at large distances, i.e. $Q(x,0) \propto -x^2$. On the other hand, in the regime $2(d-1) < \eta < 2(d+1)$ the equal-time Green's function can be calculated via bosonization, and vanishes like a *stretched exponential* at large distances. We shall refer to normal Fermi systems with this property as *exotic quantum liquids*. It is easy to show [6.5] that the stretched exponential decay of the static Debye-Waller factor implies that the momentum distribution n_k is analytic at the (non-interacting) Fermi surface, so that a sharp Fermi surface of the interacting system simply cannot be defined any more. The disappearance of a sharp Fermi surface in strongly correlated Fermi systems is certainly not a special feature of the singular interactions considered here. For example, models with correlated hopping [6.6, 6.7] show similar behavior. The various critical values for η derived in this section are summarized in Fig. 6.1. The fact that exotic quantum liquids do not have a sharp Fermi surface does *not* mean that in these systems the bosonization approach (which is based on the expansion of the energy dispersion for momenta in the vicinity of the *non-interacting* Fermi surface) is inconsistent. As already pointed out long time ago by Tomonaga [1.17], the existence of a singularity in the momentum distribution is really not necessary for the consistency of the bosonization

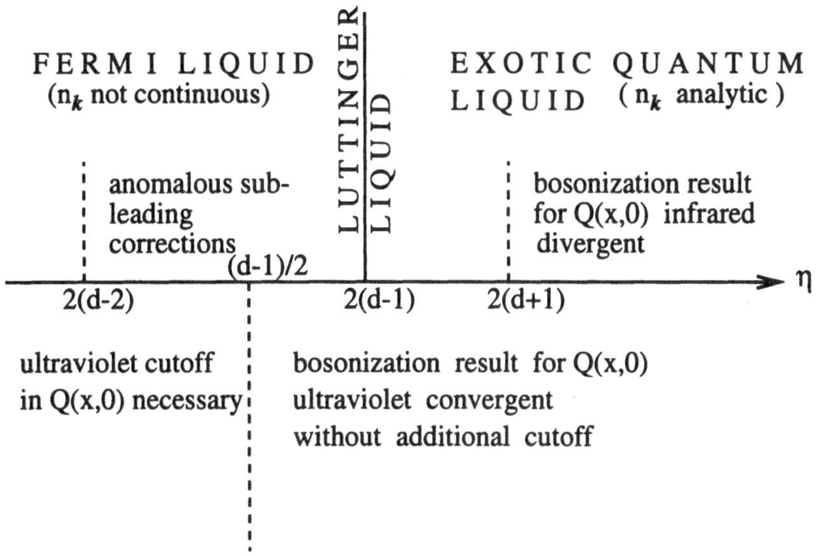

Fig. 6.1. Summary of special values of the exponent η for singular density-density interactions of the type $|q|^{-\eta}$ in d dimensions. The Luttinger liquid for $\eta = 2(d-1)$ corresponds to the marginal case, where the momentum distribution n_k is continuous but not analytic. Note that $2(d-2) < \frac{d-1}{2}$ for $d < \frac{7}{3}$.

procedure as long as (i) the thickness of the shell where the momentum distribution drops from unity to zero is small compared with the characteristic size of k_F, and (ii) the effective interaction is not too singular, so that the Debye-Waller factor $Q(x, 0)$ is mathematically well defined[5]. The condition (i) means that the smearing of the Fermi surface is small, so that *it does not matter which point k^α within the smeared shell is chosen as a reference point for the expansion of the non-interacting energy dispersion.* For singular interactions of the type discussed above the condition (ii) is satisfied as long as $\eta < 2(d+1)$.

6.3 Luttinger liquid behavior in $d = 1$

This section does not contain any new results, but shows that in $d = 1$ our formalism correctly reproduces the well-known bosonization result for the Green's function of the Tomonaga-Luttinger model.

[5] The fact that the integral defining $Q(x, 0)$ exists does not imply the existence of $Q(x, \tau)$ for $\tau \neq 0$. In fact, in $d = 1$ Eq.(6.76) tells us that the static Debye-Waller factor is mathematically ill-defined for $\eta \geq 4$, while for $\tau \neq 0$ it is easy to show from Eq.(6.84) that the integral defining $Q(x, \tau)$ does not exist as soon as $\eta \geq 1$.

In $d = 1$ we have only two Fermi points, which may be labelled by $\alpha = +, -$. The associated normal vectors are $\hat{\boldsymbol{v}}^{\alpha} = \alpha \boldsymbol{e}_x$. Obviously the matrix $\underline{f}_{\boldsymbol{q}}$ is then a 2×2-matrix[6]. The usual notation in the literature [1.13] is $[\underline{f}_{\boldsymbol{q}}]^{++} = [\underline{f}_{\boldsymbol{q}}]^{--} = g_4(\boldsymbol{q})$, and $[\underline{f}_{\boldsymbol{q}}]^{+-} = [\underline{f}_{\boldsymbol{q}}]^{-+} = g_2(\boldsymbol{q})$. Because Eqs.(6.14), (6.16) and (6.17) have been derived for the special case that all matrix elements of $\underline{f}_{\boldsymbol{q}}$ are identical, these expressions should reduce to the exact solution of the Tomonaga-Luttinger model with interaction parameters $g_4 = g_2 = f_0$, where $\lim_{\boldsymbol{q} \to 0} f_{\boldsymbol{q}} = f_0 = \text{const}$. Note that in the Tomonaga-Luttinger model the energy dispersion is linear by definition. Writing $\boldsymbol{r} = r_x \boldsymbol{e}_x$, it is clear from the general considerations of Sect. 6.2.1 that the Debye-Waller factor depends on the sector label only via $r_{\parallel}^{\alpha} = \hat{\boldsymbol{v}}^{\alpha} \cdot \boldsymbol{r} = \alpha r_x$, so that $Q^{\alpha}(r_{\parallel}^{\alpha} \hat{\boldsymbol{v}}^{\alpha}, \tau) = Q(x, \tau)$, with $x = \alpha r_x$.

To evaluate the Debye-Waller factor from Eqs.(6.14), (6.16) and (6.17), we need the RPA dynamic structure factor in $d = 1$. From Eqs.(A.30),(A.32),(A.34) and (A.36) we obtain

$$S_{\mathrm{RPA}}(\boldsymbol{q}, \omega) = Z_{\boldsymbol{q}} \delta(\omega - \omega_{\boldsymbol{q}}) \quad , \tag{6.77}$$

where the residue and the collective mode are given by

$$Z_{\boldsymbol{q}} = \frac{|\boldsymbol{q}|}{2\pi \sqrt{1 + F_{\boldsymbol{q}}}} \quad , \quad \omega_{\boldsymbol{q}} = \sqrt{1 + F_{\boldsymbol{q}}} v_{\mathrm{F}} |\boldsymbol{q}| \quad , \tag{6.78}$$

and $F_{\boldsymbol{q}} = \nu f_{\boldsymbol{q}} = f_{\boldsymbol{q}} / (\pi v_{\mathrm{F}})$ is the usual dimensionless interaction. Note that in $d = 1$ there is no single pair contribution to the RPA dynamic structure factor. Furthermore, because the Fermi surface in $d = 1$ is covered by $M = 2$ patches, we know from the considerations of Sect. 6.1.3 that $L_{\boldsymbol{q}}^{\alpha}(\tau) = 0$ in Eqs.(6.16) and (6.17). Substituting Eq.(6.77) into Eq.(6.14), we obtain

$$\begin{aligned} R &= -\frac{1}{V} \sum_{\boldsymbol{q}} f_{\boldsymbol{q}}^2 \frac{Z_{\boldsymbol{q}}}{(\omega_{\boldsymbol{q}} + |\boldsymbol{v}^{\alpha} \cdot \boldsymbol{q}|)^2} \\ &= -\frac{1}{V} \sum_{\boldsymbol{q}} \frac{\pi}{|q_x|} \frac{F_{\boldsymbol{q}}^2}{2\sqrt{1 + F_{\boldsymbol{q}}} \left[\sqrt{1 + F_{\boldsymbol{q}}} + 1 \right]^2} \quad . \end{aligned} \tag{6.79}$$

In the limit $V \to \infty$ we may replace $\frac{1}{V} \sum_{\boldsymbol{q}} f(|q_x|) \to \int_0^{\infty} \frac{dq_x}{\pi} f(q_x)$. Using the identity

$$\frac{F_{\boldsymbol{q}}^2}{2\sqrt{1 + F_{\boldsymbol{q}}} \left[\sqrt{1 + F_{\boldsymbol{q}}} + 1 \right]^2} = \frac{1 + \frac{F_{\boldsymbol{q}}}{2}}{\sqrt{1 + F_{\boldsymbol{q}}}} - 1 \quad , \tag{6.80}$$

we finally obtain

$$R = -\int_0^{\infty} \frac{dq_x}{q_x} \left[\frac{1 + \frac{F_{\boldsymbol{q}}}{2}}{\sqrt{1 + F_{\boldsymbol{q}}}} - 1 \right] \quad . \tag{6.81}$$

[6] To distinguish the wave-vector label from the collective label $q = [\boldsymbol{q}, i\omega_m]$, we shall continue to write $\boldsymbol{q}$ for the wave-vector, it being understood that $\boldsymbol{q} = q_x \boldsymbol{e}_x$, where $\boldsymbol{e}_x$ is a unit vector in the x-direction.

Similarly, we obtain from Eqs.(6.16) and (6.17)

$$\mathrm{Re}S(x,\tau) = -\int_0^\infty \frac{\mathrm{d}q_x}{q_x} \cos(q_x x) \left[\frac{1+\frac{F_q}{2}}{\sqrt{1+F_q}} \mathrm{e}^{-\sqrt{1+F_q} v_{\mathrm{F}} q_x |\tau|} - \mathrm{e}^{-v_{\mathrm{F}} q_x |\tau|} \right], \tag{6.82}$$

$$\mathrm{Im}S(x,\tau) = -\mathrm{sgn}(\tau) \int_0^\infty \frac{\mathrm{d}q_x}{q_x} \sin(q_x x) \left[\mathrm{e}^{-\sqrt{1+F_q} v_{\mathrm{F}} q_x |\tau|} - \mathrm{e}^{-v_{\mathrm{F}} q_x |\tau|} \right]. \tag{6.83}$$

Combining Eqs.(6.81) and (6.82), we can also write

$$\mathrm{Re}Q(x,\tau) = R - \mathrm{Re}S(x,\tau) = -\int_0^\infty \frac{\mathrm{d}q_x}{q_x} \times \left\{ \frac{1+\frac{F_q}{2}}{\sqrt{1+F_q}} \left[1 - \cos(q_x x) \mathrm{e}^{-\sqrt{1+F_q} v_{\mathrm{F}} q_x |\tau|} \right] - \left[1 - \cos(q_x x) \mathrm{e}^{-v_{\mathrm{F}} q_x |\tau|} \right] \right\}. \tag{6.84}$$

Eqs.(6.82)–(6.84) are identical with the well-known bosonization result for the Green's function of an interacting Fermi system with linearized energy dispersion [1.15].

Let us evaluate Eqs.(6.82)–(6.84) for interactions of the form $F_q = F_0 \mathrm{e}^{-|q|/q_{\mathrm{c}}}$, where $q_{\mathrm{c}} \ll k_{\mathrm{F}}$. From Sect. 6.2 we know that in one dimension an interaction that approaches a constant for $q \to 0$ leads to an unbounded Debye-Waller factor which grows logarithmically at large distances. The logarithmic singularity is evident in Eq.(6.81). Hence, according to Eq.(5.86) the quasi-particle residue vanishes in this case, so that the system is not a Fermi liquid. However, in the combination $R - S(x,\tau)$ the logarithmic singularity is removed, and we obtain a finite result for the Green's function. Unfortunately, for interactions of the above form the integrals in Eqs.(6.82)–(6.84) cannot be evaluated analytically. However, at length scales x large compared with the characteristic range q_{c}^{-1} of the interaction the Green's function should be independent of the precise way in which the ultraviolet cutoff is introduced. Therefore we may regularize the q_x-integrals in any convenient way. A standard regularization which leads to elementary integrals is to multiply the entire integrand by a convergence factor $\mathrm{e}^{-|q|/q_{\mathrm{c}}}$ and to replace $F_q \to F_0$ everywhere in the integrand [4.1, 6.8]. Although the cutoff q_{c} defined in this way is not identical with the cutoff in $F_0 \mathrm{e}^{-|q|/q_{\mathrm{c}}}$, it can still be identified physically with the range of the interaction in momentum space. The relevant integrals can be found in standard tables [6.3, 6.4], and we obtain

$$\mathrm{Re}Q(x,\tau) = -\frac{1+\frac{F_0}{2}}{2\sqrt{1+F_0}} \ln \left[\frac{x^2 + (\tilde{v}_{\mathrm{F}}|\tau| + q_{\mathrm{c}}^{-1})^2}{q_{\mathrm{c}}^{-2}} \right] + \frac{1}{2} \ln \left[\frac{x^2 + (v_{\mathrm{F}}|\tau| + q_{\mathrm{c}}^{-1})^2}{q_{\mathrm{c}}^{-2}} \right], \tag{6.85}$$

$$\mathrm{Im}Q(x,\tau) = \frac{\mathrm{sgn}(\tau)}{2\mathrm{i}} \left\{ -\ln\left[\frac{x + \mathrm{i}\tilde{v}_{\mathrm{F}}|\tau| + \mathrm{i}q_{\mathrm{c}}^{-1}}{x - \mathrm{i}\tilde{v}_{\mathrm{F}}|\tau| - \mathrm{i}q_{\mathrm{c}}^{-1}}\right] + \ln\left[\frac{x + \mathrm{i}v_{\mathrm{F}}|\tau| + \mathrm{i}q_{\mathrm{c}}^{-1}}{x - \mathrm{i}v_{\mathrm{F}}|\tau| - \mathrm{i}q_{\mathrm{c}}^{-1}}\right]\right\} , \tag{6.86}$$

where $\tilde{v}_{\mathrm{F}} = \sqrt{1+F_0}v_{\mathrm{F}}$ is the renormalized Fermi velocity. Combining the terms differently, the total Debye-Waller factor can also be written as

$$Q(x,\tau) = \frac{\gamma(F_0)}{2} \ln\left[\frac{q_{\mathrm{c}}^{-2}}{x^2 + (\tilde{v}_{\mathrm{F}}|\tau| + q_{\mathrm{c}}^{-1})^2}\right] + \ln\left[\frac{x + \mathrm{i}v_{\mathrm{F}}\tau + \mathrm{i}\,\mathrm{sgn}(\tau)q_{\mathrm{c}}^{-1}}{x + \mathrm{i}\tilde{v}_{\mathrm{F}}\tau + \mathrm{i}\,\mathrm{sgn}(\tau)q_{\mathrm{c}}^{-1}}\right] , \tag{6.87}$$

where the so-called *anomalous dimension* is given by

$$\gamma(F_0) \equiv \frac{1+\frac{F_0}{2}}{\sqrt{1+F_0}} - 1 = \frac{[\sqrt{1+F_0}-1]^2}{2\sqrt{1+F_0}} = \frac{F_0^2}{2\sqrt{1+F_0}\left[\sqrt{1+F_0}+1\right]^2} . \tag{6.88}$$

At $\tau = 0$ we obtain from Eq.(6.87)

$$Q(x,0) \sim -\gamma(F_0)\ln(q_{\mathrm{c}}|x|) \quad , \quad q_{\mathrm{c}}|x| \to \infty . \tag{6.89}$$

Exponentiating Eq.(6.87) and using the expression for the non-interacting real space Green's function given in Eq.(6.25), we finally obtain from Eq.(6.24) for the interacting Green's function (recall that $x = \alpha r_x$, with $\alpha = \pm$)

$$G^{\alpha}(r_x,\tau) = \left(\frac{-\mathrm{i}}{2\pi}\right)\frac{\mathrm{e}^{Q(\alpha r_x,\tau)}}{\alpha r_x + \mathrm{i}v_{\mathrm{F}}\tau} = \left(\frac{-\mathrm{i}}{2\pi}\right)\frac{1}{\alpha r_x + \mathrm{i}v_{\mathrm{F}}\tau} \times \left[\frac{\alpha r_x + \mathrm{i}v_{\mathrm{F}}\tau + \mathrm{i}\,\mathrm{sgn}(\tau)q_{\mathrm{c}}^{-1}}{\alpha r_x + \mathrm{i}\tilde{v}_{\mathrm{F}}\tau + \mathrm{i}\,\mathrm{sgn}(\tau)q_{\mathrm{c}}^{-1}}\right]\left[\frac{q_{\mathrm{c}}^{-2}}{r_x^2 + (\tilde{v}_{\mathrm{F}}|\tau| + q_{\mathrm{c}}^{-1})^2}\right]^{\gamma/2} . \tag{6.90}$$

We now observe that for times $|\tau| \gg (\tilde{v}_{\mathrm{F}}q_{\mathrm{c}})^{-1}$ or length scales $|r_x| \gg q_{\mathrm{c}}^{-1}$ we may neglect the cutoff q_{c}^{-1} when it appears in combination with $\tilde{v}_{\mathrm{F}}\tau$ or x. *In this regime* Eq.(6.90) reduces to

$$G^{\alpha}(r_x,\tau) = \left(\frac{-\mathrm{i}}{2\pi}\right)\frac{1}{\alpha r_x + \mathrm{i}\tilde{v}_{\mathrm{F}}\tau}\left[\frac{q_{\mathrm{c}}^{-2}}{r_x^2 + (\tilde{v}_{\mathrm{F}}\tau)^2}\right]^{\gamma/2} . \tag{6.91}$$

Note that this expression depends exclusively on the renormalized Fermi velocity $\tilde{v}_{\mathrm{F}}$. If we rescale both space and time by a factor of s^{-1}, then it is obvious that the interacting Green's function (6.91) satisfies

$$G^{\alpha}(r_x/s,\tau/s) = s^{1+\gamma}G^{\alpha}(r_x,\tau) . \tag{6.92}$$

Note that in d dimensions the non-interacting sector Green's function (5.48) satisfies

$$G_0^{\alpha}(r_x/s,\tau/s) = s^d G_0^{\alpha}(r_x,\tau) . \tag{6.93}$$

It is easy to see that in the asymptotic long-distance and large-time limit this scaling behavior is not changed by the interactions as long as the system is a Fermi liquid. In this case the scaling behavior of the Green's function can be determined by dimensional analysis. In the renormalization group literature [1.11], the exponent d in Eq.(6.93) is called the ***scaling dimension*** of the Green's function. Because the real space Green's function has units of inverse volume, the scaling dimension agrees with the dimensionality d of the system. The reason why the exponent γ in Eq.(6.92) is called anomalous dimension is now clear: In $d = 1$ the effect of the interactions is so drastic that the scaling behavior of the Green's function cannot be completely determined by dimensional analysis. There exists an ***anomalous*** contribution to the scaling dimension, which depends on the strength of the interaction in a non-trivial way, as given in Eq.(6.88).

6.4 Summary and outlook

In this chapter we have studied in some detail singular density-density interactions in d dimensions that diverge in the infrared limit as $|\boldsymbol{q}|^{-\eta}$. These are perhaps the simplest model systems for non-Fermi liquid behavior in higher dimensions. We have confirmed the result of Bares and Wen [6.1] that the Fermi liquid state is only stable for $\eta < 2(d-1)$. We have also identified various other special values for η, which are summarized in Fig. 6.1. Unfortunately, non-Fermi liquid behavior in $d > 1$ can only be obtained in the regime $\eta \geq 2(d-1)$, which corresponds to unphysical super-long range interactions in real space. For simplicity, we have restricted ourselves to the analysis of the static Debye-Waller factor $Q(x, 0)$, and have worked with linearized energy dispersion. As discussed at the end of Sect. 6.1.3, we expect that for the density-density interactions considered here the quadratic term in the energy dispersion will not qualitatively modify the long-distance behavior of $Q(x, 0)$. In Chap. 10 we shall show that this is *not* the case for the effective current-current interaction mediated by transverse gauge-fields.

The asymptotic long-distance behavior of the static Debye-Waller factor determines the momentum distribution $n_{\boldsymbol{k}}$ for wave-vectors close to the Fermi surface. In the regime $\eta > 2(d-1)$ we have found that $n_{\boldsymbol{k}}$ is analytic, so that the interactions completely wash out the sharpness of the Fermi surface. Our result disagrees with the works by Khodel and collaborators [6.9], who treated long-range interactions within a Hartree–Fock approach and found that the momentum distribution is completely flat in a certain finite shell around the non-interacting Fermi surface. This result has been criticized by Nozières [6.10], who argued that it is most likely an unphysical artefact of the Hartree–Fock approximation. Our non-perturbative calculation supports his arguments.

An interesting unsolved problem is the explicit evaluation of our non-perturbative result for the full **momentum- and frequency-dependence**

of the Green's function $G(\boldsymbol{k}, \omega)$ in the non-Fermi liquid regime $\eta > 2(d-1)$. As discussed by Volovik [6.11], the Green's function of non-Fermi liquids might exhibit some interesting topological structure in Fourier space, which can be used for a rather general topological definition of the Fermi surface of an interacting Fermi system. Recall that in Chap. 2.2.2 we have defined the Fermi surface via the singularity in the momentum distribution. According to this definition fermions with singular density-density interactions of the type (6.1) with $\eta > 2(d-1)$ do not have a Fermi surface. See [6.11] for an alternative topological definition of the Fermi surface, which seems to be general enough to associate a mathematically well-defined Fermi surface with a system that has an analytic momentum distribution[7]. This definition requires the knowledge of the $\boldsymbol{k}$- and ω-dependence of the Green's function $G(\boldsymbol{k}, \omega)$, and not only of the momentum distribution $n_{\boldsymbol{k}}$. As discussed in Sect. 6.1.3, for the calculation of $G(\boldsymbol{k}, \omega)$ via higher-dimensional bosonization it is most likely necessary to retain the quadratic term in the energy dispersion. On the other hand, the momentum distribution $n_{\boldsymbol{k}}$ is determined by the *static* Debye-Waller factor[8], the leading long-distance behavior of which can be calculated correctly with linearized energy dispersion

---

[7] I would like to thank G. E. Volovik for pointing this out to me, and for sending me copies of the relevant references.

[8] This is also the reason why we expect that the special values of the exponent η shown in Fig. 6.1 will not be modified by the non-linear terms in the energy dispersion.

7. Quasi-one-dimensional metals

Here comes the first experimentally relevant application of our method: The calculation of the single-particle Green's function for highly anisotropic chain-like metals. Most of the results presented in this chapter have been obtained in collaboration with V. Meden and K. Schönhammer [7.1, 7.2].

One of the main motivations for the development of the higher-dimensional bosonization approach is the fact that non-Fermi liquid behavior has been observed in the laboratory, and is therefore an experimental reality that requires theoretical explanation. The most prominent example are perhaps the normal-state properties of the high-temperature superconductors [1.24], but also experiments on quasi-one-dimensional conductors [7.3–7.6] suggest non-Fermi liquid behavior. Note that in these highly anisotropic systems the electrons interact with Coulomb forces, which for isotropic systems in $d > 1$ certainly do not destabilize the Fermi liquid state. This indicates that the experimentally seen non-Fermi liquid behavior could be due to the spatial anisotropy of these systems.

In this chapter we shall study a simple model for a quasi-one-dimensional metal, which consists of electrons moving in a periodic array of weakly coupled metallic chains embedded in three-dimensional space. The electrons are assumed to interact with realistic three-dimensional Coulomb forces, so that, even in the absence of interchain hopping, this is not a purely one-dimensional problem. Note that in $d = 1$ the logarithmic one-dimensional Fourier transform of the Coulomb potential (see Eq.(A.49)) gives rise to singularities that are stronger than in a conventional Luttinger liquid, so that the anomalous dimension diverges [7.7]. However, as will be shown in Sect. 7.1, the Coulomb forces between the chains remove this divergence, so that the long-range part of the three-dimensional Coulomb interaction in an array of chains without interchain hopping indeed leads to a Luttinger liquid. It should be kept in mind, however, that in this work we shall retain only processes with small momentum transfers, so that possible instabilities due to back-scattering or Umklapp-scattering are ignored. We are implicitly assuming that there exists a parameter regime where these processes are irrelevant.

The problem of coupled chains without interchain hopping can also be solved by means of the usual one-dimensional bosonization techniques

[7.7, 7.8]. The true power of the higher-dimensional bosonization approach becomes apparent if we switch on a finite interchain hopping $t_\perp$. In this case conventional one-dimensional bosonization cannot be used, but within our higher-dimensional bosonization approach this problem can be handled quite easily. The problem of weak interchain hopping has been discussed by Gorkov and Dzyaloshinskii [7.9] more than 20 years ago. More recently, many other authors have used various more or less systematic methods to shed more light onto this rather difficult problem [7.10–7.17].

In Sects. 7.1 and 7.2 we shall evaluate our bosonization results (6.14), (6.16) and (6.17) for the Debye-Waller factor with linearized energy dispersion in the case of an infinite array of metallic chains. However, even for finite interchain hopping $t_\perp$ the Debye-Waller factor exhibits an unphysical logarithmic nesting singularity, which is due to the fact that for linearized energy dispersion the Fermi surface is replaced by a finite number of completely flat patches. To remove this singularity, the artificial nesting symmetry of the Fermi surface has to be broken. The simplest way to do this is to work with non-linear energy dispersion. In Sect. 7.2.2 we shall use our general results (6.4)–(6.6) for the Debye-Waller factor with quadratic energy dispersion to show that the logarithmic nesting singularity is indeed removed by the curvature of the Fermi surface. Our main result is that an arbitrarily small $t_\perp$ destroys the Luttinger liquid state and leads to a finite quasi-particle residue Z^α. We explicitly calculate Z^α for small $t_\perp$ and show that there exists a large intermediate regime where the signature of characteristic Luttinger liquid properties is visible in physical observables, although the system is a Fermi liquid.

7.1 The Coulomb interaction in chains without interchain hopping

Before addressing the more interesting case of $t_\perp \neq 0$, it is useful to consider the three-dimensional Coulomb interaction in metallic chains in the absence of interchain hopping.

The Fermi surface of a periodic array of one-dimensional chains embedded in three-dimensional space without interchain hopping consists of two parallel completely flat planes, as shown in Fig. 7.1. Because the Fermi surface does not have any curvature, it will be sufficient to work with linearized energy dispersion. In this case all interaction effects are contained in the Debye-Waller factor $Q^\alpha(\boldsymbol{r}, \tau)$ given in Eqs.(6.14), (6.16) and (6.17). Note this these expressions have been derived for an arbitrary geometry of the Fermi surface, so that we simply have to substitute the parameters relevant for the case of interest here. Obviously the local Fermi velocities on the two sheets of the Fermi surface in Fig. 7.1 are exactly constant. Hence, the entire Fermi

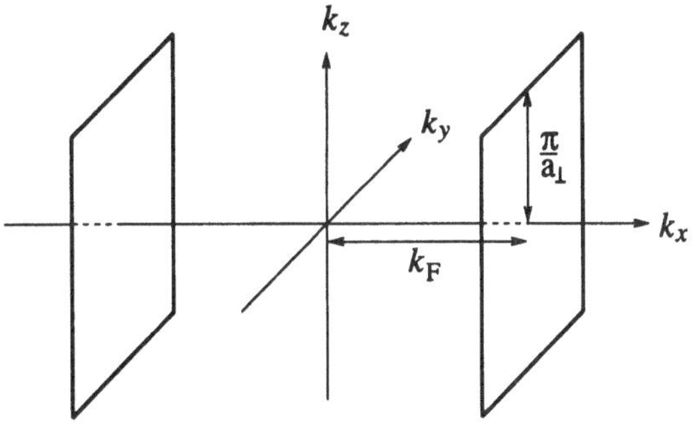

Fig. 7.1.
Fermi surface of a periodic array of one-dimensional chains embedded in three-dimensional space without interchain hopping.

surface can be covered with $M = 2$ patches, which can be identified with the two sheets of the Fermi surface. In this case the patch cutoff Λ is given by $2\pi/a_\perp$, where $a_\perp$ is the distance between the chains. Because the number of patches is finite, we know from the general analysis given in Chap. 6.1.3 that $L^\alpha_{\boldsymbol{q}}(\tau) = 0$ in Eqs.(6.16) and (6.17). Let us label the right patch in Fig. 7.1 by $\alpha = +$, and the left one by $\alpha = -$. The associated local Fermi velocities are $\boldsymbol{v}^\alpha = \alpha v_{\mathrm{F}} \boldsymbol{e}_x$. The non-interacting linearized energy dispersion close to the Fermi surface is then $\xi^\alpha_{\boldsymbol{q}} = \boldsymbol{v}^\alpha \cdot \boldsymbol{q} = \alpha v_{\mathrm{F}} q_x$, and the local density of states is

$$\nu^\alpha = \int_{-\lambda}^{\lambda} \frac{dq_x}{2\pi} \int_{-\frac{\pi}{a_\perp}}^{\frac{\pi}{a_\perp}} \frac{dq_y}{2\pi} \int_{-\frac{\pi}{a_\perp}}^{\frac{\pi}{a_\perp}} \frac{dq_z}{2\pi} \delta(\alpha v_{\mathrm{F}} q_x) = \frac{1}{2\pi v_{\mathrm{F}} a_\perp^2} \quad , \tag{7.1}$$

where λ is some radial cutoff that should be chosen large compared with the Thomas-Fermi wave-vector κ given in Eq.(7.6) below. Hence, the non-interacting polarization at long wave-lengths is

$$\Pi_0(q) = \sum_{\alpha = \pm} \nu^\alpha \frac{\boldsymbol{v}^\alpha \cdot \boldsymbol{q}}{\boldsymbol{v}^\alpha \cdot \boldsymbol{q} - \mathrm{i}\omega_m} = \nu g_1 (\frac{\mathrm{i}\omega_m}{v_{\mathrm{F}} |q_x|}) \quad , \tag{7.2}$$

where the total density of states is given by

$$\nu = \sum_{\alpha = \pm} \nu^\alpha = \frac{1}{\pi v_{\mathrm{F}} a_\perp^2} \quad , \tag{7.3}$$

and the function $g_1(\mathrm{i}y)$ is defined in Eq.(A.12). The one-dimensional form of the polarization implies that the RPA dynamic structure factor is formally identical with the one-dimensional expression (6.77), with the collective mode and the residue given by

$$\omega_{\boldsymbol{q}} = \sqrt{1 + F_{\boldsymbol{q}}} v_{\mathrm{F}} |q_x| \quad , \tag{7.4}$$

$$Z_{\boldsymbol{q}} = = \frac{\nu v_{\mathrm{F}} |q_x|}{2\sqrt{1 + F_{\boldsymbol{q}}}} = \frac{|q_x|}{2\pi a_\perp^2 \sqrt{1 + F_{\boldsymbol{q}}}} \quad . \tag{7.5}$$

Here $F_{\boldsymbol{q}} = \nu f_{\boldsymbol{q}}$ is the usual dimensionless interaction. Compared with the one-dimensional result given in Eq.(6.78), the residue (7.5) has an extra factor of $a_\perp^2$ in the denominator, because we are now dealing with a three-dimensional system. At length scales large compared with the lattice spacing $a_\perp$, we may replace the Fourier transform of the Coulomb potential by its continuum approximation, so that in this case

$$F_q = \frac{\kappa^2}{q^2} \quad , \quad \kappa^2 = \frac{4e^2}{v_F a_\perp^2} \quad , \quad \text{for } |q| a_\perp \ll 1 \quad . \tag{7.6}$$

Given the fact that q_y and q_z appear in the Debye-Waller factor only via F_q, and that the dynamic structure factor has the one-dimensional form, it is now easy to see that the frequency integration in Eqs.(6.14), (6.16) and (6.17) is exactly the same as in the one-dimensional case, so that we can simply copy the results of Chap. 6.3. From Eqs.(6.81)–(6.83) we obtain

$$R = -\int_0^\infty \frac{\mathrm{d}q_x}{q_x} \left\langle \frac{F_q^2}{2\sqrt{1+F_q}[\sqrt{1+F_q}+1]^2} \right\rangle_{\mathrm{BZ}} \quad , \tag{7.7}$$

$$\mathrm{Re}S(x,\tau) = -\int_0^\infty \frac{\mathrm{d}q_x}{q_x} \cos(q_x x) \times \left[\left\langle \frac{1+\frac{F_q}{2}}{\sqrt{1+F_q}} \mathrm{e}^{-\sqrt{1+F_q} v_F q_x |\tau|} \right\rangle_{\mathrm{BZ}} - \mathrm{e}^{-v_F q_x |\tau|} \right] \quad , \tag{7.8}$$

$$\mathrm{Im}S(x,\tau) = -\mathrm{sgn}(\tau) \int_0^\infty \frac{\mathrm{d}q_x}{q_x} \sin(q_x x) \times \left[\left\langle \mathrm{e}^{-\sqrt{1+F_q} v_F q_x |\tau|} \right\rangle_{\mathrm{BZ}} - \mathrm{e}^{-v_F q_x |\tau|} \right] \quad , \tag{7.9}$$

where for any function $f(q)$ the symbol $< f(q) >_{\mathrm{BZ}}$ denotes averaging over the transverse Brillouin zone,

$$\langle f(q) \rangle_{\mathrm{BZ}} = \frac{a_\perp^2}{(2\pi)^2} \int_{-\frac{\pi}{a_\perp}}^{\frac{\pi}{a_\perp}} \mathrm{d}q_y \int_{-\frac{\pi}{a_\perp}}^{\frac{\pi}{a_\perp}} \mathrm{d}q_z f(q) \quad . \tag{7.10}$$

The above expression for the Green's function can also be derived by means of standard one-dimensional bosonization techniques [7.7, 7.8]. However, as will be shown in Sect. 7.2, our derivation via higher-dimensional bosonization has the advantage that it can be generalized to the case of finite interchain hopping.

For $\tau = 0$ we can make progress analytically in the regime where the Thomas-Fermi screening length κ^{-1} is large compared with the transverse lattice spacing $a_\perp$, i.e. for $\kappa a_\perp \ll 1$. Because in this case all wave-vector integrals are dominated by the regime $|q| \lesssim \kappa$, it is allowed to use in Eq.(7.6) the continuum approximation for the Fourier transform of the Coulomb potential. Note that

$$\kappa a_\perp = \sqrt{\frac{4e^2}{v_F}} \quad , \tag{7.11}$$

so that the condition $\kappa a_\perp \ll 1$ means that the dimensionless coupling constant e^2/v_F should be small compared with unity. Unfortunately at experimentally relevant densities this parameter is of the order of unity, so that in this case the continuum approximation for the Fourier transform of the Coulomb potential cannot be used. To reach the experimentally relevant parameter regime, one should therefore take in F_q the discreteness of the lattice in the transverse direction into account. In [7.1] this was done by means of an Ewald summation technique [7.20]. Here we would like to restrict ourselves to the regime $\kappa a_\perp \ll 1$.

For $\tau = 0$ we need to calculate the following Brillouin zone average

$$\gamma_{\mathrm{cb}}(q_x) = \left\langle \frac{F_q^2}{2\sqrt{1+F_q}[\sqrt{1+F_q}+1]^2} \right\rangle_{\mathrm{BZ}} = \left\langle \frac{1+\frac{F_q}{2}}{\sqrt{1+F_q}} \right\rangle_{\mathrm{BZ}} - 1 \, . \tag{7.12}$$

Substituting Eq.(7.6) into Eq.(7.12) and using Eq.(7.11), the integration is elementary, and we obtain for $\kappa a_\perp \ll 1$

$$\gamma_{\mathrm{cb}}(q_x) = \frac{e^2}{2\pi v_{\mathrm{F}}} \frac{1}{\left[\frac{|q_x|}{\kappa} + \sqrt{1+(\frac{q_x}{\kappa})^2}\right]^2} \quad . \tag{7.13}$$

The asymptotic behavior for large and small $|q_x|$ is

$$\gamma_{\mathrm{cb}}(q_x) \sim \frac{e^2}{2\pi v_{\mathrm{F}}} \times \begin{cases} 1 & \text{for } |q_x| \ll \kappa \\ (\frac{\kappa}{2q_x})^2 & \text{for } |q_x| \gg \kappa \end{cases} \quad . \tag{7.14}$$

Because $\gamma_{\mathrm{cb}}(q_x)$ has a finite limit for $q_x \to 0$, the integral (7.7) defining R is logarithmically divergent, so that the system is a Luttinger liquid. Moreover, for $q_x \gg \kappa$ the function $\gamma_{\mathrm{cb}}(q_x)$ vanishes sufficiently fast to insure ultraviolet convergence of the integral defining $Q(x, 0)$. Recall that in the one-dimensional Tomonaga-Luttinger model (see Chap. 6.3) it was necessary to introduce an ultraviolet cutoff q_{c} to make the integrals convergent. The precise physical origin of this cutoff has remained somewhat obscure. In the present problem, however, the effective ultraviolet cutoff can be identified with the Thomas-Fermi screening wave-vector. To calculate the anomalous dimension, we consider the long-distance behavior of the static Debye-Waller factor. Using Eqs.(7.7),(7.8) and (7.13), and introducing the dimensionless integration variable $p = q_x/\kappa$, we obtain

$$Q(x, 0) = -\frac{e^2}{2\pi v_{\mathrm{F}}} \int_0^\infty \frac{\mathrm{d}p}{p} \frac{1-\cos(p\kappa x)}{\left[p+\sqrt{1+p^2}\right]^2} \quad . \tag{7.15}$$

To calculate the asymptotic behavior of the integral for large κx, we write

$$\frac{1-\cos(p\kappa x)}{p} = \frac{\mathrm{d}}{\mathrm{d}p}[\ln p - \mathrm{Ci}(p\kappa x)] \quad , \tag{7.16}$$

where [7.18]

$$\mathrm{Ci}(x) = -\int_x^\infty \mathrm{d}t \frac{\cos t}{t} \quad . \tag{7.17}$$

An integration by parts yields

$$Q(x,0) = -\frac{e^2}{2\pi v_\mathrm{F}} \left[\lim_{p \to 0} [\mathrm{Ci}(p\kappa|x|) - \ln p] + 2\int_0^\infty \mathrm{d}p \frac{\ln p}{\sqrt{1+p^2}[p+\sqrt{1+p^2}]^2} \right]$$
$$+ \frac{e^2}{\pi v_\mathrm{F}} \int_0^\infty \mathrm{d}p \frac{\mathrm{Ci}(p\kappa|x|)}{\sqrt{1+p^2}[p+\sqrt{1+p^2}]^2} \quad . \tag{7.18}$$

Using the fact that [7.18]

$$\lim_{p \to 0} [\mathrm{Ci}(p\kappa|x|) - \ln p] = \ln(\kappa|x|) + \gamma_\mathrm{E} \quad , \tag{7.19}$$

where $\gamma_\mathrm{E} \approx 0.577$ is the Euler constant, and noting that the last term in Eq.(7.18) vanishes as $\frac{\ln(\kappa|x|)}{\kappa|x|}$ as $x \to \infty$, we finally obtain

$$Q(x,0) \sim -\frac{e^2}{2\pi v_\mathrm{F}} \left[\ln(\kappa|x|) + b + O\left(\frac{\ln(\kappa|x|)}{\kappa|x|} \right) \right] \quad , \tag{7.20}$$

where the numerical constant b is given by

$$b = \gamma_\mathrm{E} + 2\int_0^\infty \mathrm{d}p \frac{\ln p}{\sqrt{1+p^2}[p+\sqrt{1+p^2}]^2} \quad . \tag{7.21}$$

We conclude that the interacting equal-time Green's function vanishes at large distances as

$$G^\alpha(\boldsymbol{r},0) = G_0^\alpha(\boldsymbol{r},0) \left(\frac{\mathrm{e}^{-b}}{\kappa|r_x|} \right)^{\gamma_\mathrm{cb}}$$
$$= \delta(r_y)\delta(r_z) \left(\frac{-\mathrm{i}}{2\pi} \right) \left(\frac{\mathrm{e}^{-b}}{\kappa} \right)^{\gamma_\mathrm{cb}} \frac{1}{|r_x|^{1+\gamma_\mathrm{cb}}} \quad , \tag{7.22}$$

where the anomalous dimension γ_cb is given by

$$\gamma_\mathrm{cb} \equiv \lim_{q_x \to 0} \gamma_\mathrm{cb}(q_x) = \lim_{q_x \to 0} \left\langle \frac{F_q^2}{2\sqrt{1+F_q}[\sqrt{1+F_q}+1]^2} \right\rangle_\mathrm{BZ} = \frac{e^2}{2\pi v_\mathrm{F}} \, . \tag{7.23}$$

We would like to emphasize again that this expression is only valid for $e^2/v_\mathrm{F} \ll 1$, so that it would be incorrect to extrapolate Eq.(7.23) to the experimentally relevant regime $e^2/v_\mathrm{F} = O(1)$. In this regime the simple continuum approximation for the Fourier transform of the Coulomb potential is not sufficient, and one has to use numerical methods to calculate the anomalous dimension. This numerical calculation has been performed in [7.1], with the result that in the experimentally relevant regime the anomalous dimension is indeed of the order of unity. Recent photoemission studies [7.3, 7.4, 7.6] of quasi-one-dimensional conductors suggest values for the anomalous dimension in the range 1.0 ± 0.2, which is in agreement with our result. However, the comparison of the experimental result with Eq.(7.23) is at least problematic, because our calculation was based on several idealizations which are perhaps not satisfied in the realistic experimental system. First of all, the experiments are certainly not performed on perfectly clean systems. Because any finite disorder changes the algebraic decay in Eq.(7.22) into an exponential one (see Chap. 9), the Luttinger liquid behavior is completely destroyed by impurities. Therefore one cannot exclude the possibility that the experiments do not measure the Luttinger liquid nature of the system, but are essentially determined by impurities. Another possibly unjustified idealization in our calculation is the neglect of processes with large momentum transfers, which might favour charge-density wave instabilities or other broken symmetries. The associated pseudo-gaps in the excitation spectrum will certainly lead to a further suppression of the momentum integrated spectral function in the vicinity of the Fermi energy, which competes with the suppression inherent in the Luttinger liquid state. Nevertheless, in spite of all these caveats, we believe that the large value of γ_{cb} due to long-range Coulomb forces can give rise to an important contribution to the suppression of the spectral weight seen in the experiments.

At finite τ we have not been able to calculate the integral defining $Q(x,\tau)$ analytically. In [7.1] the numerical method developed by Meden and Schönhammer [1.20] was used to calculate the full momentum- and frequency-dependent spectral function. More detailed numerical calculations can be found in the thesis by Meden [7.19]. In contrast to our present discussion, in the works [7.1,7.2,7.19] the spin degree of freedom was also taken into account, and the phenomenon of spin-charge separation was studied. The fact that the spin and the charge excitations manifest themselves with different velocities in the single-particle Green's function is one of the fundamental characteristics of a Luttinger liquid [1.15].

7.2 Finite interchain hopping

Experimentally the interchain hopping $t_{\perp}$ can never be completely turned off. Realistic Fermi surfaces of quasi-one-dimensional conductors have therefore the form shown in Fig. 7.2. The amplitude of the modulation of the Fermi surface sheets is proportional to the interchain hopping $t_{\perp}$. Because the intrachain hopping $t_{\parallel}$ is of the order of E_{F}, the relevant small dimensionless parameter which measures the quasi-one-dimensionality of the system is

$$\theta = \frac{|t_{\perp}|}{E_{\mathrm{F}}} \quad . \tag{7.24}$$

From the previous section we know that for $\theta = 0$ the system is a Luttinger liquid. We now calculate the Green's function of the system for small but finite θ, assuming transverse hopping only in the y-direction. This approximation is justified for materials where the interchain hopping $t_{\perp} = t_y$ in the y-direction is large compared with the interchain hopping in the z-direction. As discussed in [7.5], this condition is satisfied for some experimentally studied materials.

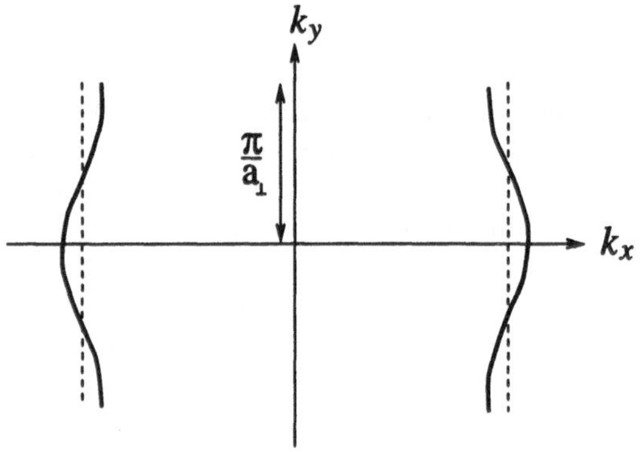

Fig. 7.2. Fermi surface of a periodic array of chains with interchain hopping. Only the intersection with the plane $k_z = 0$ is shown.

7.2.1 The 4-patch model

We now break the symmetry of the Fermi surface by deforming the flat sheets into wedges, so that we obtain a model with four patches.

For simplicity let us first assume that the Fermi surface has the shape shown in Fig. 7.3: it consists of four perfectly flat patches, which are obtained by

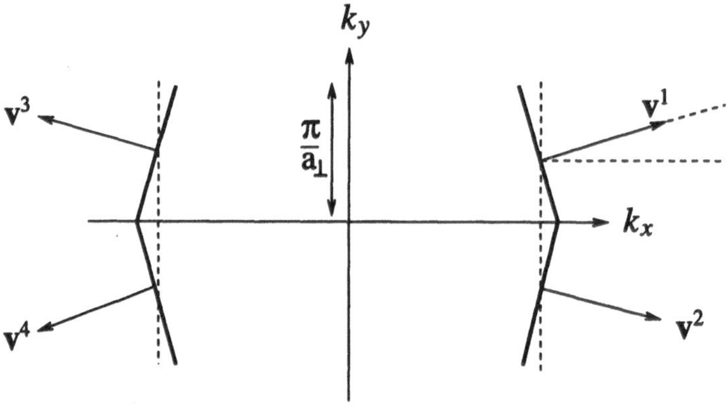

Fig. 7.3. Fermi surface of the 4-patch model. Only the intersection with a plane of constant k_z is shown.

replacing the sine modulation in Fig. 7.2 by a triangle. Because the coefficients in the Fourier decomposition of a triangular wave vanish rather slowly, the microscopic origin for such a Fermi surface is a particular superposition of long-range hoppings. We shall refer to our hopping model as the 4-*patch model.* In a sense, this is the simplest example for a non-trivial model in $d > 1$ which can be discussed within the framework of higher-dimensional bosonization. The patches are now labelled by $\alpha = 1, 2, 3, 4$. Because the curvature of the patches vanishes by construction, the local Fermi velocities are constant on a given patch. From Fig. 7.3 we see that

$$\begin{aligned} \boldsymbol{v}^1 &= (\boldsymbol{e}_x \cos\theta + \boldsymbol{e}_y \sin\theta) v_{\mathrm{F}} \\ \boldsymbol{v}^2 &= (\boldsymbol{e}_x \cos\theta - \boldsymbol{e}_y \sin\theta) v_{\mathrm{F}} \\ \boldsymbol{v}^3 &= (-\boldsymbol{e}_x \cos\theta + \boldsymbol{e}_y \sin\theta) v_{\mathrm{F}} \\ \boldsymbol{v}^4 &= (-\boldsymbol{e}_x \cos\theta - \boldsymbol{e}_y \sin\theta) v_{\mathrm{F}} \end{aligned} \quad . \tag{7.25}$$

To calculate the Green's function, we simply repeat the steps of the previous section. The non-interacting polarization is now

$$\Pi_0(q) = \frac{\nu}{4} \sum_{\alpha=1}^{4} \frac{\boldsymbol{v}^\alpha \cdot \boldsymbol{q}}{\boldsymbol{v}^\alpha \cdot \boldsymbol{q} - \mathrm{i}\omega_m} = \frac{\nu}{2} \sum_{\alpha=1}^{2} \frac{(\boldsymbol{v}^\alpha \cdot \boldsymbol{q})^2}{(\boldsymbol{v}^\alpha \cdot \boldsymbol{q})^2 + \omega_m^2} \quad , \tag{7.26}$$

where the global density of states $\nu = \sum_{\alpha=1}^{4} \nu^\alpha$ is for small θ identical with Eq.(7.3). As usual, the collective modes are obtained by solving Eq.(A.70), which for $M = 4$ patches leads to the bi-quadratic equation

$$z^4 - \left(1 + \frac{F_{\boldsymbol{q}}}{2}\right)\left(\xi_{\boldsymbol{q}}^2 + \tilde{\xi}_{\boldsymbol{q}}^2\right) z^2 + (1 + F_{\boldsymbol{q}}) \xi_{\boldsymbol{q}}^2 \tilde{\xi}_{\boldsymbol{q}}^2 = 0 \quad , \tag{7.27}$$

where we have introduced the notation

$$\xi_{\boldsymbol{q}} = \boldsymbol{v}^1 \cdot \boldsymbol{q} = v_{\mathrm{F}}(q_x \cos\theta + q_y \sin\theta) \quad , \tag{7.28}$$

$$\tilde{\xi}_{\boldsymbol{q}} = \boldsymbol{v}^2 \cdot \boldsymbol{q} = v_{\mathrm{F}}(q_x \cos\theta - q_y \sin\theta) \quad . \tag{7.29}$$

The bi-quadratic equation (7.27) is easily solved. The two solutions are

$$\begin{aligned} \omega_{\boldsymbol{q}}^2 = & \left(1 + \frac{F_{\boldsymbol{q}}}{2}\right) \frac{\xi_{\boldsymbol{q}}^2 + \tilde{\xi}_{\boldsymbol{q}}^2}{2} \\ & + \frac{1}{2}\left[F_{\boldsymbol{q}}^2 \left(\frac{\xi_{\boldsymbol{q}}^2 + \tilde{\xi}_{\boldsymbol{q}}^2}{2}\right)^2 + (1 + F_{\boldsymbol{q}})(\xi_{\boldsymbol{q}}^2 - \tilde{\xi}_{\boldsymbol{q}}^2)^2\right]^{1/2} \quad , \end{aligned} \tag{7.30}$$

$$\begin{aligned} \tilde{\omega}_{\boldsymbol{q}}^2 = & \left(1 + \frac{F_{\boldsymbol{q}}}{2}\right) \frac{\xi_{\boldsymbol{q}}^2 + \tilde{\xi}_{\boldsymbol{q}}^2}{2} \\ & - \frac{1}{2}\left[F_{\boldsymbol{q}}^2 \left(\frac{\xi_{\boldsymbol{q}}^2 + \tilde{\xi}_{\boldsymbol{q}}^2}{2}\right)^2 + (1 + F_{\boldsymbol{q}})(\xi_{\boldsymbol{q}}^2 - \tilde{\xi}_{\boldsymbol{q}}^2)^2\right]^{1/2} \quad . \end{aligned} \tag{7.31}$$

Note that for small θ

$$\frac{\xi_{\boldsymbol{q}}^2 + \tilde{\xi}_{\boldsymbol{q}}^2}{2} \approx v_{\mathrm{F}}^2 (q_x^2 + \theta^2 q_y^2) \quad , \tag{7.32}$$

$$\xi_{\boldsymbol{q}}^2 - \tilde{\xi}_{\boldsymbol{q}}^2 \approx 4\theta v_{\mathrm{F}}^2 q_x q_y \quad , \tag{7.33}$$

$$\xi_{\boldsymbol{q}} \tilde{\xi}_{\boldsymbol{q}} \approx v_{\mathrm{F}}^2 (q_x^2 - \theta^2 q_y^2) \quad , \tag{7.34}$$

and that

$$\omega_q^2 - \tilde{\omega}_q^2 = \left[F_q^2 \left(\frac{\xi_q^2 + \tilde{\xi}_q^2}{2} \right)^2 + (1 + F_q)(\xi_q^2 - \tilde{\xi}_q^2)^2 \right]^{1/2} . \tag{7.35}$$

The right-hand side of Eq.(7.31) is non-negative because

$$\left[\left(1 + \frac{F_q}{2} \right) \frac{\xi_q^2 + \tilde{\xi}_q^2}{2} \right]^2 - \frac{1}{4} \left[F_q^2 \left(\frac{\xi_q^2 + \tilde{\xi}_q^2}{2} \right)^2 + (1 + F_q)(\xi_q^2 - \tilde{\xi}_q^2)^2 \right]$$
$$= (1 + F_q)\xi_q^2 \tilde{\xi}_q^2 \geq 0 . \tag{7.36}$$

Therefore both modes ω_q and $\tilde{\omega}_q$ are not damped and give rise to δ-function peaks in the dynamic structure factor. The dielectric function is then given by

$$\epsilon_{\mathrm{RPA}}(\boldsymbol{q}, \omega) \equiv 1 + F_q \Pi_0(\boldsymbol{q}, \omega) = \frac{(\omega^2 - \omega_q^2)(\omega^2 - \tilde{\omega}_q^2)}{(\omega^2 - \xi_q^2)(\omega^2 - \tilde{\xi}_q^2)} , \tag{7.37}$$

so that the RPA polarization is simply

$$\Pi_{\mathrm{RPA}}(\boldsymbol{q}, \omega) = \nu \frac{\xi_q^2 \tilde{\xi}_q^2 - \omega^2 \frac{\xi_q^2 + \tilde{\xi}_q^2}{2}}{(\omega^2 - \omega_q^2)(\omega^2 - \tilde{\omega}_q^2)} . \tag{7.38}$$

Note that $\epsilon_{\mathrm{RPA}}(\boldsymbol{q}, \xi_q) = \epsilon_{\mathrm{RPA}}(\boldsymbol{q}, \tilde{\xi}_q) = \infty$, in agreement with Eq.(6.23). The RPA dynamic structure factor is then easily calculated from Eq.(2.45). For $\omega > 0$ we obtain

$$S_{\mathrm{RPA}}(\boldsymbol{q}, \omega) = Z_q \delta(\omega - \omega_q) + \tilde{Z}_q \delta(\omega - \tilde{\omega}_q) , \tag{7.39}$$

with the residues given by

$$Z_q = \frac{\nu}{2\omega_q} \frac{\omega_q^2 \frac{\xi_q^2 + \tilde{\xi}_q^2}{2} - \xi_q^2 \tilde{\xi}_q^2}{\omega_q^2 - \tilde{\omega}_q^2} , \tag{7.40}$$

$$\tilde{Z}_q = \frac{\nu}{2\tilde{\omega}_q} \frac{\xi_q^2 \tilde{\xi}_q^2 - \tilde{\omega}_q^2 \frac{\xi_q^2 + \tilde{\xi}_q^2}{2}}{\omega_q^2 - \tilde{\omega}_q^2} . \tag{7.41}$$

In the limit $\theta \to 0$ we have $\tilde{\xi}_q \to \xi_q \to v_{\mathrm{F}} q_x$, so that $\omega_q \to \sqrt{1 + F_q} v_{\mathrm{F}} |q_x|$ and $\tilde{\omega}_q \to v_{\mathrm{F}} |q_x|$. It is also easy to see that the residue Z_q reduces in this limit to the result (7.5) without interchain hopping, while the residue $\tilde{Z}_q$ vanishes.

To calculate the Green's function, we substitute Eq.(7.39) into Eqs.(6.14), (6.16) and (6.17). Because the dynamic structure factor consists of a sum of two δ-functions, the frequency integration is trivial. As before $L_q^\alpha(\tau) = 0$, because we have covered the Fermi surface with a finite number of patches. To see whether the interchain hopping destroys the Luttinger liquid state, it is sufficient to calculate the static Debye-Waller factor. Substituting Eq.(7.39) into Eqs.(6.14) and (6.16), we obtain

$$Q^\alpha(r_\parallel^\alpha \hat{v}^\alpha, 0) = R^\alpha - S^\alpha(r_\parallel^\alpha \hat{v}^\alpha, 0)$$
$$= -\frac{1}{V}\sum_{q}\left[1-\cos(\hat{v}^\alpha \cdot q r_\parallel^\alpha)\right] f_q^2 \left[\frac{Z_q}{(\omega_q + |v^\alpha \cdot q|)^2} + \frac{\tilde{Z}_q}{(\tilde{\omega}_q + |v^\alpha \cdot q|)^2}\right] . \tag{7.42}$$

To evaluate Eq.(7.42), we need to simplify the above expressions for the collective modes and the associated residues. Depending on the relative order of magnitude of F_q and θ, three regimes have to be distinguished,

$$\begin{array}{l} (a) : \theta \ll 1 \ll F_q \\ (b) : \theta \ll F_q \ll 1 \\ (c) : F_q \ll \theta \ll 1 \end{array} . \tag{7.43}$$

Note that in the weak coupling regime (b) the energy scale set by the interaction is smaller than the intrachain hopping energy $t_\parallel$, but still large compared with the interchain hopping $t_\perp$. In the second weak coupling regime (c) the interaction is even smaller than the kinetic energy associated with transverse hopping. Because for $|q| \lesssim \kappa$ the dimensionless Coulomb interaction F_q is large compared with unity, in the present problem only the strong coupling regime (a) is of interest.

We begin with the evaluation of the first term in Eq.(7.42) involving the mode ω_q. We then show that the contribution of the second term, which involves the other mode $\tilde{\omega}_q$, grows actually logarithmically for $r_\parallel^\alpha \to \infty$, signaling Luttinger liquid behavior. However, this is due to an unphysical nesting symmetry inherent in our 4-patch model with flat patches; in Sect. 7.2.2 we shall slightly deform our patches so that they have a finite curvature, and show that in this case the contribution from the second mode remains bounded for all $r_\parallel^\alpha$ and is negligible compared with the contribution from the first mode.

The plasmon mode

From Eqs.(7.30) and (7.31) it is easy to see that, up to higher orders in θ/F_q, the collective density mode ω_q in the strong coupling regime can be approximated by

$$\omega_q \approx v_F \sqrt{1+F_q}\sqrt{q_x^2 + \theta^2 q_y^2} . \tag{7.44}$$

Note that for $\theta \to 0$ this mode reduces to the plasmon mode (7.4) in the absence of interchain hopping. Therefore we shall refer to the collective mode ω_q as the plasmon mode. Substituting Eq.(7.44) into Eq.(7.40), we obtain for the associated residue

$$Z_q \approx \frac{\nu v_F \sqrt{q_x^2 + \theta^2 q_y^2}}{2\sqrt{1+F_q}} , \tag{7.45}$$

which should be compared with Eq.(7.5). Note that the only effect of the interchain hopping is the replacement $|q_x| \to \sqrt{q_x^2 + \theta^2 q_y^2}$. The contribution

$R^{\alpha}_{\rm pl}$ of the plasmon mode to the constant part R^{α} of the Debye-Waller factor (7.42) is then for small θ given by

$$R^{\alpha}_{\rm pl} = - \int_0^{\infty} {\rm d}q_x \left\langle \frac{1}{\sqrt{q_x^2 + \theta^2 q_y^2}} \frac{F_{\boldsymbol{q}}^2}{2\sqrt{1+F_{\boldsymbol{q}}}\left[\sqrt{1+F_{\boldsymbol{q}}}+1\right]^2} \right\rangle_{\rm BZ} . \quad (7.46)$$

Although $R^{\alpha}_{\rm pl}$ is to this order in θ independent of α, we shall keep the patch index here. If we set $\theta = 0$ in this expression, we recover the previous result (7.7) in the absence of interchain hopping, which is logarithmically divergent. This divergence is due to the fact that for $\theta = 0$ the first factor in Eq.(7.46) can be pulled out of the averaging bracket. However, for any finite θ the q_x- and q_y-integrations are correlated, so that it is not possible to factorize the integrations. Hence, any non-zero value of θ couples the phase space of the $\boldsymbol{q}$-integration. Because for $\theta \to 0$ the integral in Eq.(7.46) is logarithmically divergent, the coefficient of the leading logarithmic term can be extracted by ignoring the q_x-dependence of the second factor in the averaging bracket. Then we obtain to leading logarithmic order

$$\begin{aligned} R^{\alpha}_{\rm pl} &\sim - \int_0^{\kappa} {\rm d}q_x \left\langle \frac{1}{\sqrt{q_x^2 + \theta^2 q_y^2}} \lim_{q_x \to 0} \left[\frac{F_{\boldsymbol{q}}^2}{2\sqrt{1+F_{\boldsymbol{q}}}\left[\sqrt{1+F_{\boldsymbol{q}}}+1\right]^2} \right] \right\rangle_{\rm BZ} \\ &= - \left\langle \ln\left(\frac{\kappa}{\theta |q_y|}\right) \lim_{q_x \to 0} \left[\frac{F_{\boldsymbol{q}}^2}{2\sqrt{1+F_{\boldsymbol{q}}}\left[\sqrt{1+F_{\boldsymbol{q}}}+1\right]^2} \right] \right\rangle_{\rm BZ} \\ &= -\gamma_{\rm cb} \left[\ln\left(\frac{1}{\theta}\right) + b_1 \right] , \end{aligned} \quad (7.47)$$

where $\gamma_{\rm cb}$ is given in Eq.(7.23), and b_1 is a numerical constant of the order of unity.

The contribution $S^{\alpha}_{\rm pl}(r^{\alpha}_{\parallel}\hat{\boldsymbol{v}}^{\alpha}, 0)$ of the plasmon mode to the spatially varying part of the Debye-Waller factor at equal times can be calculated analogously. Note that $r^{\alpha}_{\parallel} = \hat{\boldsymbol{v}}^{\alpha} \cdot \boldsymbol{r} = \pm r_x \cos\theta \pm r_y \sin\theta$. Repeating the steps leading to Eq.(7.46), we obtain

$$\begin{aligned} S^{\alpha}_{\rm pl}(r^{\alpha}_{\parallel}\hat{\boldsymbol{v}}^{\alpha}, 0) = & - \int_0^{\infty} {\rm d}q_x \cos(q_x r^{\alpha}_{\parallel}) \\ & \times \left\langle \frac{\cos(\theta q_y r^{\alpha}_{\parallel})}{\sqrt{q_x^2 + \theta^2 q_y^2}} \frac{F_{\boldsymbol{q}}^2}{2\sqrt{1+F_{\boldsymbol{q}}}\left[\sqrt{1+F_{\boldsymbol{q}}}+1\right]^2} \right\rangle_{\rm BZ} . \end{aligned} \quad (7.48)$$

Because the Thomas-Fermi wave-vector κ acts as an effective ultraviolet cut-off, the value of the integral (7.48) is determined by the regime $|\boldsymbol{q}| \lesssim \kappa$. For $\theta\kappa |r^{\alpha}_{\parallel}| \ll 1$ we may approximate in this regime $\cos(\theta q_y r^{\alpha}_{\parallel}) \approx 1$ under the integral sign. Furthermore, for $\kappa |r^{\alpha}_{\parallel}| \gg 1$ the oscillating factor $\cos(q_x r^{\alpha}_{\parallel})$ effectively replaces κ by $|r^{\alpha}_{\parallel}|^{-1}$ as relevant ultraviolet cutoff. We conclude that in the parametrically large intermediate regime

$$\kappa^{-1} \ll |r^{\alpha}_{\|}| \ll (\theta\kappa)^{-1} \tag{7.49}$$

we have to leading logarithmic order

$$S^{\alpha}_{\mathrm{pl}}(r^{\alpha}_{\|}\hat{\boldsymbol{v}}^{\alpha}, 0) \sim -\gamma_{\mathrm{cb}} \left[\ln \left(\frac{1}{\theta\kappa |r^{\alpha}_{\|}|} \right) + b_2 \right] \quad , \tag{7.50}$$

where b_2 is another numerical constant.

The nesting mode

Let us now focus on the contribution from the second term in Eq.(7.42), which involves the collective mode $\tilde{\omega}_{\boldsymbol{q}}$. With the help of Eq.(7.36) the dispersion of this mode can also be written as

$$\tilde{\omega}^2_{\boldsymbol{q}} = \left(1 + \frac{F_{\boldsymbol{q}}}{2} \right) \frac{\xi^2_{\boldsymbol{q}} + \tilde{\xi}^2_{\boldsymbol{q}}}{2} \left[1 - (1 - G_{\boldsymbol{q}})^{1/2} \right] \, , \tag{7.51}$$

$$G_{\boldsymbol{q}} \equiv \frac{(1 + F_{\boldsymbol{q}})}{(1 + \frac{F_{\boldsymbol{q}}}{2})^2} \frac{4 \xi^2_{\boldsymbol{q}} \tilde{\xi}^2_{\boldsymbol{q}}}{\left(\xi^2_{\boldsymbol{q}} + \tilde{\xi}^2_{\boldsymbol{q}} \right)^2} \quad . \tag{7.52}$$

For $|G_{\boldsymbol{q}}| \ll 1$ this implies to leading order

$$\tilde{\omega}_{\boldsymbol{q}} \approx \sqrt{ \frac{1 + F_{\boldsymbol{q}}}{1 + \frac{F_{\boldsymbol{q}}}{2}} \frac{|\xi_{\boldsymbol{q}}| |\tilde{\xi}_{\boldsymbol{q}}|}{\sqrt{\xi^2_{\boldsymbol{q}} + \tilde{\xi}^2_{\boldsymbol{q}}}} } \quad . \tag{7.53}$$

From this expression it is obvious that $\tilde{\omega}_{\boldsymbol{q}}$ vanishes on the planes defined by $\xi_{\boldsymbol{q}} = 0$ or $\tilde{\xi}_{\boldsymbol{q}} = 0$. These equations define precisely the set of points on the Fermi surface. The vanishing of the collective mode $\tilde{\omega}_{\boldsymbol{q}}$ on the Fermi surface is due to the fact that by construction the curvature of the patches is exactly zero, so that the Fermi surface has a nesting symmetry: patches 1 and 4 (or 2 and 3) in Fig. 7.3 can be connected by vectors in the directions of $\boldsymbol{v}^1$ (or $\boldsymbol{v}^2$) that can be attached to an arbitrary point on the patches. For realistic Fermi surfaces of the type shown in Fig. 7.2 this nesting symmetry is absent, so that the associated zero modes do not exist. The vanishing of the mode $\tilde{\omega}_{\boldsymbol{q}}$ gives rise to an unphysical singularity in Eq.(7.42). To see this more clearly, it is necessary to calculate the residue $\tilde{Z}_{\boldsymbol{q}}$ in the regime $|G_{\boldsymbol{q}}| \ll 1$. Expanding the square root in Eq.(7.51) to second order in $G_{\boldsymbol{q}}$, we obtain

$$\tilde{\omega}^2_{\boldsymbol{q}} = \frac{1 + F_{\boldsymbol{q}}}{1 + \frac{F_{\boldsymbol{q}}}{2}} \frac{\xi^2_{\boldsymbol{q}} \tilde{\xi}^2_{\boldsymbol{q}}}{\xi^2_{\boldsymbol{q}} + \tilde{\xi}^2_{\boldsymbol{q}}} \left[1 + \frac{G_{\boldsymbol{q}}}{4} + O(G^2_{\boldsymbol{q}}) \right] \quad . \tag{7.54}$$

The numerator in the expression for the associated residue $\tilde{Z}_{\boldsymbol{q}}$ (see Eq.(7.41)) can then be written as

$$\xi^2_{\boldsymbol{q}} \tilde{\xi}^2_{\boldsymbol{q}} - \tilde{\omega}^2_{\boldsymbol{q}} \frac{\xi^2_{\boldsymbol{q}} + \tilde{\xi}^2_{\boldsymbol{q}}}{2} \approx \frac{\xi^2_{\boldsymbol{q}} \tilde{\xi}^2_{\boldsymbol{q}}}{2 + F_{\boldsymbol{q}}} \left[1 - (1 + F_{\boldsymbol{q}}) \frac{G_{\boldsymbol{q}}}{4} \right] \quad . \tag{7.55}$$

For simplicity let us first consider the regime $F_q \gg 1$. From the definition (7.52) it is clear that in this case the condition $|G_q| \ll 1$ is valid for all values of ξ_q and $\tilde{\xi}_q$. Because the terms of order G_q^2 that have been ignored in Eq.(7.54) are proportional to F_q^{-2}, it is consistent to expand the right-hand side of Eq.(7.55) to first order in F_q^{-1}, in which case we obtain

$$\xi_q^2 \tilde{\xi}_q^2 - \tilde{\omega}_q^2 \frac{\xi_q^2 + \tilde{\xi}_q^2}{2} \approx \frac{\xi_q^2 \tilde{\xi}_q^2}{F_q} \left[\frac{(\xi_q^2 - \tilde{\xi}_q^2)^2}{(\xi_q^2 + \tilde{\xi}_q^2)^2} + O(F_q^{-1}) \right] \quad . \tag{7.56}$$

Substituting this expression into Eq.(7.41) and using Eqs.(7.53) and (7.44), we obtain for $F_q \gg 1$

$$\tilde{Z}_q \approx \frac{\nu}{\sqrt{2} F_q^2} \frac{|\xi_q| |\tilde{\xi}_q|}{\sqrt{\xi_q^2 + \tilde{\xi}_q^2}} \frac{(\xi_q^2 - \tilde{\xi}_q^2)^2}{(\xi_q^2 + \tilde{\xi}_q^2)^2} \quad . \tag{7.57}$$

Note that this expression correctly vanishes if we set $\theta = 0$. We conclude that for $F_q \gg 1$ the second term in Eq.(7.42) involves the integrand

$$\frac{\tilde{Z}_q}{(\tilde{\omega}_q + |\boldsymbol{v}^\alpha \cdot \boldsymbol{q}|)^2} \approx \frac{\nu |\tilde{\xi}_q|}{\sqrt{2} F_q^2 |\xi_q| \sqrt{\xi_q^2 + \tilde{\xi}_q^2}} \frac{(\xi_q^2 - \tilde{\xi}_q^2)^2}{(\xi_q^2 + \tilde{\xi}_q^2)^2} \left[\frac{\sqrt{2} |\tilde{\xi}_q|}{\sqrt{\xi_q^2 + \tilde{\xi}_q^2}} + 1 \right]^{-2} , \tag{7.58}$$

where, without loss of generality, we have set $\boldsymbol{v}^\alpha \cdot \boldsymbol{q} = \xi_q$. To discuss the singularities of this integrand, it is convenient to choose the integration variables $q_{\parallel} = \hat{\boldsymbol{v}}^1 \cdot \boldsymbol{q} = q_x \cos\theta + q_y \sin\theta$ and $q_{\perp} = -q_x \sin\theta + q_y \cos\theta$. Then $\xi_q = v_{\mathrm{F}} q_{\parallel}$ and $\tilde{\xi}_q = v_{\mathrm{F}} (q_{\parallel} - 2\theta q_{\perp})$ to leading order in θ. Hence,

$$\frac{(\xi_q^2 - \tilde{\xi}_q^2)^2}{(\xi_q^2 + \tilde{\xi}_q^2)^2} \sim \begin{cases} \theta^2 q_{\perp}^2 / q_{\parallel}^2 & \text{for } |q_{\parallel}| \gtrsim \theta |q_{\perp}| \\ 1 & \text{for } |q_{\parallel}| \lesssim \theta |q_{\perp}| \end{cases} \quad . \tag{7.59}$$

Note that the condition $|q_{\parallel}| \lesssim \theta |q_{\perp}|$ is equivalent with $|\xi_q| \lesssim |\tilde{\xi}_q|$. Geometrically this means that the wave-vector $\boldsymbol{q}$ is almost parallel to the surface of the first and fourth patch, so that its projection $q_{\parallel}$ onto the local normals $\hat{\boldsymbol{v}}^1$ and $\hat{\boldsymbol{v}}^4$ is much smaller than the projection onto the normals $\hat{\boldsymbol{v}}^2$ and $\hat{\boldsymbol{v}}^3$ of the other two patches, see Fig. 7.4. The contribution from the regime $|q_{\parallel}| \gtrsim \theta |q_{\perp}|$ to Eq.(7.42) is finite and small in the strong coupling limit of interest here. On the other hand, in the regime $|q_{\parallel}| \lesssim \theta |q_{\perp}|$ we have

$$\frac{\tilde{Z}_q}{(\tilde{\omega}_q + |\boldsymbol{v}^\alpha \cdot \boldsymbol{q}|)^2} \approx \frac{\nu}{v_{\mathrm{F}} |q_{\parallel}|} \frac{1}{\sqrt{2}(\sqrt{2}+1)^2 F_q^2} \quad , \quad \text{for } F_q \gg 1 \quad . \tag{7.60}$$

Substituting this expression into Eq.(7.42), we see that the contribution of the nesting mode to the constant part R^α of our Debye-Waller factor leads to the logarithmically divergent integral $\int_0^{\theta |q_{\perp}|} \frac{d q_{\parallel}}{q_{\parallel}}$. Of course, in the combination $R^\alpha - S^\alpha (r_{\parallel}^\alpha \hat{\boldsymbol{v}}^\alpha, \tau)$ this divergence is removed, and we obtain a Debye-Waller factor that grows logarithmically at large distances. This is precisely the Luttinger liquid behavior discussed in Chap. 6.3, so that our 4-patch model is a

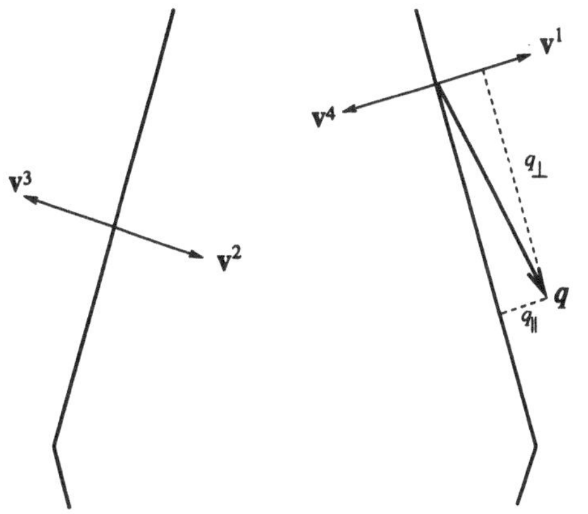

Fig. 7.4. Wave-vector $\boldsymbol{q}$ that contributes to the nesting instability in the 4-patch model. The direction of $\boldsymbol{q}$ is almost perpendicular to $\boldsymbol{v}^1$ and $\boldsymbol{v}^4$, so that $|q_{\parallel}| \lesssim \theta |q_{\perp}|$ and hence $|\xi_{\boldsymbol{q}}| \lesssim |\tilde{\xi}_{\boldsymbol{q}}|$.

higher-dimensional example for a Luttinger liquid. However, the logarithmic growth of the static Debye-Waller factor at large distances is *not* due to the collective mode $\omega_{\boldsymbol{q}}$ which in the limit $\theta \to 0$ can be identified with the plasmon without interchain hopping; instead, in our 4-patch model the Luttinger liquid behavior is generated by the new nesting mode $\tilde{\omega}_{\boldsymbol{q}}$, which disappears at $\theta = 0$.

Clearly, the non-Fermi liquid behavior of our 4-patch model is due to the artificial nesting symmetry of the Fermi surface, which manifests itself for $|q_{\parallel}| \lesssim \theta |q_{\perp}|$. In this regime the dimensionless parameter $G_{\boldsymbol{q}}$ in Eq.(7.51) is small compared with unity for all $F_{\boldsymbol{q}}$, so that it is easy to repeat the above calculations for arbitrary $F_{\boldsymbol{q}}$. We obtain from Eq.(7.53) to leading order

$$\tilde{\omega}_{\boldsymbol{q}} \approx \sqrt{\frac{1+F_{\boldsymbol{q}}}{1+\frac{F_{\boldsymbol{q}}}{2}}} v_{\mathrm{F}} |q_{\parallel}| \quad , \quad |q_{\parallel}| \lesssim \theta |q_{\perp}| \quad , \tag{7.61}$$

and from Eqs.(7.41) and (7.55)

$$\tilde{Z}_{\boldsymbol{q}} \approx \frac{\nu v_{\mathrm{F}} |q_{\parallel}|}{4[1+\frac{F_{\boldsymbol{q}}}{2}]^{\frac{3}{2}}[1+F_{\boldsymbol{q}}]^{\frac{1}{2}}} \quad , \quad |q_{\parallel}| \lesssim \theta |q_{\perp}| \quad . \tag{7.62}$$

In the limit $F_{\boldsymbol{q}} \gg 1$ this expression agrees with Eq.(7.57) if we restrict ourselves to the regime $|\xi_{\boldsymbol{q}}| \lesssim |\tilde{\xi}_{\boldsymbol{q}}|$. We conclude that for $|q_{\parallel}| \lesssim \theta |q_{\perp}|$

$$\frac{\tilde{Z}_{\boldsymbol{q}}}{(\tilde{\omega}_{\boldsymbol{q}} + |\boldsymbol{v}^{\alpha} \cdot \boldsymbol{q}|)^2} \approx \frac{\nu}{4 v_{\mathrm{F}} |q_{\parallel}| [1+\frac{F_{\boldsymbol{q}}}{2}]^{\frac{3}{2}} [1+F_{\boldsymbol{q}}]^{\frac{1}{2}} \left[\sqrt{\frac{1+F_{\boldsymbol{q}}}{1+\frac{F_{\boldsymbol{q}}}{2}}} + 1 \right]^2} \quad . \tag{7.63}$$

It is now obvious that the nesting singularity exists for arbitrary coupling strength. However, this singularity is an unphysical feature of our 4-patch model, and does not exist for realistic Fermi surfaces shown in Fig. 7.2. We shall now refine our model by giving the patches a finite curvature. We then use our bosonization results for non-linear energy dispersion derived in Chap. 5.2

to show that the contribution from the nesting mode becomes negligible compared with the contribution from the plasmon mode $\omega_{\boldsymbol{q}}$.

7.2.2 How curvature kills the nesting singularity

We consider a generalized 4-patch model with curved patches, and first give a simple intuitive argument why the quadratic term in the energy dispersion removes the nesting singularity. We then use our result (6.4) for the Debye-Waller factor with non-linear energy dispersion to confirm this argument by explicit calculation.

It is physically clear that any finite curvature of the patches will destroy the nesting symmetry and hence remove the logarithmic divergence in the Debye-Waller factor. Let us therefore replace the completely flat patches of Fig. 7.3 by the slightly curved patches shown in Fig. 7.5. The corresponding energy

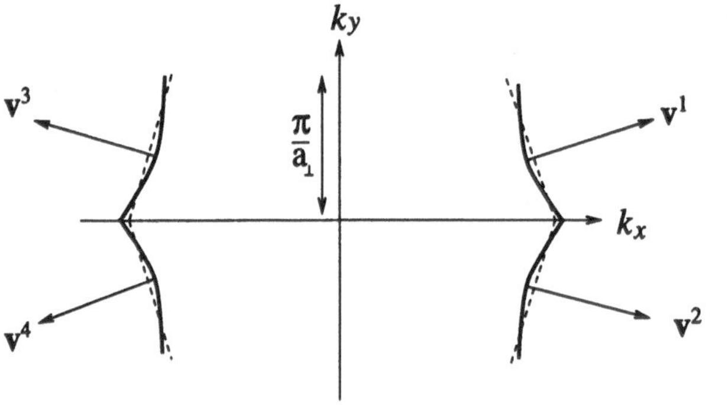

Fig. 7.5. Fermi surface of the 4-patch model with curved patches. If the component of $\boldsymbol{q}$ perpendicular to $\boldsymbol{v}^{\alpha}$ is denoted by $q_{\perp}$, the patches can be described by energy dispersions of the form (7.64) with negative effective mass $m_{\perp}$.

dispersions are

$$\xi_{\boldsymbol{q}}^{\alpha} = \boldsymbol{v}^{\alpha} \cdot \boldsymbol{q} + \frac{q_{\perp}^2}{2m_{\perp}} \quad , \quad q_{\perp} = \hat{\boldsymbol{v}}_{\perp}^{\alpha} \cdot \boldsymbol{q} \quad , \tag{7.64}$$

where $\hat{\boldsymbol{v}}_{\perp}^{\alpha}$ is a unit vector perpendicular to $\boldsymbol{v}^{\alpha}$, and the effective mass $m_{\perp}$ is negative. Note that terms of the form $q_{\parallel}^2/(2m_{\parallel})$ do not describe the curvature of the patches and can be ignored for our purpose (recall the discussion after Eq.(5.101) in Chap. 5.2). Let us first estimate the effect of curvature in a simple qualitative way, which leads to exactly the same result as the explicit evaluation of the bosonization expression for curved Fermi surfaces. Obviously, the curvature term in the energy dispersion becomes important for $v_{\mathrm{F}}|q_{\parallel}| \lesssim q_{\perp}^2/(2|m_{\perp}|)$. Therefore we expect that for curved patches the lower limit for the $q_{\parallel}$-integral will effectively be replaced by $q_{\perp}^2/(2|m_{\perp}|v_{\mathrm{F}})$. We conclude that the effect of curvature can be qualitatively taken into account by substituting

$$\int_0^{\theta|q_\perp|} \frac{dq_\parallel}{q_\parallel} \to \int_{\frac{q_\perp^2}{2|m_\perp|v_F}}^{\theta|q_\perp|} \frac{dq_\parallel}{q_\parallel} = \ln\left(\frac{2|m_\perp|v_F\theta}{|q_\perp|}\right) \quad . \tag{7.65}$$

In physically relevant cases we expect $|m_\perp| \approx m_\parallel/\theta = k_F/(v_F\theta)$, so that the right-hand side of Eq.(7.65) reduces to the integrable logarithmic factor $\ln(2k_F/|q_\perp|)$. Note that the above argument is only consistent if $q_\perp^2/(2|m_\perp|v_F) \ll \theta|q_\perp|$, even for the largest relevant $q_\perp$. Keeping in mind that the effective ultraviolet-cutoff for the $q_\perp$-integral is the Thomas-Fermi wave-vector κ (see Eq.(7.11)), this condition reduces to $\kappa \ll k_F$. Combining Eqs.(7.60) and (7.65), it is easy to see that the (regularized) nesting mode simply renormalizes the numerical constant b_1 in Eq.(7.47). We therefore conclude that the leading small-θ behavior of R^α is entirely due to the plasmon mode $\omega_{\boldsymbol{q}}$.

We now confirm the above argument with the help of the bosonization result for the Green's function for non-linear energy dispersion derived in Chap. 5.2. With finite curvature we should replace the expression (7.42) for the constant part of the Debye-Waller factor of the 4-patch model by R_1^α given in Eq.(6.4). Of course, we should now use the dynamic structure factor $S_{\mathrm{RPA}}(\boldsymbol{q},\omega)$ corresponding to the Fermi surface shown in Fig. 7.5. Due to the curvature of patches, $S_{\mathrm{RPA}}(\boldsymbol{q},\omega)$ is now more complicated than in Eq.(7.39). Apart from a δ-function peak representing the physical plasmon mode, we expect that, due to Landau damping, the peak associated with the nesting mode $\tilde{\omega}_{\boldsymbol{q}}$ is now spread out into a continuum in a finite frequency interval. However, in order to estimate the fate of the nesting mode in the presence of curvature, it is sufficient to substitute the dynamic structure factor (7.39) for flat patches into Eq.(6.4). Certainly, if in this approximation the nesting singularity is removed, then more accurate approximations for $S_{\mathrm{RPA}}(\boldsymbol{q},\omega)$ will lead to the same result, because the curvature terms smooth out the sharpness of the nesting mode. Combining then Eqs.(7.39) and (6.4), we see that the contribution of the nesting mode to R_1^α is given by

$$R_{1,\mathrm{nest}}^\alpha \approx -\frac{1}{V}\sum_{\boldsymbol{q}} f_{\boldsymbol{q}}^2 \tilde{Z}_{\boldsymbol{q}} \frac{2\mathrm{sgn}(\xi_{\boldsymbol{q}}^\alpha)}{\frac{q_\perp^2}{|m_\perp|}(\tilde{\omega}_{\boldsymbol{q}} + |\xi_{\boldsymbol{q}}^\alpha|)} \quad , \tag{7.66}$$

with $\tilde{Z}_{\boldsymbol{q}}$ and $\tilde{\omega}_{\boldsymbol{q}}$ given in Eqs.(7.41) and (7.51). From Sect. 7.2.1 we know that possible nesting singularities are due to the regime $|q_\parallel| \lesssim \theta|q_\perp|$. Thus, restricting the integral in Eq.(7.66) to this regime, we have from Eqs.(7.57) and (7.53) in the strong coupling limit $f_{\boldsymbol{q}}^2\tilde{Z}_{\boldsymbol{q}} \approx v_F|q_\parallel|/(\sqrt{2}\nu)$ and $\tilde{\omega}_{\boldsymbol{q}} \approx \sqrt{2}v_F|q_\parallel|$. Recall that for the three-dimensional Coulomb interaction the strong-coupling condition $\nu f_{\boldsymbol{q}} \gg 1$ is equivalent with $|\boldsymbol{q}| \ll \kappa$, where the Thomas-Fermi wave-vector κ is given in Eq.(7.11). It is useful to introduce again the integration variables $q_\parallel = \hat{\boldsymbol{v}}^\alpha \cdot \boldsymbol{q}$ and $q_\perp = \hat{\boldsymbol{v}}_\perp^\alpha \cdot \boldsymbol{q}$. Putting everything together, we find that the contribution from the critical regime $|q_\parallel| \lesssim \theta|q_\perp|$ to Eq.(6.4) can be written as

$$R^{\alpha}_{1,\text{nest}} \approx -\frac{\sqrt{2}\kappa}{\pi^3 \nu} \int_0^{\kappa} \mathrm{d}q_{\perp} \frac{|m_{\perp}|}{q_{\perp}^2} \int_{-\theta|q_{\perp}|}^{\theta|q_{\perp}|} \mathrm{d}q_{\parallel} \frac{|q_{\parallel}| \mathrm{sgn}(q_{\parallel} - \frac{q_{\perp}^2}{2|m_{\perp}|v_{\mathrm{F}}})}{\sqrt{2}|q_{\parallel}| + \left| q_{\parallel} - \frac{q_{\perp}^2}{2|m_{\perp}|v_{\mathrm{F}}} \right|} \ . \tag{7.67}$$

The $q_{\parallel}$-integration can now be performed analytically. The integral is proportional to $q_{\perp}^2/|m_{\perp}|$, which cancels the singular factor of $|m_{\perp}|/q_{\perp}^2$ in Eq.(7.67). We obtain

$$R^{\alpha}_{1,\text{nest}} \approx -\frac{\sqrt{2}\kappa}{2(\sqrt{2}+1)^2 \pi^3 \nu v_{\mathrm{F}}} \int_0^{\kappa} \mathrm{d}q_{\perp} \left[\ln \left(\frac{2|m_{\perp}|v_{\mathrm{F}}\theta}{q_{\perp}} \right) + b_3 \right] \ , \tag{7.68}$$

where b_3 is a numerical constant of the order of unity. This is the same type of integral as in Eq.(7.65), so that our simple intuitive arguments given above are now put on a more solid basis. As already mentioned, in physically relevant cases we expect $|m_{\perp}|v_{\mathrm{F}}\theta \approx k_{\mathrm{F}}$. Using then Eqs.(7.3) and (7.11), we finally obtain

$$R^{\alpha}_{1,\text{nest}} \approx -\gamma_{\text{cb}} b_4 \ , \tag{7.69}$$

where $\gamma_{\text{cb}} = e^2/(2\pi v_{\mathrm{F}})$ (see Eq.(7.23)), and b_4 is another numerical constant of the order of unity. Thus, for patches with finite curvature the contribution of the nesting mode is finite. It is also easy to see that the curvature terms do *not* modify the logarithmic small-θ behavior of R^{α} given in Eq.(7.47). This is so because the leading $\ln(1/\theta)$-term in Eq.(7.47) is generated by the energy scale $v_{\mathrm{F}}\theta|q_{\perp}|$, which is by assumption larger than the curvature energy $q_{\perp}^2/(2|m_{\perp}|)$.

7.2.3 Anomalous scaling in a Fermi liquid

Now comes the important conclusion about the physical system of interest (to be distinguished from the 4-patch model discussed in Sect. 7.2.1).

Comparing Eq.(7.69) with the corresponding contribution (7.47) to the Debye-Waller factor that is due to the plasmon mode $\omega_{\boldsymbol{q}}$, we conclude that for small θ the Debye-Waller factor is dominated by the plasmon mode. In particular, for realistic Fermi surfaces of the form shown in Fig. 7.2 the constant part R^{α} of the Debye-Waller factor is finite. To leading logarithmic order for small θ we may therefore approximate $R^{\alpha} \approx R^{\alpha}_{\text{pl}}$, where R^{α}_{pl} is given in Eq.(7.47). We conclude that for any non-zero θ the system is a Fermi liquid, with quasi-particle residue

$$Z^{\alpha} = \mathrm{e}^{R^{\alpha}} \propto \theta^{\gamma_{\text{cb}}} \ , \tag{7.70}$$

where γ_{cb} is given in Eq.(7.23). Thus, for $\theta \to 0$ the quasi-particle residue vanishes with a non-universal power of θ, which *can be identified with the anomalous dimension of the corresponding Luttinger liquid that would exist for $\theta = 0$ at the same value of the dimensionless coupling constant e^2/v_{F}.*

Combining Eqs.(7.47) and (7.50), we obtain for the total static Debye-Waller factor to leading logarithmic order in θ

$$Q^{\alpha}(r_{\parallel}^{\alpha}\hat{v}^{\alpha},0) = R^{\alpha} - S^{\alpha}(r_{\parallel}^{\alpha}\hat{v}^{\alpha},0)$$
$$= -\gamma_{\text{cb}}\left[\ln(\kappa|r_{\parallel}^{\alpha}|) + O(1)\right] \quad , \quad \kappa^{-1} \ll |r_{\parallel}^{\alpha}| \ll (\theta\kappa)^{-1} \quad . \tag{7.71}$$

Exponentiating this expression, we see that the interacting Green's function satisfies the anomalous scaling relation,

$$G^{\alpha}(\boldsymbol{r}/s,0) = s^{3+\gamma_{\text{cb}}}G^{\alpha}(\boldsymbol{r},0) \quad , \quad \kappa^{-1} \ll |r_{\parallel}^{\alpha}| \, , \, |r_{\parallel}^{\alpha}|/s \ll (\theta\kappa)^{-1} \quad . \tag{7.72}$$

Thus, in spite of the fact that the system is a Fermi liquid, there exists for small θ a parametrically large intermediate regime where the interacting Green's function satisfies the anomalous scaling law typical for Luttinger liquids, as discussed in Chap. 6.3. Moreover, the effective anomalous exponent *is precisely given by the anomalous dimension of the Luttinger liquid that would exist for $\theta = 0$*. This is a very important result, because in realistic experimental systems the interchain hopping $t_{\perp}$ is never exactly zero. We thus arrive at the important conclusion that for small θ the anomalous dimension of the Luttinger liquid is in principle measurable, although strictly speaking the system is a Fermi liquid. The relevance of $t_{\perp}$ in an infinite array of weakly coupled chains has also been discussed in [7.11, 7.15] by means of a perturbative expansion to lowest order in $t_{\perp}$. In contrast, our approach is non-perturbative in $t_{\perp}$.

7.2.4 The nesting singularity for general Fermi surfaces

We show that quite generally the nesting symmetries introduced via the patching construction give rise to logarithmic singularities and hence to unphysical Luttinger liquid behavior.

The nesting singularity discussed in Sect. 7.2.1 is not a special feature of our 4-patch model. Singularities of this type will appear in any model where the Fermi surface is covered by a finite number M of flat patches, such that at least some of the patches have a nesting symmetry. The simplest analytically tractable case is perhaps a square Fermi surface ($M = 4$) in two dimensions, which has first been discussed by Mattis [7.21], and more recently by Hlubina [7.22] and by Luther [7.23]. However, unless there exists a real physical nesting symmetry in the problem, these nesting singularities are artificially generated by approximating a curved Fermi surface by a collection of completely flat patches.

There are several ways to cure this problem. The simplest one is perhaps to choose the patches such that nesting symmetries do not exist. For example, in the case of a circular Fermi surface in $d = 2$ we avoid artificial nesting symmetries by choosing an odd number of identical patches (see Fig. 5.1 for

$M = 5$). The disadvantage of this construction is that it explicitly breaks the inversion symmetry of the Fermi surface, so that the negative frequency part of the dynamic structure factor has to be treated separately[1].

The second possibility is to take the limit $M \to \infty$ at some intermediate point in the calculation, for example in the Debye-Waller factor given in Eqs.(5.31)–(5.33). Because for finite M the residue of the nesting mode in the dynamic structure factor is proportional to M^{-1}, its contribution vanishes in the limit $M \to \infty$. To see this, suppose that we approximate a spherical Fermi surface with an even number M of identical patches (see Fig. 2.3 for $M = 12$ in two dimensions). The corresponding non-interacting polarization $\Pi_0(\boldsymbol{q}, z)$ is given in Eq.(A.66). From the discussion of the nesting mode in the 4-patch model in Sect. 7.2.1 we expect that for some directions of $\boldsymbol{q}$ there will exist one particular patch P^{μ}_{Λ} such that the energy $|\boldsymbol{v}^{\mu} \cdot \boldsymbol{q}|$ is much smaller than all the other energies $|\boldsymbol{v}^{\alpha} \cdot \boldsymbol{q}|$, $\alpha \neq \mu$. Furthermore, we expect that for sufficiently small $q^{\mu}_{\parallel} \equiv \hat{\boldsymbol{v}}^{\mu} \cdot \boldsymbol{q}$ the nesting mode $\omega^{\mu}_{\boldsymbol{q}}$ gives rise to a δ-function peak in the dynamic structure factor with $\omega^{\mu}_{\boldsymbol{q}} \propto |\boldsymbol{v}^{\mu} \cdot \boldsymbol{q}|$. Because for sufficiently small $q^{\mu}_{\parallel}$ this energy is much smaller than $|\boldsymbol{v}^{\alpha} \cdot \boldsymbol{q}|$ with $\alpha \neq \mu$, the energy dispersion of the nesting mode can be approximately calculated by setting $z^2 = 0$ in all terms with $\alpha \neq \mu$ in the expression (A.66) for the non-interacting polarization $\Pi_0(\boldsymbol{q}, z)$ for finite patch number. This yields for the polarization in the regime of wave-vectors $\boldsymbol{q}$ satisfying $|\boldsymbol{v}^{\mu} \cdot \boldsymbol{q}| \ll |\boldsymbol{v}^{\alpha} \cdot \boldsymbol{q}|$ for $\alpha \neq \mu$

$$\Pi_0(\boldsymbol{q}, z) \approx \frac{\nu}{M} \left[M - 2 + \frac{2(\boldsymbol{v}^{\mu} \cdot \boldsymbol{q})^2}{(\boldsymbol{v}^{\mu} \cdot \boldsymbol{q})^2 - z^2} \right] \quad . \tag{7.73}$$

The collective mode equation (A.70) is then easily solved, with the result that the dispersion of the nesting mode is given by

$$\omega^{\mu}_{\boldsymbol{q}} = \sqrt{\frac{1 + F_{\boldsymbol{q}}}{1 + \frac{M-2}{M} F_{\boldsymbol{q}}}} |\boldsymbol{v}^{\mu} \cdot \boldsymbol{q}| \quad . \tag{7.74}$$

For the associated residue we obtain with the help of Eq.(A.75)

$$Z^{\mu}_{\boldsymbol{q}} \equiv \frac{\nu^2}{F^2_{\boldsymbol{q}} \frac{\partial}{\partial z} \Pi_0(\boldsymbol{q}, z)\big|_{z=\omega^{\mu}_{\boldsymbol{q}}}} = \frac{\nu |\boldsymbol{v}^{\mu} \cdot \boldsymbol{q}|}{M \left[1 + \frac{M-2}{M} F_{\boldsymbol{q}}\right]^{\frac{3}{2}} \left[1 + F_{\boldsymbol{q}}\right]^{\frac{1}{2}}} \quad . \tag{7.75}$$

If we set $M = 4$ we recover the corresponding expressions (7.61) and (7.62) for the nesting mode in the 4-patch model. From Eq.(6.14) it is now obvious that in the thermodynamic limit the constant part R^{μ} of the Debye-Waller factor is proportional to $\frac{1}{M} \int_0 \frac{dq_{\parallel}}{q_{\parallel}}$. Clearly, the logarithmic divergence for finite M is removed if the take the limit $M \to \infty$. It should be kept in mind, however, that taking the limit $M \to \infty$ at intermediate stages of the calculation is not

[1] As already mentioned in the first footnote of Chap. 6, in this case the relation (2.42) between the imaginary part of the polarization and the dynamic structure factor (and all equations derived from Eq.(2.42)) are not correct. In particular, the expressions derived in Chap. 6.1 cannot be used.

quite satisfactory, because $M \to \infty$ implies that the patch cutoff Λ vanishes. As discussed in Chap. 6.1.3, in this case it is difficult to formally justify our derivation of the Green's function with linearized energy dispersion given in Chap. 5.1, because the condition $q_c \ll \Lambda$ is violated in this limit (see Fig. 2.5). In Sect. 7.2.2 we have solved this problem with the help of our background field method developed in Chap. 5.2, which leads to a simple way for including the effect of the curvature of the Fermi surface into the bosonization procedure.

7.3 Summary and outlook

In this chapter we have used our non-perturbative higher-dimensional bosonization approach to calculate the single-particle Green's function of weakly coupled metallic chains. This problem is not only of current experimental interest, but its solution via higher-dimensional bosonization also nicely illustrates the approximations inherent in this approach. In particular, we have shown that the replacement of a curved Fermi surface by a finite number of flat patches leads to unphysical logarithmic singularities and to Luttinger behavior in the Green's function when at least two opposite patches are parallel. In this respect we agree with the works by Mattis [7.21] and by Hlubina [7.22], who studied this problem in the special case of a square Fermi surface. However, we have also shown that for more realistic curved Fermi surfaces these logarithmic singularities disappear. Any finite value of the interchain hopping $t_\perp$ leads then to a bounded Debye-Waller factor, signalling Fermi liquid behavior.

Very recently the singularities generated by flat regions on opposite sides of a two-dimensional Fermi surface have been analyzed by Zheleznyak *et al.* [7.24] with the help of the parquet approximation [7.9,7.25]. These authors obtained results which are, at least at the first sight, at variance with our finding (as well as with [7.21,7.22]). Note, however, that in our approach we have ignored the spin degree of freedom as well as scattering processes involving large momentum transfers. In particular, we have not taken into account the instabilities towards charge- or spin-density wave order, which according to the authors of [7.24] become essential at sufficiently low temperatures. It is therefore not surprising that we obtain a different result than Zheleznyak *et al.* [7.24]. Our calculation is restricted to a parameter regime where the low energy physics is dominated by forward scattering. The existence of such a regime is by no means obvious [7.9], and we have assumed that for some range of temperature, interchain hopping, and interaction strength the instabilities mentioned above can indeed be ignored.

Our finding that any finite value of the interchain hopping leads to a Fermi liquid is supported by lowest order perturbation theory [7.15]. However, there have been recent claims in the literature [7.26,7.27] that coupled chains with finite $t_\perp$ can remain Luttinger liquids if the interaction is sufficiently strong, so that the anomalous dimension characterizing the Luttinger liquid at $t_\perp = 0$ exceeds a certain critical value. It is important to realize that this result can only be obtained within an approach that *allows for a change in the shape of the Fermi surface as the interaction is turned on*[2]. Unfortunately, higher-dimensional bosonization with linearized energy dispersion cannot describe the renormalization of the shape of the Fermi surface due to the interactions, because after the linearization the relative position of the flat patches on the Fermi surface remains completely rigid[3]. On the other hand, our more general bosonization result for the Green's function with non-linear energy dispersion derived in Chap. 5.2 certainly incorporates also the renormalization of the shape of the Fermi surface due to the interactions. Thus, an extremely interesting open problem is the **full analysis of the higher-dimensional bosonization result for the Green's function of coupled chains with non-linear energy dispersion.** Note that within the Gaussian approximation one should not only calculate the Debye-Waller factor $Q_1^{\alpha}(\boldsymbol{r}, \tau)$ in Eqs.(5.151)–(5.153), but also the prefactor self-energy $\Sigma_1^{\alpha}(\tilde{q})$ and the vertex function $Y^{\alpha}(\tilde{q})$ given in Eqs.(5.159) and (5.162). Furthermore, although for $t_{\perp} = 0$ and for linearized energy dispersion the Gaussian approximation is exact and correctly reproduces the solution of the Tomonaga-Luttinger model (see Chap. 6.3), it is not clear whether for finite $t_{\perp}$ the Gaussian approximation is still sufficient, so that it might be necessary to include at least certain sub-classes of the non-Gaussian corrections discussed in Chap. 5.2. Obviously, the problem of coupled chains is far from being solved. We hope that the methods developed in this book will help to shed more light onto this very interesting problem.

Finally, we would like to point out that the problem of calculating the Green's function of an *infinite array* of coupled chains is very different from the problem of *two* coupled chains [7.13,7.14,7.16,7.17]. The two-chain problem is it not so easy to solve by means of higher-dimensional bosonization, because in this case the Fermi surface consists of four isolated points, which evidently cannot be treated as a simple higher-dimensional surface. It turns out that even for long-range Coulomb interactions it is impossible to map the two-chain system onto a pure forward scattering problem. In fact, the calculation of the Green's function in the two-chain system can be mapped onto an effective back-scattering problem in one dimension [7.28,7.29], which in general cannot be solved exactly. However, if one assumes certain special values of the interchain and intrachain interaction, the Green's function of an arbitrary number of coupled chains can be calculated exactly [7.28]. Although these special interactions are perhaps unphysical, it is interesting to note that in these models Luttinger liquid behavior coexists with coherent interchain hopping [7.29]. This seems to disagree with the result of Clarke, Strong, and Anderson [7.26, 7.30], who claim that Luttinger liquid behavior necessarily destroys coherent interchain hopping.

[2] I would like to thank Steven Strong for his detailed explanations of this point.

[3] Recall in this context our discussion at the end of Chap. 5.1.5. concerning the absence of effective mass renormalizations for linearized energy dispersion.

8. Electron-phonon interactions

We couple electrons to phonons via Coulomb forces, and show that for isotropic three-dimensional systems the long-range part of the Coulomb interaction cannot destabilize the Fermi liquid state. However, Luttinger liquid behavior in three dimensions can be due to quasi-one-dimensional anisotropy in the electronic band structure or in the phonon frequencies. A brief account of the results presented in this chapter has been published in [8.1].

The interplay between the vibrations of the ionic lattice in a solid and the interactions between the conduction electrons still lacks a complete understanding [8.2–8.4]. Conventionally this problem is approached perturbatively, which is possible as long as the mass M_{i} of the ions is much larger than the effective mass m of the electrons. In this case a theorem due to Migdal [8.5] guarantees that, to leading order in $\sqrt{m/M_{\mathrm{i}}}$, the electron-phonon vertex is not renormalized by phonon corrections. However, in heavy fermion systems the parameter $\sqrt{m/M_{\mathrm{i}}}$ is not necessarily small, so that Migdal's theorem may not be valid. Then the self-consistent renormalization of the phonon energies due to the coupling to the electrons cannot be neglected [1.7, 8.6]. In diagrammatic approaches it is often tacitly assumed that the phonons remain well defined collective modes [1.6, 8.7]. Moreover, an implicit assumption in the proof of Migdal's theorem is that the electronic system is a Fermi liquid. In view of the experimental evidence of non-Fermi liquid behavior in the normal state of some of the high-temperature superconductors [1.24], it is desirable to study the coupled electron-phonon system by means of a method which does not assume *a priori* a Fermi liquid. Our functional bosonization approach fulfills this requirement, so that it offers a new non-perturbative way to study coupled electron-phonon systems in $d > 1$. In one dimension the problem of electron-phonon interactions has recently been analyzed via bosonization in the works [8.8, 8.9].

We would like to emphasize, however, that we shall retain only processes involving small momentum transfers and neglect superconducting instabilities. Recall that in BCS superconductors the phonons mediate an effective attractive interaction between the electrons, which at low enough temperatures overcomes the repulsive Coulomb interaction and leads to superconductivity [8.10]. Thus, the analysis presented below is restricted to the parameter regime where the electronic system is in the normal metallic state. However, we do ***not*** assume that the electronic system is a Fermi liquid.

Throughout this chapter we shall work with linearized energy dispersion, because we shall focus on the calculation of the ***static*** Debye-Waller factor $Q^{\alpha}(\boldsymbol{r}, 0)$. As discussed in Chap. 6.1.3, the long-distance behavior of this quantity should only be weakly affected by the non-linear terms in the energy dispersion. Note that this approximation is most likely not sufficient for the calculation of $Q^{\alpha}(\boldsymbol{r}, \tau)$ for $\tau \neq 0$, because in this case the double pole that appears in the bosonization result for the Debye-Waller factor with linearized energy dispersion leads to some unphysical features (see the discussion at the beginning of Chap. 5.2 and in Chap. 6.1.3). In this case one should retain the non-linear terms in the energy dispersion.

This chapter is subdivided into four main sections. In Sect. 8.1 we define the coupled electron-phonon system in the language of functional integrals. By integrating over the phonon degrees of freedom, we then derive the effective action for the electrons, and determine the precise form of the effective retarded density-density interaction between the electrons mediated by the phonons. Because this interaction is of the density-density type discussed in Chap. 5, we obtain in Sect. 8.2 a non-perturbative expression for the electronic Green's function by simply substituting the proper effective interaction $f_{q}^{\mathrm{RPA},\alpha}$ into Eqs.(5.32) and (5.33). In Sect. 8.3 we show that our approach takes also the renormalization of the phonon spectrum due to the coupling to the electrons into account. Finally, in Sect. 8.4 we shall explicitly calculate the quasi-particle residue and examine the conditions under which the residue can become small or even vanishes. In particular, we discuss one-dimensional phonons with dispersion $\Omega_{\boldsymbol{q}} = c_{\mathrm{s}}|q_x|$ that are coupled to three-dimensional electrons with a spherical Fermi surface. We show that in this case the quasi-particle residue vanishes at the points $\boldsymbol{k}^{\alpha} = \pm k_{\mathrm{F}} \boldsymbol{e}_x$ on the Fermi sphere, and that close to these special points the single-particle Green's function exhibits Luttinger liquid behavior.

8.1 The effective interaction

We introduce a simple model for electrons that are coupled to longitudinal acoustic phonons and derive the associated retarded electron-electron interaction by means of functional integration.

8.1.1 The Debye model

Following the classic textbook by Fetter and Walecka [1.6], we use the Debye model to describe the interaction between electrons and longitudinal acoustic (LA) phonons. In this model the ionic background charge is approximated by a homogeneous elastic medium. Although the ions in real solids form a lattice, the discrete lattice structure is unimportant for LA phonons with wave-vectors $|\boldsymbol{q}| \ll k_{\mathrm{F}}$. For a detailed description of this model and its physical justification see chapter 12 of the book by Fetter and Walecka [1.6]. However, some subtleties concerning screening and phonon energy renormalization have been ignored in [1.6]. To clarify these points, we first give a careful derivation of the effective electron-electron interaction in this model via functional integration.

In our Euclidean functional integral approach, the dynamics of the isolated phonon system is described via the action

$$S_{\mathrm{ph}}\{b\} = \beta \sum_{q} [-\mathrm{i}\omega_m + \Omega_{\boldsymbol{q}}] b_q^{\dagger} b_q \quad , \tag{8.1}$$

where b_q is a complex field representing the phonons in the coherent state functional integral. For simplicity let us first assume isotropic acoustic phonons, with dispersion relation $\Omega_{\boldsymbol{q}} = c_{\mathrm{s}}|\boldsymbol{q}|$, where c_{s} is the *bare* velocity of sound, which is determined by the *short-range part* of the Coulomb potential and all other non-universal forces between the ions. The *long-range* part of the Coulomb potential will be treated explicitly[1]. In Eq.(8.1) and all subsequent expressions involving phonon variables it is understood that wave-vector summations are cut off when the phonon frequency reaches the Debye frequency [8.11]. As before, the electronic degrees of freedom are represented by a Grassmann field ψ, so that the total action of the interacting electron-phonon system is

$$S\{\psi, b\} = S_0\{\psi\} + S_{\mathrm{ph}}\{b\} + S_{\mathrm{int}}\{\psi, b\} \quad . \tag{8.2}$$

Here $S_0\{\psi\}$ describes the dynamics of the non-interacting electron system (see Eq.(3.3)), and $S_{\mathrm{int}}\{\psi, b\}$ represents the Coulomb energy associated with all charge fluctuations in the system,

$$S_{\mathrm{int}}\{\psi, b\} = \frac{e^2}{2} \int_0^{\beta} \mathrm{d}\tau \int \mathrm{d}\boldsymbol{r} \int \mathrm{d}\boldsymbol{r}' \frac{\rho^{\mathrm{tot}}(\boldsymbol{r}, \tau) \rho^{\mathrm{tot}}(\boldsymbol{r}', \tau)}{|\boldsymbol{r} - \boldsymbol{r}'|} \quad , \tag{8.3}$$

where

$$\rho^{\mathrm{tot}}(\boldsymbol{r}, \tau) = \psi^{\dagger}(\boldsymbol{r}, \tau)\psi(\boldsymbol{r}, \tau) - \rho^{\mathrm{ion}}(\boldsymbol{r}, \tau) \tag{8.4}$$

represents the total density of charged particles at point $\boldsymbol{r}$ and imaginary time τ. The ionic density $\rho^{\mathrm{ion}}(\boldsymbol{r}, \tau)$ is of the form

$$\rho^{\mathrm{ion}}(\boldsymbol{r}, \tau) = z\frac{N}{V} + \delta\rho^{\mathrm{ion}}(\boldsymbol{r}, \tau) \quad , \tag{8.5}$$

where the first term represents the charge density of the uniform background charge, which in the absence of phonons exactly compensates the total charge of the conduction electrons. Here $z \geq 1$ is the valence of the ions and zN is the total number of conduction electrons. The fluctuating component of the ionic charge density is related to the bosonic field b_q via

---

[1] From Appendix A.3.1 it is clear that the boundary between the long- and short-wavelength regimes is defined by the Thomas-Fermi wave-vector $\kappa = (4\pi e^2 \nu)^{1/2}$.

$$\delta\rho^{\text{ion}}(\boldsymbol{r},\tau) = -z\frac{N}{V}\nabla\cdot\boldsymbol{d}(\boldsymbol{r},\tau) \quad , \tag{8.6}$$

where the displacement field $\boldsymbol{d}(\boldsymbol{r},\tau)$ is given by [1.6]

$$\boldsymbol{d}(\boldsymbol{r},\tau) = \frac{-\mathrm{i}}{\sqrt{N}}\sum_{\boldsymbol{q}}\frac{\hat{\boldsymbol{q}}}{\sqrt{2M_{\mathrm{i}}\Omega_{\boldsymbol{q}}}}\left[b_{\boldsymbol{q}}\mathrm{e}^{\mathrm{i}(\boldsymbol{q}\cdot\boldsymbol{r}-\omega_m\tau)} - b_{\boldsymbol{q}}^{\dagger}\mathrm{e}^{-\mathrm{i}(\boldsymbol{q}\cdot\boldsymbol{r}-\omega_m\tau)}\right] \quad , \tag{8.7}$$

so that

$$\nabla\cdot\boldsymbol{d}(\boldsymbol{r},\tau) = \frac{1}{\sqrt{N}}\sum_{\boldsymbol{q}}\frac{|\boldsymbol{q}|}{\sqrt{2M_{\mathrm{i}}\Omega_{\boldsymbol{q}}}}\left[b_{\boldsymbol{q}}\mathrm{e}^{\mathrm{i}(\boldsymbol{q}\cdot\boldsymbol{r}-\omega_m\tau)} + b_{\boldsymbol{q}}^{\dagger}\mathrm{e}^{-\mathrm{i}(\boldsymbol{q}\cdot\boldsymbol{r}-\omega_m\tau)}\right] \quad . \tag{8.8}$$

Substituting Eq.(8.4) into Eq.(8.3), we obtain three contributions, which after Fourier transformation can be written as

$$S_{\text{int}}\{\psi,b\} = S_{\text{int}}^{\text{el}}\{\psi\} + S_{\text{int}}^{\text{el-ph}}\{\psi,b\} + S_{\text{int}}^{\text{ph}}\{b\} \quad , \tag{8.9}$$

with

$$S_{\text{int}}^{\text{el}}\{\psi\} = \frac{\beta}{2V}\sum_{\boldsymbol{q}} f_{\boldsymbol{q}}^{\text{cb}}\rho_{-\boldsymbol{q}}\rho_{\boldsymbol{q}} \quad , \tag{8.10}$$

$$S_{\text{int}}^{\text{el-ph}}\{\psi,b\} = -\frac{\beta}{2V}\sum_{\boldsymbol{q}} f_{\boldsymbol{q}}^{\text{cb}}\left[\rho_{-\boldsymbol{q}}\rho_{\boldsymbol{q}}^{\text{ion}} + \rho_{-\boldsymbol{q}}^{\text{ion}}\rho_{\boldsymbol{q}}\right] \quad , \tag{8.11}$$

$$S_{\text{int}}^{\text{ph}}\{b\} = \frac{\beta}{2V}\sum_{\boldsymbol{q}} f_{\boldsymbol{q}}^{\text{cb}}\rho_{-\boldsymbol{q}}^{\text{ion}}\rho_{\boldsymbol{q}}^{\text{ion}} \quad , \tag{8.12}$$

where we have defined

$$f_{\boldsymbol{q}}^{\text{cb}} = \begin{cases} \frac{4\pi e^2}{\boldsymbol{q}^2} & \text{for } \boldsymbol{q}\neq 0 \\ 0 & \text{for } \boldsymbol{q}=0 \end{cases} \quad . \tag{8.13}$$

The Fourier coefficients of the densities can be expressed in terms of the Fourier coefficients $\psi_{\boldsymbol{k}}$ and $b_{\boldsymbol{q}}$ of the electron and phonon fields,

$$\rho_{\boldsymbol{q}} = \sum_{\boldsymbol{k}}\psi_{\boldsymbol{k}}^{\dagger}\psi_{\boldsymbol{k}+\boldsymbol{q}} \quad , \tag{8.14}$$

$$\rho_{\boldsymbol{q}}^{\text{ion}} = -z\sqrt{N}\frac{|\boldsymbol{q}|}{\sqrt{2M_{\mathrm{i}}\Omega_{\boldsymbol{q}}}}\left[b_{\boldsymbol{q}} + b_{-\boldsymbol{q}}^{\dagger}\right] \quad . \tag{8.15}$$

The part of the action involving the phonon degrees of freedom can then be written as

$$\begin{aligned} S_{\text{ph}}\{b\} + S_{\text{int}}^{\text{ph}}\{b\} + S_{\text{int}}^{\text{el-ph}}\{b,\psi\} &= \beta\sum_{\boldsymbol{q}}\Big[(-\mathrm{i}\omega_m + \Omega_{\boldsymbol{q}})b_{\boldsymbol{q}}^{\dagger}b_{\boldsymbol{q}} \\ &+\frac{W_{\boldsymbol{q}}}{4}(b_{\boldsymbol{q}} + b_{-\boldsymbol{q}}^{\dagger})(b_{-\boldsymbol{q}} + b_{\boldsymbol{q}}^{\dagger}) + g_{\boldsymbol{q}}\rho_{-\boldsymbol{q}}(b_{\boldsymbol{q}} + b_{-\boldsymbol{q}}^{\dagger})\Big] \quad , \end{aligned} \tag{8.16}$$

with

$$W_q = \left[\frac{z^2 N}{V} \frac{q^2}{M_{\rm i}} \right] \frac{f_q^{\rm cb}}{\Omega_q} \quad , \tag{8.17}$$

$$g_q = \left[\frac{z^2 N}{V} \frac{q^2}{M_{\rm i}} \right]^{1/2} \frac{f_q^{\rm cb}}{\sqrt{2V\Omega_q}} \quad . \tag{8.18}$$

At this point Fetter and Walecka [1.6] make the following two approximations: (a) the bare Coulomb interaction $f_q^{\rm cb}$ in $S_{\rm int}^{\rm el-ph}\{\psi, b\}$ is replaced by the static screened interaction, $4\pi e^2/q^2 \to 4\pi e^2/\kappa^2$, and (b) the contribution $S_{\rm int}^{\rm ph}\{b\}$ is simply dropped. We shall see shortly that the approximation (b) amounts to ignoring the self-consistent renormalization of the phonon frequencies [1.7,8.6]. Although Fetter and Walecka [1.6] argue that these approximations correctly describe the physics of screening, it is not quite satisfactory that one has to rely here on words and not on calculations. Because in our bosonization method screening can be derived from first principles, we do not follow the "screening by hand" procedure of [1.6], and retain at this point all terms in Eqs.(8.10)–(8.12) with the bare Coulomb interaction.

8.1.2 Integration over the phonons

In this way we obtain the effective electron-electron interaction mediated by the phonons.

We are interested in the exact electronic Green's function of the interacting many-body system. The Matsubara Green's function can be written as a functional integral average

$$G(k) = -\beta \frac{\int \mathcal{D}\{\psi\} \mathcal{D}\{b\} \, {\rm e}^{-S\{\psi,b\}} \psi_k \psi_k^\dagger}{\int \mathcal{D}\{\psi\} \mathcal{D}\{b\} \, {\rm e}^{-S\{\psi,b\}}} \quad . \tag{8.19}$$

Evidently the b-integration in Eq.(8.19) is Gaussian, and can therefore be carried out *exactly*. After a straightforward integration we obtain the following exact expression for the interacting Green's function

$$G(k) = -\beta \frac{\int \mathcal{D}\{\psi\} \, {\rm e}^{-S_{\rm eff}\{\psi\}} \psi_k \psi_k^\dagger}{\int \mathcal{D}\{\psi\} \, {\rm e}^{-S_{\rm eff}\{\psi\}}} \quad , \tag{8.20}$$

with

$$S_{\rm eff}\{\psi\} = S_0\{\psi\} + S_{\rm int}^{\rm el}\{\psi\} - \beta \sum_q \left[\frac{g_q^2 \Omega_q}{\omega_m^2 + \Omega_q^2 + \Omega_q W_q} \right] \rho_{-q} \rho_q \, . \tag{8.21}$$

The last term is the effective interaction between the electrons mediated by the phonons. Combining the last two terms in Eq.(8.21) and using the above definitions of W_q and g_q, we finally arrive at

$$S_{\rm eff}\{\psi\} = S_0\{\psi\} + \frac{\beta}{2V} \sum_q f_q \rho_{-q} \rho_q \quad , \tag{8.22}$$

where the total effective interaction is given by

$$f_q = f_q^{\mathrm{cb}} \left[1 - \frac{f_q^{\mathrm{cb}} \frac{z^2 N q^2}{V M_{\mathrm{i}}}}{\omega_m^2 + \Omega_q^2 + f_q^{\mathrm{cb}} \frac{z^2 N q^2}{V M_{\mathrm{i}}}} \right] = f_q^{\mathrm{cb}} \frac{\omega_m^2 + \Omega_q^2}{\omega_m^2 + \Omega_q^2 + f_q^{\mathrm{cb}} \frac{z^2 N q^2}{V M_{\mathrm{i}}}} \ . \tag{8.23}$$

Defining the electron-phonon coupling constant γ via

$$\frac{z^2 N q^2}{V M_{\mathrm{i}}} \equiv \nu^2 \gamma^2 \Omega_q^2 \ , \tag{8.24}$$

where ν is the density of states, we see that Eq.(8.23) can also be written as

$$f_q = \frac{f_q^{\mathrm{cb}}}{1 + \nu^2 \gamma^2 f_q^{\mathrm{cb}} \frac{\Omega_q^2}{\omega_m^2 + \Omega_q^2}} \ . \tag{8.25}$$

It is instructive to compare Eq.(8.25) with the expression that would result from the "screening by hand" procedure described above. The approximation (a) amounts to the replacement

$$g_q^2 \Omega_q \to \frac{z^2 N q^2}{V M_{\mathrm{i}}} \left(\frac{4\pi e^2}{\kappa^2} \right)^2 \frac{1}{2V} \tag{8.26}$$

in Eq.(8.21), while (b) is equivalent with $W_q \to 0$. Using the fact that $\kappa^2 = 4\pi e^2 \nu$, it is easy to see that in this approximation the effective interaction f_q in Eq.(8.23) is replaced by

$$f_q \to f_q^{\mathrm{cb}} - \gamma^2 \frac{\Omega_q^2}{\omega_m^2 + \Omega_q^2} \ . \tag{8.27}$$

For consistency, we should also replace $4\pi e^2/q^2 \to 4\pi e^2/\kappa^2$ in the direct Coulomb interaction, which amounts to setting $f_q^{\mathrm{cb}} \to 1/\nu$ in the first term of Eq.(8.27). Evidently the phonon contribution in Eq.(8.27) can be obtained from an expansion of the exact result (8.25) *to first order* in γ^2 and the subsequent replacement $f_q^{cb} \to \frac{1}{\nu}$ in the phonon part. By performing these replacements one implicitly neglects the self-consistent renormalization of the phonon frequencies [1.7]. Therefore one should also replace in Eq.(8.27) $\Omega_q \to \tilde{\Omega}_q$, where the *renormalized* phonon frequencies $\tilde{\Omega}_q$ include by definition the effect of the electronic degrees of freedom on the phonon dynamics. In this way one arrives at the usual form of the electron-phonon interaction that is frequently used in the literature. Evidently in the conventional "screening by hand" approach the renormalization of the phonon dispersion due to interactions with the electrons remains unknown, and it is implicitly assumed that a *self-consistent* calculation would lead to an effective interaction of the form (8.27), with well-defined phonon modes. Note that the coupling to the electronic system will certainly lead to a finite damping of the phonon mode, which is not properly described by Eq.(8.27). In contrast, the effective interaction in Eq.(8.25) is an *exact* consequence of the microscopic model defined in Eqs.(8.1)–(8.3). In fact, as will be shown in Sect. 8.3, the phonon energy shift and damping *can be derived* from this expression!

8.2 The Debye-Waller factor

Given the effective frequency-dependent density-density interaction (8.25), it is now easy to obtain a non-perturbative expression for the single-particle Green's function, which is valid even if the system is not a Fermi liquid.

Because the phonons simply modify the effective density-density interaction, we can obtain a non-perturbative expression for the interacting Green's function by substituting the interaction (8.25) into our general bosonization formula for linearized energy dispersion given in Eqs.(5.31)–(5.33) and (5.37)–(5.39). Because the interaction f_q in Eq.(8.25) does not depend on the patch indices, the effective interaction $f_q^{\mathrm{RPA},\alpha}$ in Eqs.(5.32) and (5.33) is the usual RPA interaction, so that the Debye-Waller factor associated with patch α is given by

$$Q^\alpha(\boldsymbol{r},\tau) = \frac{1}{\beta V}\sum_q f_q^{\mathrm{RPA}} \frac{1-\cos(\boldsymbol{q}\cdot\boldsymbol{r}-\omega_m\tau)}{(\mathrm{i}\omega_m - \boldsymbol{v}^\alpha\cdot\boldsymbol{q})^2} \quad , \tag{8.28}$$

with

$$f_q^{\mathrm{RPA}} = \frac{f_q}{1+f_q\Pi_0(q)} = \frac{f_q^{\mathrm{cb}}}{1+f_q^{\mathrm{cb}}\Pi_{\mathrm{ph}}(q)} \quad , \tag{8.29}$$

where

$$\Pi_{\mathrm{ph}}(q) = \Pi_0(q) + \nu\tilde{\gamma}^2\frac{1}{1+\omega_m^2/\Omega_q^2} \tag{8.30}$$

is the *dressed inverse phonon propagator* [8.7]. Here $\tilde{\gamma}^2$ is the dimensionless measure for the strength of the electron-phonon coupling,

$$\tilde{\gamma}^2 = \nu\gamma^2 = \frac{z^2N}{VM_{\mathrm{i}}\nu c_{\mathrm{s}}^2} \quad . \tag{8.31}$$

Using Eq.(A.5), this reduces for a spherical three-dimensional Fermi surface to[2]

$$\tilde{\gamma}^2 = \frac{z}{3}\frac{m}{M_{\mathrm{i}}}\left(\frac{v_{\mathrm{F}}}{c_{\mathrm{s}}}\right)^2 \quad . \tag{8.32}$$

We conclude that the phonons simply give rise to an additive contribution to the non-interacting polarization. Assuming that the Fermi surface is spherically symmetric, we can also write

$$\Pi_{\mathrm{ph}}(q) = \nu g_{\mathrm{ph}}(\boldsymbol{q},\mathrm{i}\omega_m) \quad , \tag{8.33}$$

where the dimensionless function $g_{\mathrm{ph}}(\boldsymbol{q},\mathrm{i}\omega_m)$ is given by

[2] Note that the total number of conduction electrons is now zN, so that we should replace in Eq.(A.5) $N \to zN$.

$$g_{\rm ph}(\boldsymbol{q}, {\rm i}\omega_m) = g_3(\frac{{\rm i}\omega_m}{v_{\rm F}|\boldsymbol{q}|}) + \tilde{\gamma}^2 g_1(\frac{{\rm i}\omega_m}{\Omega_{\boldsymbol{q}}}) \quad , \tag{8.34}$$

and the functions $g_1({\rm i}y)$ and $g_3({\rm i}y)$ are defined in Eqs.(A.12) and (A.14). Note that the phonon part of Eq.(8.34) involves the dimensionless function $g_1({\rm i}y)$ that appears in the polarization of the one-dimensional electron gas, see Eq.(7.2). Of course, here the origin for this function is the coupling of the electron system to ***another well defined collective mode***, whereas in the chain-model it was essentially due to the ***shape of the Fermi surface***. However, the appearance of the one-dimensional polarization function in Eq.(8.34) suggests the possibility that a quasi-one-dimensional phonon dispersion $\Omega_{\boldsymbol{q}}$ might lead to Luttinger behavior even if the electron dispersion is three-dimensional. We shall confirm this expectation in Sect. 8.4.2.

Because all effects due to the phonons are contained in the function $g_{\rm ph}(\boldsymbol{q}, {\rm i}\omega_m)$, the general expressions for the various contributions to the Debye-Waller factor derived in Chap. 6.1 remain valid. We simply have to use the corresponding RPA dynamic structure factor,

$$S_{\rm RPA}(\boldsymbol{q}, \omega) = \frac{\nu}{\pi} {\rm Im} \left\{ \frac{g_{\rm ph}(\boldsymbol{q}, \omega + {\rm i}0^+)}{1 + \left(\frac{\kappa}{q}\right)^2 g_{\rm ph}(\boldsymbol{q}, \omega + {\rm i}0^+)} \right\} \quad . \tag{8.35}$$

In the following section we shall discuss the form of $S_{\rm RPA}(\boldsymbol{q}, \omega)$ in some detail.

8.3 Phonon energy shift and phonon damping

We show that the dynamic structure factor (8.35) contains the the self-consistent renormalization of the phonon dynamics due to the coupling to the electronic system.

The renormalization of the phonon spectrum due to the coupling to the electrons can be obtained from the phonon peak of $S_{\rm RPA}(\boldsymbol{q}, \omega)$. The qualitative behavior of the dynamic structure factor can be determined from simple physical considerations [1.7], and is shown in Fig. 8.1. In the absence of phonons, $S_{\rm RPA}(\boldsymbol{q}, \omega)$ consists of a sum of two terms, which are discussed in detail in Appendix A.2. The first term $S^{\rm col}_{\rm RPA}(\boldsymbol{q}, \omega)$ is a δ-function peak due to the collective plasmon mode. From Eq.(A.57) we see that for the Coulomb interaction in $d = 3$ the plasmon approaches at long wavelengths a finite value, the plasma frequency $\omega_{\rm pl} = v_{\rm F}\kappa/\sqrt{3}$. Within the RPA this mode is not damped, so that its contribution to the dynamic structure factor is

$$S^{\rm col}_{\rm RPA}(\boldsymbol{q}, \omega) = Z_{\boldsymbol{q}} \delta(\omega - \omega_{\rm pl}) \quad , \tag{8.36}$$

with the residue $Z_{\boldsymbol{q}}$ given in Eq.(A.60). For small $zm/M_{\rm i}$ this contribution is only weakly affected by phonons. This follows from the fact that at the plasma

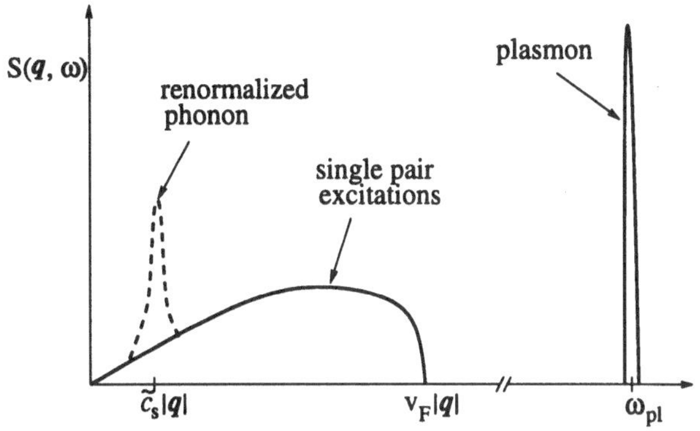

Fig. 8.1. A rough sketch of the various contributions to the RPA dynamic structure factor (8.35) in the regime where the phonon mode is well defined. Here $\tilde{c}_s$ is the renormalized phonon velocity, see Eq.(8.45).

frequency the ratio of the phonon to the electron contribution in Eq.(8.34) is for $|\boldsymbol{q}| \ll \kappa$ given by

$$\frac{\tilde{\gamma}^2 g_1(\frac{\omega_{pl}}{\Omega_q})}{g_3(\frac{\omega_{pl}}{v_F|\boldsymbol{q}|})} \approx 3\tilde{\gamma}^2 \left(\frac{\Omega_q}{v_F q}\right)^2 = 3\tilde{\gamma}^2 \left(\frac{c_s}{v_F}\right)^2 = z\frac{m}{M_i} \quad . \tag{8.37}$$

Evidently we may ignore the effect of the phonon on the plasmon mode provided

$$\tilde{\gamma}\frac{c_s}{v_F} = \sqrt{\frac{zm}{3M_i}} \ll 1 \quad . \tag{8.38}$$

Note that the validity of the Migdal theorem in a Fermi liquid is based on precisely this condition. In addition to the plasmon mode, the dynamic structure factor is non-zero in the regime $\omega \leq v_F|\boldsymbol{q}|$. In the absence of phonons, $S_{RPA}(\boldsymbol{q}, \omega)$ is here a rather featureless function, representing the decay of collective density fluctuations into particle hole pairs, i.e. Landau damping (see Eq.(A.27)). Mathematically, the Landau damping arises from the finite imaginary part of the function $g_3(\frac{\omega}{v_F|\boldsymbol{q}|} + i0^+)$ for $\omega < v_F|\boldsymbol{q}|$. As long as the *renormalized* phonon velocity is small compared with v_F and phonon damping is negligible, we expect that phonons give rise to an additional narrow peak that sticks out of the smooth background due to Landau damping. This is the renormalized phonon mode.

We now confirm this picture by explicitly calculating the approximate form of the dynamic structure factor in the vicinity of the phonon peak. To determine the *renormalized* phonon frequency, we look for solutions of the collective mode equation

$$1 + \left(\frac{\kappa}{q}\right)^2 g_{ph}(\boldsymbol{q}, z) = 0 \quad . \tag{8.39}$$

Anticipating that this equation has a solution with $|z| \ll v_F|\boldsymbol{q}|$, we may approximate the function $g_3(z)$ in Eq.(8.34) by the expansion of $g_3(x + i0^+)$ for small x, which is according to Eqs.(A.11) and (A.19) given by

$$g_3(x + \mathrm{i}0^+) \approx 1 + \mathrm{i}\frac{\pi}{2}x \quad . \tag{8.40}$$

Substituting this approximation for g_3 into Eq.(8.34) and using Eq.(A.15), we find the following cubic equation for the dressed phonon frequency,

$$z^2 - \Omega_q^2 \left[1 + \frac{\tilde{\gamma}^2}{1 + (\frac{q}{\kappa})^2}\right] + \mathrm{i}\frac{\pi}{2}\frac{1}{[1 + (\frac{q}{\kappa})^2]}\frac{z}{v_\mathrm{F}|\boldsymbol{q}|}\left[z^2 - \Omega_q^2\right] = 0 \quad . \tag{8.41}$$

If we ignore the damping term, this equation has a solution at $z = \tilde{\Omega}_q$, where the renormalized phonon frequency is

$$\tilde{\Omega}_q = \Omega_q \sqrt{1 + \frac{\tilde{\gamma}^2}{1 + \left(\frac{q}{\kappa}\right)^2}} \quad . \tag{8.42}$$

For $\tilde{\Omega}_q \ll v_\mathrm{F}|\boldsymbol{q}|$ the cubic term in Eq.(8.41) can be treated perturbatively. This term shifts the solution to $z = \tilde{\Omega}_q - \mathrm{i}\tilde{\Gamma}_q$, with the damping given by

$$\tilde{\Gamma}_q = \frac{\pi}{4}\frac{\Omega_q^2}{v_\mathrm{F}|\boldsymbol{q}|}\frac{\tilde{\gamma}^2}{[1 + (\frac{q}{\kappa})^2]^2} \quad . \tag{8.43}$$

Note that

$$\begin{aligned} \frac{\tilde{\Gamma}_q}{\tilde{\Omega}_q} &= \frac{\pi}{4}\frac{c_\mathrm{s}}{v_\mathrm{F}}\frac{\tilde{\gamma}^2}{[1 + (\frac{q}{\kappa})^2]^{\frac{3}{2}}[1 + \tilde{\gamma}^2 + (\frac{q}{\kappa})^2]^{\frac{1}{2}}} \\ &\approx \frac{\pi}{4}\frac{c_\mathrm{s}}{v_\mathrm{F}}\frac{\tilde{\gamma}^2}{\sqrt{1 + \tilde{\gamma}^2}} \quad , \quad \text{for } |\boldsymbol{q}| \ll \kappa \quad , \end{aligned} \tag{8.44}$$

so that the collective phonon mode is always well defined as long as the condition (8.38) is satisfied. Thus, in the regime $\tilde{\gamma} \ll v_\mathrm{F}/c_\mathrm{s}$ there is a well defined narrow peak with frequency $\tilde{\Omega}_q$ and width $\tilde{\Gamma}_q$ in the dynamic structure factor, which sticks out of the smooth background due to the particle hole continuum (see Fig. 8.1). This corresponds to the renormalized phonon mode. Using Eq.(8.42) we may define a wave-vector-dependent phonon velocity

$$\tilde{\Omega}_q = \tilde{c}_\mathrm{s}(\boldsymbol{q})|\boldsymbol{q}| \quad , \quad \tilde{c}_\mathrm{s}(\boldsymbol{q}) = c_\mathrm{s}\sqrt{1 + \frac{\tilde{\gamma}^2}{1 + \left(\frac{q}{\kappa}\right)^2}} \quad . \tag{8.45}$$

The renormalization of the phonon velocity is obviously a screening effect. At short length scales there is no screening charge around the phonon, so that it propagates with the bare velocity. At long wavelengths, however, the phonon has to drag along the screening cloud, so that its velocity is modified. For large $\tilde{\gamma}$ the renormalized phonon velocity reduces at long wavelengths to $\tilde{c}_\mathrm{s}(0) \approx c_\mathrm{s}\tilde{\gamma}$. For a spherical three-dimensional Fermi surface we may use Eq.(8.32) to rewrite this as

$$\tilde{c}_\mathrm{s}(0) \approx \sqrt{\frac{z}{3}\frac{m}{M_\mathrm{i}}}v_\mathrm{F} \quad . \tag{8.46}$$

This well-known result is called the Bohm-Staver relation [8.11, 8.12]. Note that the renormalized phonon velocity (8.46) is independent of the bare velocity c_s.

To calculate the dynamic structure factor in the vicinity of the phonon peak, we also need the height of the peak. Expanding the denominator in Eq.(8.35) around $\omega = \tilde{\Omega}_{\boldsymbol{q}}$, we obtain for the residue associated with the phonon peak

$$Z_{\boldsymbol{q}}^{\mathrm{ph}} = \frac{\nu}{(\frac{\kappa}{|\boldsymbol{q}|})^4 \frac{\partial}{\partial \omega} g_{\mathrm{ph}}(\boldsymbol{q}, \omega + \mathrm{i}0^+)\big|_{\omega=\tilde{\Omega}_{\boldsymbol{q}}}} = \frac{\nu}{2} \tilde{\Omega}_{\boldsymbol{q}} \left(\frac{|\boldsymbol{q}|}{\kappa} \right)^4 \frac{\tilde{\gamma}^2}{[1 + (\frac{q}{\kappa})^2][1 + \tilde{\gamma}^2 + (\frac{q}{\kappa})^2]} \quad . \tag{8.47}$$

Compared with the residue of the plasmon peak in Eq.(A.60), the phonon residue is at long wavelengths smaller by a factor of

$$\left(\frac{q}{\kappa} \right)^2 \frac{\Omega_{\boldsymbol{q}}}{\omega_{\mathrm{pl}}} \frac{\tilde{\gamma}^2}{\sqrt{1 + \tilde{\gamma}^2}} \quad . \tag{8.48}$$

Note that this is a small parameter even at $\boldsymbol{q}^2 \approx \kappa^2$ provided Eq.(8.38) is satisfied. In summary, for $\tilde{\gamma} \ll v_{\mathrm{F}} / c_s$ the total dynamic structure factor can be approximated by

$$S_{\mathrm{RPA}}(\boldsymbol{q}, \omega) = S_{\mathrm{RPA}}^{\mathrm{col}}(\boldsymbol{q}, \omega) + S_{\mathrm{RPA}}^{\mathrm{sp}}(\boldsymbol{q}, \omega) + \frac{Z_{\boldsymbol{q}}^{\mathrm{ph}}}{\pi} \frac{\tilde{\Gamma}_{\boldsymbol{q}}}{(\omega - \tilde{\Omega}_{\boldsymbol{q}})^2 + \tilde{\Gamma}_{\boldsymbol{q}}^2} \quad , \tag{8.49}$$

with $\tilde{\Omega}_{\boldsymbol{q}}$, $\tilde{\Gamma}_{\boldsymbol{q}}$ and $Z_{\boldsymbol{q}}^{\mathrm{ph}}$ given in Eqs.(8.42), (8.43) and (8.47). The plasmon contribution $S_{\mathrm{RPA}}^{\mathrm{col}}(\boldsymbol{q}, \omega)$ is given in Eq.(8.36), while the single pair contribution $S_{\mathrm{RPA}}^{\mathrm{sp}}(\boldsymbol{q}, \omega)$ is given in Eq.(A.27).

8.4 The quasi-particle residue

We now calculate the quasi-particle residue Z^α and determine the conditions under which Z^α becomes small or even vanishes.

According to Eqs.(5.86) and (6.14), the quasi-particle residue associated with patch P_Λ^α on the Fermi surface is $Z^\alpha = \mathrm{e}^{R^\alpha}$, where the constant part R^α of the Debye-Waller can be written as

$$R^\alpha = - \int \frac{\mathrm{d}\boldsymbol{q}}{(2\pi)^3} (f_{\boldsymbol{q}}^{\mathrm{cb}})^2 \int_0^\infty \mathrm{d}\omega \frac{S_{\mathrm{RPA}}(\boldsymbol{q}, \omega)}{(\omega + |\boldsymbol{v}^\alpha \cdot \boldsymbol{q}|)^2} \quad , \tag{8.50}$$

with $S_{\mathrm{RPA}}(\boldsymbol{q}, \omega)$ given in Eq.(8.35). We would like to emphasize that this expression is valid for arbitrary strength of the electron-phonon interaction. In particular, it is valid for $\tilde{\gamma} \gtrsim v_{\mathrm{F}} / c_s$, where the phonon mixes with the plasmon and the decomposition (8.49) of the dynamic structure factor is not valid. In this case we should use Eq.(8.35). It is not difficult to see that the integral exists for arbitrary values of $\tilde{\gamma}$ provided neither the electron dispersion nor the phonon dispersion is one-dimensional. Therefore phonons that couple to electrons via long-range Coulomb forces cannot destabilize the Fermi liquid state.

In order to make progress analytically, we shall restrict ourselves from now on to the regime $\tilde{\gamma} \ll v_F/c_s$. Then the phonons can be considered as well defined collective modes, so that the dynamic structure factor can be approximated by Eq.(8.49). As shown in Chap. 6.2.3 (see Eq.(6.56)), the contribution of the first two terms in Eq.(8.49) to R^α can be written as $-(\frac{\kappa}{k_F})^2 \frac{\tilde{r}_3}{2}$, where the numerical constant $\tilde{r}_3$ is given in Eq.(6.57). Because by assumption $\tilde{\Gamma}_{\boldsymbol{q}} \ll \tilde{\Omega}_{\boldsymbol{q}}$, the last term in Eq.(8.49) acts under the integral in Eq.(8.50) like a δ-function, so that

$$R^\alpha = -\left(\frac{\kappa}{k_F}\right)^2 \frac{\tilde{r}_3}{2} + R^\alpha_{\mathrm{ph}} \quad , \tag{8.51}$$

with

$$R^\alpha_{\mathrm{ph}} = -\frac{\tilde{\gamma}^2}{2\nu} \int \frac{d\boldsymbol{q}}{(2\pi)^3} \frac{\tilde{\Omega}_{\boldsymbol{q}}}{\left[\tilde{\Omega}_{\boldsymbol{q}} + |\boldsymbol{v}^\alpha \cdot \boldsymbol{q}|\right]^2 \left[1 + (\frac{q}{\kappa})^2\right] \left[1 + \tilde{\gamma}^2 + (\frac{q}{\kappa})^2\right]} \ . \tag{8.52}$$

8.4.1 Isotropic phonon dispersion

Let us first evaluate Eq.(8.52) for the isotropic phonon dispersion $\Omega_{\boldsymbol{q}} = c_s|\boldsymbol{q}|$. Using Eq.(A.5) we obtain

$$R^\alpha_{\mathrm{ph}} = -\frac{\tilde{\gamma}^2}{2k_F^2} \frac{c_s}{v_F} \int \frac{d\boldsymbol{q}}{4\pi} \frac{|\boldsymbol{q}|}{\left[\frac{\tilde{c}_s(\boldsymbol{q})}{v_F}|\boldsymbol{q}| + |\hat{\boldsymbol{v}}^\alpha \cdot \boldsymbol{q}|\right]^2} \frac{1}{\left[1 + (\frac{q}{\kappa})^2\right]^{\frac{3}{2}} \left[1 + \tilde{\gamma}^2 + (\frac{q}{\kappa})^2\right]^{\frac{1}{2}}} \ . \tag{8.53}$$

Because according to Eq.(8.45) the renormalized phonon velocity $\tilde{c}_s(\boldsymbol{q})$ depends only on $|\boldsymbol{q}|$, the angular integration can now be done exactly. The relevant integral is just the function $h_3(x)$ given in Eq.(6.34). After a simple rescaling we obtain

$$R^\alpha_{\mathrm{ph}} = -\frac{\tilde{\gamma}^2}{4} \left(\frac{\kappa}{k_F}\right)^2 \int_0^\infty dx \frac{1}{\left[1 + \tilde{\gamma}^2 + x\right] \left[1 + x + \frac{c_s}{v_F}(1+x)^{\frac{1}{2}}(1 + \tilde{\gamma}^2 + x)^{\frac{1}{2}}\right]} \ . \tag{8.54}$$

Clearly, in the regime (8.38) we may ignore the term proportional to c_s/v_F in the denominator of Eq.(8.54). The integral is then elementary,

$$\int_0^\infty dx \frac{1}{[1 + \tilde{\gamma}^2 + x][1 + x]} = \frac{1}{\tilde{\gamma}^2} \ln(1 + \tilde{\gamma}^2) \quad , \tag{8.55}$$

so that we finally obtain

$$R^{\alpha}_{\rm ph} = -\frac{1}{4}\left(\frac{\kappa}{k_{\rm F}}\right)^2 \ln(1+\tilde{\gamma}^2) \quad . \tag{8.56}$$

Note that the small parameter $c_{\rm s}/v_{\rm F}$ has disappeared in the prefactor, so that the final result depends only on the dimensionless strength of the electron-phonon coupling $\tilde{\gamma}^2$. Combining Eqs.(8.56) and (8.51), and using the fact that $(\kappa/k_{\rm F})^2 = 2e^2/(\pi v_{\rm F})$ (see Appendix A.3.1), we obtain

$$R^{\alpha} = -\frac{e^2}{\pi v_{\rm F}}\left[\tilde{r}_3 + \frac{1}{2}\ln(1+\tilde{\gamma}^2)\right] \quad . \tag{8.57}$$

In the regime $\kappa \ll k_{\rm F}$ where our bosonization approach is most accurate, the prefactor $e^2/(\pi v_{\rm F})$ in Eq.(8.57) is a small number, see Eq.(A.51). For weak electron-phonon coupling $\tilde{\gamma}^2$ we may expand $\ln(1+\tilde{\gamma}^2) \approx \tilde{\gamma}^2$. Because $\tilde{r}_3$ is a number of the order of unity, the phonon contribution to the quasi-particle residue is then negligible. On the other hand, for large $\tilde{\gamma}^2$ the phonon contribution is dominant. Exponentiating Eq.(8.57) we obtain for the quasi-particle residue

$$Z^{\alpha} = \left[\frac{{\rm e}^{-\tilde{r}_3}}{\sqrt{1+\tilde{\gamma}^2}}\right]^{\frac{e^2}{\pi v_{\rm F}}} \quad . \tag{8.58}$$

If we take the high-density limit $v_{\rm F} \to \infty$ at fixed $\tilde{\gamma}$, the quasi-particle residue approaches unity. On the other hand, if we keep the density fixed but increase the electron-phonon coupling $\tilde{\gamma}$, we obtain

$$Z^{\alpha} = \left[\frac{{\rm e}^{-\tilde{r}_3}}{\tilde{\gamma}}\right]^{\frac{e^2}{\pi v_{\rm F}}} \quad , \quad 1 \ll \tilde{\gamma}^2 \ll \left(\frac{v_{\rm F}}{c_{\rm s}}\right)^2 \quad . \tag{8.59}$$

8.4.2 Quasi-one-dimensional electrons or phonons

It is straightforward to generalize our results for anisotropic systems. For example, for strictly one-dimensional electron dispersion the polarization in Eq.(8.30) is given by

$$\Pi_0(q) = \nu \frac{(v_{\rm F} q_x)^2}{\omega_m^2 + (v_{\rm F} q_x)^2} = \nu g_1\left(\frac{{\rm i}\omega_m}{v_{\rm F}|q_x|}\right) \quad . \tag{8.60}$$

In this case it is not difficult to show that Eq.(8.52) gives rise to Luttinger liquid behavior *even if the phonon dispersion is three-dimensional.*

Alternatively, we may couple one-dimensional phonons to three-dimensional electrons. Then we should set $\Omega_q = c_{\rm s}|q_x|$ in Eqs.(8.42) and (8.52), while choosing for $\Pi_0(q)$ the usual three-dimensional polarization. Let us examine this possibility more closely. From Eq.(8.52) we obtain in this case

$$R_{\mathrm{ph}}^{\alpha} = -\frac{\tilde{\gamma}^2}{2k_{\mathrm{F}}^2} \frac{c_{\mathrm{s}}}{v_{\mathrm{F}}} \int \frac{d\boldsymbol{q}}{4\pi} \frac{|q_x|}{\left[\frac{c_{\mathrm{s}}(\boldsymbol{q})}{v_{\mathrm{F}}}|q_x| + |\hat{\boldsymbol{v}}^{\alpha} \cdot \boldsymbol{q}|\right]^2 \left[1 + (\frac{\boldsymbol{q}}{\kappa})^2\right]^{\frac{3}{2}} \left[1 + \tilde{\gamma}^2 + (\frac{\boldsymbol{q}}{\kappa})^2\right]^{\frac{1}{2}}} . \tag{8.61}$$

The crucial observation is now that for $\hat{\boldsymbol{v}}^{\alpha} = \pm \boldsymbol{e}_x$ we have $|\hat{\boldsymbol{v}}^{\alpha} \cdot \boldsymbol{q}| = |q_x|$, so that the ***phase space for the q_x-integration is decoupled from the remaining phase space and the integral is logarithmically divergent***[3]. For all other directions $\hat{\boldsymbol{v}}^{\alpha} \neq \pm \boldsymbol{e}_x$, the phase space for the $\boldsymbol{q}$-integration is coupled, so that the logarithmic divergence is cut off and the quasi-particle residue is finite. Although for $\hat{\boldsymbol{v}}^{\alpha} = \pm \boldsymbol{e}_x$ the integral in Eq.(8.61) is logarithmically divergent, the total Debye-Waller factor $Q^x(r_x \boldsymbol{e}_x, \tau)$ is finite[4]. Because the divergence in R_{ph}^x is logarithmic, we expect Luttinger liquid behavior. To obtain the anomalous dimension, it is sufficient to calculate the leading logarithmic term in the large-distance expansion of $Q^x(r_x \boldsymbol{e}_x, 0)$. Introducing the dimensionless integration variable $\boldsymbol{p} = \boldsymbol{q}/\kappa$, we obtain from Eq.(8.61)

$$Q^x(r_x \boldsymbol{e}_x, 0) = -\frac{e^2}{2\pi^2 v_{\mathrm{F}}} \frac{c_{\mathrm{s}}}{v_{\mathrm{F}}} \tilde{\gamma}^2 \int_0^{\infty} \mathrm{d}p_x \frac{1 - \cos(p_x \kappa r_x)}{p_x} \int_{-\infty}^{\infty} \mathrm{d}p_y \int_{-\infty}^{\infty} \mathrm{d}p_z$$
$$\times \frac{1}{\left[1 + \frac{c_{\mathrm{s}}}{v_{\mathrm{F}}} \sqrt{1 + \frac{\tilde{\gamma}^2}{1+\boldsymbol{p}^2}}\right]^2 \left[1 + \boldsymbol{p}^2\right]^{\frac{3}{2}} \left[1 + \tilde{\gamma}^2 + \boldsymbol{p}^2\right]^{\frac{1}{2}}} . \tag{8.62}$$

In the regime (8.38) where the phonon mode is well defined we may again ignore the term proportional to $c_{\mathrm{s}}/v_{\mathrm{F}}$ in the first factor of the second line in Eq.(8.62). Furthermore, to extract the leading logarithmic term, we may set $p_x = 0$ in the second line of Eq.(8.62). The p_y- and p_z-integrations can then easily be performed in circular coordinates, so that we finally obtain

$$Q^x(r_x \boldsymbol{e}_x, 0) \sim -\gamma_{\mathrm{ph}} \ln(\kappa |r_x|) \quad , \quad \kappa |r_x| \to \infty \quad , \tag{8.63}$$

where the anomalous dimension is

$$\gamma_{\mathrm{ph}} = \frac{e^2}{\pi v_{\mathrm{F}}} \frac{c_{\mathrm{s}}}{v_{\mathrm{F}}} \left[\sqrt{1 + \tilde{\gamma}^2} - 1\right] . \tag{8.64}$$

Note that for weak electron-phonon coupling the anomalous dimension γ_{ph} is proportional to $\tilde{\gamma}^2$, while in the strong coupling limit it is of order $\tilde{\gamma}$. However, one should keep in mind that Eq.(8.64) has been derived for $\tilde{\gamma} \ll v_{\mathrm{F}}/c_{\mathrm{s}}$ (see Eq.(8.38)), so that in the regime of validity of Eq.(8.64) the anomalous dimension is always small compared with unity.

It is also interesting to calculate the quasi-particle residue in the vicinity of the Luttinger liquid points $\boldsymbol{k}^{\alpha} = \pm k_{\mathrm{F}} \boldsymbol{e}_x$ on the Fermi surface. A quantitative measure for the vicinity to these points is the parameter $\delta = 1 - |\hat{\boldsymbol{v}}^{\alpha} \cdot \boldsymbol{e}_x|$. Obviously $\delta = 0$ corresponds to the Luttinger liquid points, so that for small

---

[3] We have encountered precisely the same situation before in our analysis of metallic chains without interchain hopping, see Chap. 7.1.

[4] We use the label $\alpha = x$ for the patch with $\boldsymbol{k}^{\alpha} = k_{\mathrm{F}} \boldsymbol{e}_x$.

enough δ we should obtain a Fermi liquid with small quasi-particle residue. A simple calculation shows that for $0 < \delta \ll c_s/v_F$ the constant part R^{α}_{ph} of the Debye-Waller factor is finite, and behaves as

$$R^{\alpha}_{\text{ph}} \sim -\gamma_{\text{ph}} \left[\ln \left(\frac{c_s}{v_F \delta} \right) + c + O(\frac{v_F \delta}{c_s}) \right] \quad , \quad \delta \ll \frac{c_s}{v_F} \quad , \tag{8.65}$$

where $c = O(1)$ is a numerical constant. Hence, for $\delta \to 0$ the quasi-particle residue vanishes as

$$Z^{\alpha}_{\text{ph}} \propto \left[\frac{v_F \delta}{c_s} \right]^{\gamma_{\text{ph}}} \quad , \quad \delta \ll \frac{c_s}{v_F} \quad . \tag{8.66}$$

Note that the exponent is given by the anomalous dimension of the Luttinger liquid that would exist for $\delta = 0$. Recall that the quasi-particle residue of weakly coupled chains discussed in Chap. 7.2 shows a very similar behavior. Obviously the parameter θ in Eq.(7.70), which measures the closeness of the coupled chain system to one-dimensionality, corresponds to $v_F \delta / c_s$ in Eq.(8.66). Both parameters are a dimensionless measure for the distance to the Luttinger liquid points in a suitably defined parameter space. From Eq.(8.65) it is also clear that in the present problem the vicinity to the Luttinger liquid points $\boldsymbol{k}^{\alpha} = \pm k_F \boldsymbol{e}_x$ becomes only apparent in the regime $\delta \ll c_s/v_F$. For $\delta \gtrsim c_s/v_F$ the correction term of order $v_F \delta / c_s$ in Eq.(8.65) cannot be ignored. In the extreme case $\delta = 1$ the integration in Eq.(8.61) gives rise to a factor

$$\int_0^{\kappa} \mathrm{d}q_y \frac{|q_x|}{\left[\frac{c_s}{v_F} |q_x| + q_y \right]^2} \propto \frac{v_F}{c_s} \quad , \tag{8.67}$$

so that outside a small neighborhood of the points $\boldsymbol{k}^{\alpha} = \pm k_F \boldsymbol{e}_x$ the prefactor of R^{α}_{ph} has the same order of magnitude as in the isotropic case, see Eq.(8.56).

8.5 Summary and outlook

In this chapter we have studied the Debye-model for electron-phonon interactions with the help of our non-perturbative bosonization approach. The Debye-model has been discussed and physically motivated in the classic textbook by Fetter and Walecka [1.6]. However, these authors did not treat the screening problem in a formally convincing manner (although the physical content of their "screening-by-hand" approach is correct). In Sect. 8.1 we have shown by means of functional integration that the screening of the Coulomb interaction in the Debye-model can be derived in a very simply way from first principles.

Higher-dimensional bosonization predicts that long-wavelength isotropic LA phonons that couple to the electrons via long-range Coulomb forces can never destabilize the Fermi liquid state in $d > 1$. On the other hand, anisotropy in the phonon dispersion can lead to small quasi-particle residues at corresponding patches of the Fermi surface, while the shape of the Fermi surface remains spherical. Of course, in realistic materials the phonon dispersion cannot be strictly one-dimensional on general grounds[5], but we know from Chap. 7.2 that the vicinity to the Luttinger liquid point in a suitably defined parameter space is sufficient to lead to characteristic Luttinger liquid features in the spectral function of a Fermi liquid. More generally, our calculation suggests that the coupling between electrons and ***any well defined quasi-one-dimensional collective mode can lead to Luttinger liquid behavior in three-dimensional Fermi systems.***

Finally, let us again point out some open research problems. So far we have explicitly evaluated the static Debye-Waller factor $Q^{\alpha}(\boldsymbol{r},0)$ in the regime $\tilde{\gamma} \ll v_{\mathrm{F}}/c_{\mathrm{s}}$ (see Eq.(8.38)) where phonons and plasmons involve different energy scales. Although we have convinced ourselves that the Fermi liquid remains stable in the strong coupling regime $\tilde{\gamma} \gtrsim v_{\mathrm{F}}/c_{\mathrm{s}}$ (where Migdal's theorem does not apply), the calculation of the **Debye-Waller factor for strong electron-phonon coupling** still remains to be done. Let us emphasize that our non-perturbative result for the Green's function is also valid in this case, but its explicit evaluation most likely requires considerable numerical work. An even more interesting (but also more difficult) problem is the evaluation of our non-perturbative result for the Green's function of electrons with non-linear energy dispersion given in Eqs.(5.181)–(5.187) for our coupled electron-phonon system.

Another direction for further research is based on the expectation that, at sufficiently low temperatures, the retarded interaction mediated by the phonons will drive the Fermi system into a **superconducting state**. As already mentioned in Chap. 5.4, with the help of a Hubbard-Stratonovich field that couples to the relevant order parameter [5.14] it should not be too difficult to incorporate superconductivity into our functional bosonization formalism. In this way our approach might offer a non-perturbative way to study superconducting symmetry breaking in correlated Fermi systems.

---

[5] Even at $T = 0$ a one-dimensional harmonic crystal is not stable. For example, the mean square displacement of a given site diverges logarithmically with the size of the system. At $T > 0$ the divergence is even linear. I would like to thank Roland Zeyher for pointing this out to me.

9. Fermions in a stochastic medium

We use our background field method to calculate the disorder averaged single-particle Green's function of fermions subject to a time-dependent random potential with long-range spatial correlations. We show that bosonization provides a microscopic basis for the description of the quantum dynamics of an interacting many-body system via an effective stochastic model with Gaussian probability distribution. In the limit of static disorder our method is equivalent with conventional perturbation theory based on the lowest order Born approximation. We also critically discuss the linearization of the energy dispersion, and give a simple example where this approximation leads to an unphysical result. Some of the calculations described in this chapter have been published in [9.1].

The complicated quantum dynamics of a many-body system of interacting electrons can sometimes by modeled by an effective non-interacting system that is coupled to a dynamic random potential with a suitably defined probability distribution [9.2]. Although the precise form of the probability distribution is in principle completely determined by the nature of the degrees of freedom that couple to the electrons (for example photons, phonons, or magnons), one usually has to rely on perturbation theory to characterize the random potential of the effective stochastic model. In this chapter we shall show that for random potentials with sufficiently long-range spatial correlations our bosonization approach allows us to relate the probability distribution of the effective stochastic model in a very direct and essentially non-perturbative way to the underlying many-body system.

The dynamic random potential could also be due to some non-equilibrium external forces. In this case the identification with an underlying many-body system is meaningless. The motion of a single isolated electron in an externally given time-dependent random potential has recently been discussed by many authors [9.3–9.9]. Here we would like to focus on the problem of calculating the average Green's function of electrons in the presence of a filled Fermi sea. We shall show that our functional integral formulation of higher-dimensional bosonization offers a new non-perturbative approach to this problem in arbitrary dimensions.

Although within the conventional operator approach this connection between bosonization and random systems seems rather surprising, it is obvious within our functional bosonization approach: In Chap. 5 the calculation of the Green's function of the interacting system has been mapped via a Hubbard-Stratonovich transformation onto the problem of calculating the average Green's function of an effective non-interacting system in a dynamic random potential $V^{\alpha}(\boldsymbol{r}, \tau)$, see Eqs.(5.1), (5.14), and (5.103). As shown in Chap. 5.1, for linearized energy dispersion and for sufficiently long-range potentials $V^{\alpha}(\boldsymbol{r}, \tau)$ it is possible to calculate the Green's function $\mathcal{G}^{\alpha}(\boldsymbol{r}, \boldsymbol{r}', \tau, \tau')$ for a given realization of the random potential without resorting to perturbation theory. The translationally invariant Green's function of the many-body system is then obtained by averaging $\mathcal{G}^{\alpha}(\boldsymbol{r}, \boldsymbol{r}', \tau, \tau')$ over all realizations of the random potential $V^{\alpha}(\boldsymbol{r}, \tau)$. Of course, in the interacting many-body system the probability distribution for this averaging is determined by the nature of the interaction and the kinetic energy (see Eqs.(3.34)–(3.37)), while in the stochastic model the probability distribution of the random potential has to be specified externally. However, in our calculation of the Green's function $\mathcal{G}^{\alpha}(\boldsymbol{r}, \boldsymbol{r}', \tau, \tau')$ for frozen random potential the nature of the probability distribution is irrelevant, so that the method described in Chap. 5 can be directly applied to disordered systems.

9.1 The average Green's function

We introduce a model of non-interacting fermions subject to a general dynamic random potential and derive a non-perturbative expression for the average Green's function by translating the results of Chap. 5 into the language of disordered systems.

9.1.1 Non-interacting disordered fermions

The Green's function $\mathcal{G}(\boldsymbol{r}, \boldsymbol{r}', \tau, \tau')$ of non-interacting fermions moving under the influence of an imaginary time random potential $U(\boldsymbol{r}, \tau)$ is defined via the usual equation

$$\left[-\partial_{\tau} - \frac{(-\mathrm{i}\nabla_{\boldsymbol{r}})^2}{2m} + \mu - U(\boldsymbol{r}, \tau) \right] \mathcal{G}(\boldsymbol{r}, \boldsymbol{r}', \tau, \tau') = \delta(\boldsymbol{r} - \boldsymbol{r}') \delta^{*}(\tau - \tau') \ . \quad (9.1)$$

We assume that the random potential has a Gaussian probability distribution with zero average and general covariance function $C(\boldsymbol{r} - \boldsymbol{r}', \tau - \tau')$, i.e.

$$\overline{U(\boldsymbol{r}, \tau)} = 0 \ , \quad (9.2)$$

$$\overline{U(\boldsymbol{r}, \tau) U(\boldsymbol{r}', \tau')} = C(\boldsymbol{r} - \boldsymbol{r}', \tau - \tau') \ , \quad (9.3)$$

where the over-bar denotes averaging over the probability distribution $\mathcal{P}\{U\}$ of the random potential U. Explicitly, the probability distribution is given by

$$\mathcal{P}\{U\} = \frac{\mathrm{e}^{-\frac{1}{2\beta V}\sum_q C_q^{-1} U_{-q} U_q}}{\int \mathcal{D}\{U\} \mathrm{e}^{-\frac{1}{2\beta V}\sum_q C_q^{-1} U_{-q} U_q}} \quad , \tag{9.4}$$

where the Fourier components of the random potential and the covariance function are

$$U_q = \int_0^\beta \mathrm{d}\tau \int \mathrm{d}\boldsymbol{r} \mathrm{e}^{-\mathrm{i}(\boldsymbol{q}\cdot\boldsymbol{r} - \omega_m \tau)} U(\boldsymbol{r}, \tau) \quad , \tag{9.5}$$

$$C_q = \int_0^\beta \mathrm{d}\tau \int \mathrm{d}\boldsymbol{r} \mathrm{e}^{-\mathrm{i}(\boldsymbol{q}\cdot\boldsymbol{r} - \omega_m \tau)} C(\boldsymbol{r}, \tau) \quad . \tag{9.6}$$

Hence,

$$\overline{U_q U_{-q}} \equiv \int \mathcal{D}\{U\} \mathcal{P}\{U\} U_q U_{-q} = \beta V C_q \quad . \tag{9.7}$$

All statistical properties of our model are contained in the covariance function $C_q = C_{\boldsymbol{q},\mathrm{i}\omega_m}$. If we would like to describe an underlying many-body system in thermal equilibrium [9.2], then it is (at least in principle) possible to continue the covariance function to real frequencies, so that the average real time dynamics corresponding to Eq.(9.1) can be obtained by analytic continuation. On the other hand, for an externally specified non-equilibrium potential $U(\boldsymbol{r}, \tau)$ there is in general no simple relation between real and imaginary time dynamics[1].

We are interested in the average Green's function

$$G(\boldsymbol{r} - \boldsymbol{r}', \tau - \tau') = \overline{\mathcal{G}(\boldsymbol{r}, \boldsymbol{r}', \tau, \tau')} \quad . \tag{9.8}$$

For an exact calculation of the average Green's function one should first solve the differential equation (9.1) for an arbitrary realization of the random potential, and then average the result with the probability distribution (9.4). Usually this an impossible task, so that one has to use some approximate method. A widely used perturbative approach, which works very well for *time-independent* random potentials, is based on the impurity diagram technique [1.3]. In the metallic regime it is often sufficient to calculate the self-energy in lowest order Born approximation. For static disorder the average Green's function is then found to vanish at distances large compared with the correlation range of the covariance function as [1.3, 9.10–9.12]

$$G(\boldsymbol{r} - \boldsymbol{r}', \tau - \tau') = G_0(\boldsymbol{r} - \boldsymbol{r}', \tau - \tau') \mathrm{e}^{-\frac{|\boldsymbol{r} - \boldsymbol{r}'|}{2\ell}} \quad , \tag{9.9}$$

where G_0 is the Green's function of the clean system, and the length ℓ is called the elastic mean free path. In Fourier space Eq.(9.9) becomes

$$G(k) = \frac{1}{\mathrm{i}\tilde{\omega}_n - (\frac{\boldsymbol{k}^2}{2m} - \mu) + \mathrm{sgn}(\tilde{\omega}_n)\frac{\mathrm{i}}{2\tau}} \quad , \tag{9.10}$$

[1] However, for some special cases the analytic continuation is certainly possible. For example, in Sect. 9.3.1 we shall discuss the Gaussian white noise limit, where C_q is a frequency-independent constant, so that the analytic continuation is trivial.

where $\tau = \ell/v_F$ is the elastic lifetime. The extra factor of $e^{-\frac{|r-r'|}{2\ell}}$ in Eq.(9.9) is nothing but the usual Debye-Waller factor that arises in the Gaussian averaging procedure. Below we shall show that this factor can also be obtained as a special case of the Debye-Waller that is generated via bosonization.

Within our bosonization approach the average Green's function is calculated in the most direct way: First we obtain the exact Green's function for a given realization of the random potential, and then this expression is averaged. As explained in detail in Chap. 5.1, our approach is most accurate if there exists a cutoff $q_c \ll k_F$ such that for $|q| \gtrsim q_c$ the Fourier components U_q of the random potential (and hence also the Fourier components C_q of the covariance function) become negligibly small. In other words, we should restrict ourselves to random potentials with sufficiently long-range spatial correlations. Evidently the most popular model of static δ-function correlated disorder does not fall into this category. This would correspond to $C_q = \gamma_0 \beta \delta_{\omega_m,0}$, where the parameter γ_0 is related to the elastic lifetime τ via $\gamma_0 = (2\pi\tau\nu)^{-1}$. However, in view of the fact that a random potential with a finite correlation range q_c^{-1} is expected to lead for distances $|r| \gg q_c^{-1}$ to qualitatively identical results for single-particle properties as a δ-function correlated random potential, the restriction to long-range correlations seems not to be very serious.

To model the disorder, we simply add the term

$$S_{\rm dis}\{\psi, U\} = \beta \sum_q \sum_\alpha U_{-q} \rho_q^\alpha \tag{9.11}$$

to the action (3.25) in our Grassmannian functional integral (3.24). Here ρ_q^α is the sector density defined in Eq.(3.5). The average Green's function can now be calculated by repeating the steps described in Chap. 5.1. For simplicity, in this chapter we shall work with linearized energy dispersion. In Sect. 9.4 we shall further comment on the accuracy of this approximation in the present context. Thus, after subdividing the Fermi surface into patches as described in Chap. 2.4, we linearize the energy dispersion locally and thus replace Eq.(9.1) by a *linear* partial differential equation for the sector Green's function $\mathcal{G}^\alpha(r, r', \tau, \tau')$ (see Eq.(5.14))

$$[-\partial_\tau + i v^\alpha \cdot \nabla_r - U(r,\tau)]\, \mathcal{G}^\alpha(r, r', \tau, \tau') = \delta(r - r')\delta^*(\tau - \tau') \quad . \tag{9.12}$$

As shown in Chap. 5.1.1, the exact solution of this linear differential equation is given by Schwinger's ansatz [5.1], and can be written as (see Eqs.(5.17), (5.22) and (5.23))

$$\mathcal{G}^\alpha(r, r', \tau, \tau') = G_0^\alpha(r - r', \tau - \tau') \times \exp\left[\frac{1}{\beta V}\sum_q U_q \frac{e^{i(q\cdot r - \omega_m \tau)} - e^{i(q\cdot r' - \omega_m \tau')}}{i\omega_m - v^\alpha \cdot q}\right] \quad . \tag{9.13}$$

The Gaussian average of this expression is now trivial and yields the usual Debye-Waller factor,

$$\overline{\mathcal{G}^{\alpha}(\boldsymbol{r},\boldsymbol{r}',\tau,\tau')} \equiv G^{\alpha}(\boldsymbol{r}-\boldsymbol{r}',\tau-\tau') = G_0^{\alpha}(\boldsymbol{r}-\boldsymbol{r}',\tau-\tau')e^{Q^{\alpha}_{\text{dis}}(\boldsymbol{r}-\boldsymbol{r}',\tau-\tau')} \quad , \tag{9.14}$$

with

$$Q^{\alpha}_{\text{dis}}(\boldsymbol{r}-\boldsymbol{r}',\tau-\tau') = -\frac{1}{2(\beta V)^2}\sum_{\boldsymbol{q}}\overline{U_{\boldsymbol{q}}U_{-\boldsymbol{q}}}\frac{\left|e^{i(\boldsymbol{q}\cdot\boldsymbol{r}-\omega_m\tau)}-e^{i(\boldsymbol{q}\cdot\boldsymbol{r}'-\omega_m\tau')}\right|^2}{(i\omega_m-\boldsymbol{v}^{\alpha}\cdot\boldsymbol{q})^2} = -\frac{1}{\beta V}\sum_{\boldsymbol{q}}C_{\boldsymbol{q}}\frac{1-\cos[\boldsymbol{q}\cdot(\boldsymbol{r}-\boldsymbol{r}')-\omega_m(\tau-\tau')]}{(i\omega_m-\boldsymbol{v}^{\alpha}\cdot\boldsymbol{q})^2} \quad . \tag{9.15}$$

The average Matsubara Green's function can then be written as (see Eqs.(5.37) (5.39))

$$G(k) = \sum_{\alpha}\Theta^{\alpha}(\boldsymbol{k})\int d\boldsymbol{r}\int_0^{\beta} d\tau e^{-i[(\boldsymbol{k}-\boldsymbol{k}^{\alpha})\cdot\boldsymbol{r}-\tilde{\omega}_n\tau]}G_0^{\alpha}(\boldsymbol{r},\tau)e^{Q^{\alpha}_{\text{dis}}(\boldsymbol{r},\tau)} \quad . \tag{9.16}$$

This completes the solution of the non-interacting problem.

9.1.2 Interacting disordered fermions

Disorder and interactions are treated on equal footing in our bosonization approach, so that it is easy to include electron-electron interactions into the above calculation. Eq.(9.12) should then be replaced by

$$[-\partial_{\tau} + i\boldsymbol{v}^{\alpha}\cdot\nabla_{\boldsymbol{r}} - U(\boldsymbol{r},\tau) - V^{\alpha}(\boldsymbol{r},\tau)]\,\mathcal{G}^{\alpha}(\boldsymbol{r},\boldsymbol{r}',\tau,\tau') = \delta(\boldsymbol{r}-\boldsymbol{r}')\delta^{*}(\tau-\tau') \quad , \tag{9.17}$$

where $V^{\alpha}(\boldsymbol{r},\tau)$ is the same Hubbard-Stratonovich field as in Eq.(5.14). The solution of this equation is again of the form (9.13), with $U_{\boldsymbol{q}}$ replaced by $U_{\boldsymbol{q}} + (\beta V)V^{\alpha}_{q}$, where V^{α}_{q} are the Fourier components[2] of $V^{\alpha}(\boldsymbol{r},\tau)$. Given the exact solution of Eq.(9.17), we obtain the translationally invariant average Green's function of the interacting many-body system by averaging over the disorder and over the Hubbard-Stratonovich field. Explicitly,

$$G^{\alpha}(\boldsymbol{r}-\boldsymbol{r}',\tau-\tau') = \int\mathcal{D}\{U\}\mathcal{P}\{U\}\int\mathcal{D}\{\phi^{\alpha}\}\mathcal{P}\{\phi^{\alpha},U\}\,\mathcal{G}^{\alpha}(\boldsymbol{r},\boldsymbol{r}',\tau,\tau') \quad , \tag{9.18}$$

with $\mathcal{P}\{U\}$ given in Eq.(9.4). The probability distribution $\mathcal{P}\{\phi^{\alpha},U\}$ has exactly the same form as in Eqs.(3.35)–(3.37), the only modification being that the elements of the infinite matrix $\hat{V}$ in Eq.(3.37) are now given by

[2] The additional factor of βV is due to the different normalizations of the Fourier transformations, compare Eqs.(5.13) and (9.5).

$$[\hat{V}]_{kk'} = \frac{\mathrm{i}}{\beta} \sum_{\alpha} \Theta^{\alpha}(\boldsymbol{k}) \left[\phi^{\alpha}_{\boldsymbol{k}-\boldsymbol{k}'} - \frac{\mathrm{i}}{V} U_{\boldsymbol{k}-\boldsymbol{k}'} \right] \quad . \tag{9.19}$$

Recall that according to Eq.(3.31) the Fourier components V^{α}_{q} of the potential $V^{\alpha}(\boldsymbol{r}, \tau)$ in Eq.(9.17) are related to the Fourier components ϕ^{α}_{q} of our Hubbard-Stratonovich field via $V^{\alpha}_{q} = \frac{\mathrm{i}}{\beta}\phi^{\alpha}_{q}$. For long-range random-potentials the closed loop theorem guarantees that the Gaussian approximation is very accurate, so that we may approximate

$$\mathcal{P}\{\phi^{\alpha}, U\} \approx \frac{\exp\left[-S_{\mathrm{eff},2}\{\phi^{\alpha}\} - \frac{\mathrm{i}}{V} \sum_{q\alpha\alpha'} [\tilde{\underline{\Pi}}_0(q)]^{\alpha\alpha'} \phi^{\alpha}_{-q} U_q \right]}{\int \mathcal{D}\{\phi^{\alpha}\} \exp\left[-S_{\mathrm{eff},2}\{\phi^{\alpha}\} - \frac{\mathrm{i}}{V} \sum_{q\alpha\alpha'} [\tilde{\underline{\Pi}}_0(q)]^{\alpha\alpha'} \phi^{\alpha}_{-q} U_q \right]} \quad , \tag{9.20}$$

where the Gaussian action $S_{\mathrm{eff},2}\{\phi^{\alpha}\}$ is given in Eq.(4.30), and the matrix elements $[\tilde{\underline{\Pi}}_0(q)]^{\alpha\alpha'}$ of the rescaled sector polarization are defined in Eq.(4.22). Note that by construction

$$\int \mathcal{D}\{\phi^{\alpha}\} \mathcal{P}\{\phi^{\alpha}, U\} = 1 \quad , \tag{9.21}$$

i.e. for any given realization of the random potential U the distribution $\mathcal{P}\{\phi^{\alpha}, U\}$ is properly normalized. Because the random potential U in Eq.(9.20) appears also in the denominator, it seems at the first sight that the averaging over $\mathcal{P}\{U\}$ in Eq.(9.18) cannot be directly performed, so that one has to use the replica approach [9.13]. Fortunately this is not the case, because we have the freedom of integrating first over the ϕ^{α}-field before averaging over the disorder. Then it is easy to see that the U-dependence of the denominator in Eq.(9.20) is exactly cancelled by a corresponding factor in the numerator, so that the averaging can be carried out exactly, *without resorting to the replica approach*. Thus, after performing the trivial Gaussian integrations we obtain for the average sector Green's function of the interacting many-body system

$$G^{\alpha}(\boldsymbol{r}, \tau) = G^{\alpha}_{0}(\boldsymbol{r}, \tau) \exp\left[Q^{\alpha}(\boldsymbol{r}, \tau) + \tilde{Q}^{\alpha}_{\mathrm{dis}}(\boldsymbol{r}, \tau) \right] \quad , \tag{9.22}$$

where the Debye-Waller factor $Q^{\alpha}(\boldsymbol{r}, \tau)$ due to the interactions is given in Eqs.(5.31)–(5.33), and the modified Debye-Waller factor $\tilde{Q}^{\alpha}_{\mathrm{dis}}(\boldsymbol{r}, \tau)$ due to disorder is obtained from $Q^{\alpha}_{\mathrm{dis}}(\boldsymbol{r}, \tau)$ in Eq.(9.15) by replacing the bare covariance function C_q by the *screened* covariance function

$$C^{\mathrm{RPA}}_{q} = \frac{C_q}{\left[1 + \Pi_0(q) f_q\right]^2} \quad . \tag{9.23}$$

Diagrammatically this expression describes the screening of the impurity potential by the electron-electron interaction. The corresponding Feynman diagrams are shown in Fig. 9.1. Note that in Fourier space the screening correction in Fig. 9.1 is $-U_q \Pi_0(q) f^{\mathrm{RPA}}_{q}$, which should be added to the bare disorder potential U_q. Hence, the total screened disorder potential has the Fourier components

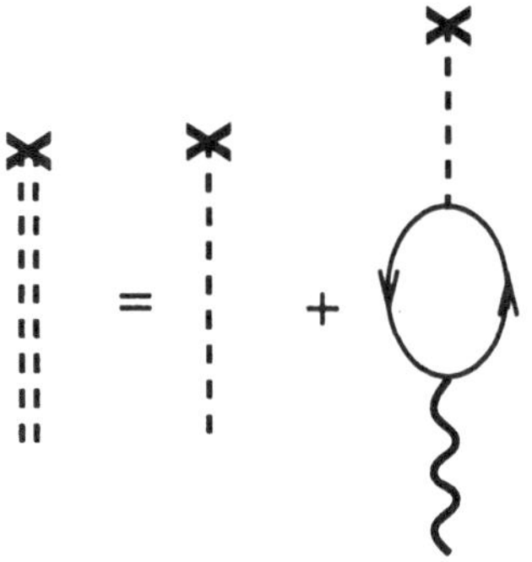

Fig. 9.1. Screening of the impurity potential. The bare impurity potential is denoted by a dashed line with a cross, and the thick wavy line represents the RPA interaction (see Fig. 4.3 (d)). The effective screened disorder potential U_q^{RPA} is denoted by a double dashed line with cross. In [9.15] we have discussed these diagrams in a different context.

$$U_q^{\mathrm{RPA}} = U_q - U_q \frac{\Pi_0(q) f_q}{1 + \Pi_0(q) f_q} = \frac{U_q}{1 + \Pi_0(q) f_q} \quad . \tag{9.24}$$

In $d = 1$ a result similar to Eq.(9.22) has also been obtained by Kleinert [9.13], and by Hu and Das Sarma [9.14]. However, Kleinert has obtained his result by combining functional bosonization [1.42] with the replica approach to treat the disorder averaging. As shown in this section, there is no need for introducing replicas if one integrates over the Hubbard-Stratonovich field *before* averaging over the disorder. In the expression derived by Hu and Das Sarma [9.14] the screening of the random potential is not explicitly taken into account.

9.2 Static disorder

We show that for static random potentials with sufficiently long-range correlations Eq.(9.16) agrees precisely with the usual perturbative result.

According to Eq.(9.5) the Fourier coefficients U_q of a time-independent random potential $U(\boldsymbol{r})$ are

$$U_q = \beta \delta_{\omega_m,0} U_{\boldsymbol{q}} \quad , \quad U_{\boldsymbol{q}} = \int \mathrm{d}\boldsymbol{r} \mathrm{e}^{-\mathrm{i}\boldsymbol{q}\cdot\boldsymbol{r}} U(\boldsymbol{r}) \quad . \tag{9.25}$$

For simplicity let us assume that the Fourier transform of the static correlator has a simple separable form[3],

$$C_q = \frac{\beta}{V} \overline{U_q U_{-q}} = \beta \delta_{\omega_m,0} \gamma_{\boldsymbol{q}} \quad , \quad \gamma_{\boldsymbol{q}} = \gamma_0 \mathrm{e}^{-|\boldsymbol{q}|_1 / q_c} \quad , \tag{9.26}$$

where $|\boldsymbol{q}|_1 = \sum_{i=1}^{d} |q_i|$. As discussed in Chap. 5.1.3, for linearized energy dispersion we may set $\boldsymbol{r} = r_{\parallel}^{\alpha} \hat{\boldsymbol{v}}^{\alpha}$ in the argument of the Debye-Waller factor,

---

[3] Any other cutoff function (for example $\mathrm{e}^{-\boldsymbol{q}^2/q_c^2}$) yields qualitatively identical results. Our choice leads to particularly simple integrals.

because the function $G_0^\alpha(\boldsymbol{r},\tau)$ is proportional to $\delta^{(d-1)}(\boldsymbol{r}_\perp^\alpha)$, see Eq.(5.48). Then we obtain from Eq.(9.15) for $V \to \infty$

$$Q_{\mathrm{dis}}^\alpha(r_\parallel^\alpha \hat{\boldsymbol{v}}^\alpha, \tau) = -\frac{\gamma_0}{|\boldsymbol{v}^\alpha|^2} \int \frac{\mathrm{d}\boldsymbol{q}}{(2\pi)^d} \mathrm{e}^{-|\boldsymbol{q}|_1/q_c} \frac{1-\cos(\hat{\boldsymbol{v}}^\alpha \cdot \boldsymbol{q} r_\parallel^\alpha)}{(\hat{\boldsymbol{v}}^\alpha \cdot \boldsymbol{q})^2} \quad . \tag{9.27}$$

Note that for a spherical Fermi surface $|\boldsymbol{v}^\alpha| = v_{\mathrm{F}}$ is independent of the patch index, but in general $\boldsymbol{v}^\alpha$ depends on α. The Debye-Waller factor is independent of τ because we have assumed a static random potential. For $|r_\parallel^\alpha q_c| \gg 1$ the integral in Eq.(9.27) is easily done and yields

$$Q_{\mathrm{dis}}^\alpha(r_\parallel^\alpha \hat{\boldsymbol{v}}^\alpha, \tau) \sim -\frac{|r_\parallel^\alpha|}{2\ell^\alpha} \quad , \quad |r_\parallel^\alpha q_c| \gg 1 \quad , \tag{9.28}$$

where the inverse elastic mean free path ℓ^α is given by

$$\frac{1}{\ell^\alpha} = \left(\frac{q_c}{\pi}\right)^{d-1} \frac{\gamma_0}{|\boldsymbol{v}^\alpha|^2} \quad . \tag{9.29}$$

We conclude that at large distances

$$G^\alpha(\boldsymbol{r},\tau) = G_0^\alpha(\boldsymbol{r},\tau) \exp\left[-\frac{|\hat{\boldsymbol{v}}^\alpha \cdot \boldsymbol{r}|}{2\ell^\alpha}\right] \quad . \tag{9.30}$$

The complete averaged real space Green's function is then according to Eq.(5.43) given by

$$G(\boldsymbol{r},\tau) = \sum_\alpha \mathrm{e}^{i\boldsymbol{k}^\alpha \cdot \boldsymbol{r}} G_0^\alpha(\boldsymbol{r},\tau) \exp\left[-\frac{|\hat{\boldsymbol{v}}^\alpha \cdot \boldsymbol{r}|}{2\ell^\alpha}\right] \quad . \tag{9.31}$$

From Eq.(9.28) it is evident that in $d = 1$ any finite static disorder destroys the Luttinger liquid features in the momentum distribution [9.14]. Recall that regular interactions in one-dimensional Fermi systems give rise to a contribution to the Debye-Waller factor that grows only logarithmically as $r_\parallel^\alpha \to \infty$, see Eq.(6.89). At sufficiently large distances this logarithmic divergence is completely negligible compared with the linear divergence due to disorder in Eq.(9.28). Note also that the linear growth of the Debye-Waller factor in Eq.(9.28) is independent of the dimensionality of the system, and implies that the momentum distribution $n_{\boldsymbol{k}^\alpha+\boldsymbol{q}}$ is for small $\boldsymbol{q}$ an analytic function of $\boldsymbol{q}$. Thus, any finite disorder washes out the singularities in the momentum distribution.

For a comparison with the usual perturbative result, let us also calculate the Fourier transform of Eq.(9.30). Shifting the coordinate origin to point $\boldsymbol{k}^\alpha$ on the Fermi surface by setting $\boldsymbol{k} = \boldsymbol{k}^\alpha + \boldsymbol{q}$, and choosing $|\boldsymbol{q}| \ll q_c$, it is easy to show that Eq.(9.30) implies for the averaged Matsubara Green's function

$$G(\boldsymbol{k}^\alpha + \boldsymbol{q}, \mathrm{i}\tilde{\omega}_n) = G^\alpha(\boldsymbol{q}, \mathrm{i}\tilde{\omega}_n) = \frac{1}{\mathrm{i}\tilde{\omega}_n - \boldsymbol{v}^\alpha \cdot \boldsymbol{q} + \mathrm{sgn}(\tilde{\omega}_n)\frac{\mathrm{i}}{2\tau^\alpha}} \quad , \tag{9.32}$$

where the inverse elastic lifetime associated with sector α is given by

$$\frac{1}{\tau^\alpha} = \frac{|\boldsymbol{v}^\alpha|}{\ell^\alpha} = \left(\frac{q_c}{\pi}\right)^{d-1} \frac{\gamma_0}{|\boldsymbol{v}^\alpha|} \quad . \tag{9.33}$$

Eqs.(9.32) and (9.33) agree with the usual perturbative result of the lowest order Born approximation for the average self-energy. The relevant diagram is shown in Fig. 9.2, and yields for the imaginary part of the self-energy

$$\begin{aligned} \mathrm{Im}\Sigma(\boldsymbol{k}) &= \frac{1}{V^2} \sum_{\boldsymbol{q}} \overline{U_{\boldsymbol{q}} U_{-\boldsymbol{q}}} \mathrm{Im}G(\boldsymbol{k}+\boldsymbol{q}, -\mathrm{i}0^+) \\ &= \frac{\gamma_0}{V} \sum_{\boldsymbol{q}} \mathrm{e}^{-|\boldsymbol{q}|_1/q_c} \mathrm{Im}G(\boldsymbol{k}+\boldsymbol{q}, -\mathrm{i}0^+) \quad . \end{aligned} \tag{9.34}$$

Because the random potential is static, the self-energy does not depend on the frequency. Note that for $q_c = \infty$, corresponding to a random potential with

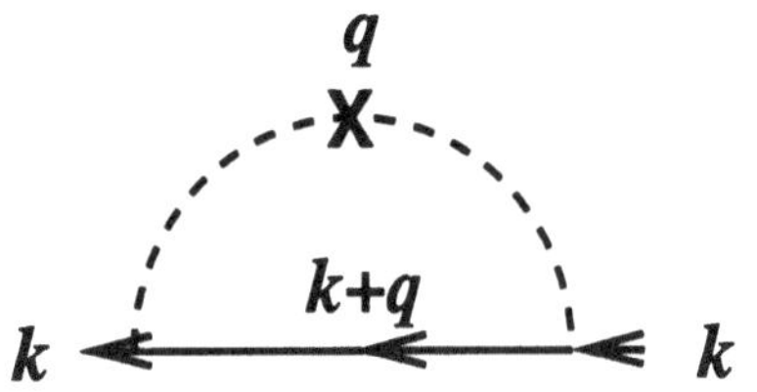

Fig. 9.2. Lowest order Born approximation for the average self-energy of non-interacting fermions in a static random potential. The dashed line with the cross denotes the average $\overline{U_{\boldsymbol{q}} U_{-\boldsymbol{q}}}$.

δ-function correlation in real space, we may shift $\boldsymbol{q}+\boldsymbol{k} \to \boldsymbol{q}$ in Eq.(9.34), so that the self-energy is independent of $\boldsymbol{k}$. Then Eq.(9.34) reduces to the usual result $\frac{1}{2\tau} = \mathrm{Im}\Sigma = \pi\gamma_0\nu$. As already mentioned, the approximations leading to Eq.(9.31) are not accurate in this case, because the correlator involves also large momentum transfers. On the other hand, for $q_c \ll k_F$ only wave-vectors $|\boldsymbol{q}| \ll k_F$ contribute in Eq.(9.34), so that we may linearize the energy dispersion. Then we obtain

$$\begin{aligned} \frac{1}{\tau^\alpha} &\equiv 2\mathrm{Im}\Sigma(\boldsymbol{k}^\alpha) = 2\pi\gamma_0 \int \frac{\mathrm{d}\boldsymbol{q}}{(2\pi)^d} \mathrm{e}^{-|\boldsymbol{q}|_1/q_c} \delta(\boldsymbol{v}^\alpha \cdot \boldsymbol{q}) \\ &= \frac{\gamma_0}{(2\pi)^{d-1}} (2q_c)^{d-1} \frac{1}{|\boldsymbol{v}^\alpha|} \quad , \end{aligned} \tag{9.35}$$

which agrees precisely with Eq.(9.33). We conclude that for static disorder with long-range correlations our bosonization approach reproduces the lowest order Born approximation for the elastic lifetime.

9.3 Dynamic disorder

We first derive a strikingly simple relation between interacting Fermi systems and effective stochastic models with time-dependent disorder. We then explicitly evaluate the average Green's function in some simple cases.

The case of a static random potential is not very exciting, because we have simply reproduced the perturbative result. New interesting physics emerges if we consider a general dynamic random potential. To calculate the average Green's function, we should specify the dynamic covariance function C_q in Eq.(9.7) and then evaluate the Debye-Waller factor (9.15). If we would like to describe with our stochastic model an underlying interacting many-body system in thermal equilibrium, then the form of C_q is determined by the nature of the interaction. In the case of the coupled electron-phonon system at high temperatures an explicit microscopic calculation of C_q has been given by Girvin and Mahan [9.2], who found that the disorder can be modeled by a white noise dynamic random potential, corresponding to a frequency-independent C_q. The identification of C_q with the parameters of the underlying many-body system given in [9.2] is based on a perturbative calculation of the self-energy at high temperatures.

In contrast, our functional bosonization approach allows us to relate the covariance function C_q of the random system at low temperatures in a direct and essentially non-perturbative way to the underlying many-body system. Evidently, the requirement that the average Green's function of the random system should be identical with the Green's function of the interacting many-body system without disorder is equivalent with the postulate that the corresponding Debye-Waller factors should be identical. Comparing then $Q^{\alpha}_{\rm dis}(\boldsymbol{r},\tau)$ in Eq.(9.15) with the Debye-Waller factor $Q^{\alpha}(\boldsymbol{r},\tau)$ due to a general density-density interaction given in Eqs.(5.31)–(5.33), we conclude that we should identify

$$C_q = -f_q^{\rm RPA} = -f_q + f_q^2 \int_0^\infty {\rm d}\omega S_{\rm RPA}(\boldsymbol{q},\omega)\frac{2\omega}{\omega^2+\omega_m^2} \quad , \tag{9.36}$$

where f_q is the bare interaction of the underlying many-body system, and we have used Eq.(6.3) to express $f_q^{\rm RPA}$ in terms of the dynamic structure factor. Eq.(9.36) is the link between the phenomenological stochastic model and the microscopic many-body system. In spite of its apparent simplicity, Eq.(9.36) is a highly non-trivial result, because it is based on a non-perturbative re-summation of the entire perturbation series of the many-body problem.

9.3.1 Gaussian white noise

Even if the random potential is determined by some non-equilibrium external forces, it is useful to decompose the covariance function C_q as in Eq.(9.36), because then we can simply use the results of Chap. 6.1 to evaluate the Debye-Waller factor. Let us first consider the case of Gaussian white noise random potential with covariance given by[4]

$$C_q = C_0 {\rm e}^{-|\boldsymbol{q}|_1/q_c} \quad . \tag{9.37}$$

[4] Note that the constant C_0 has units of volume × energy, just like the usual Landau interaction parameters.

Because a white noise random potential involves fluctuations on all energy scales with equal weight, the covariance function C_q is independent of the frequency. Comparing Eq.(9.37) with Eq.(9.36), it is clear that the corresponding Debye-Waller factor can be simply obtained from Eqs.(6.14), (6.16), and (6.17) by setting $f_q = -C_0 e^{-|q|_1/q_c}$ and $S_{\text{RPA}}(q, \omega) = 0$. From Eq.(6.14) it is then obvious that in this case the constant part of the Debye-Waller factor,

$$R^{\alpha}_{\text{dis}} = -\frac{1}{\beta V} \sum_{q} \frac{C_q}{(i\omega_m - v^{\alpha} \cdot q)^2} \quad , \tag{9.38}$$

vanishes for $\beta \to \infty$. This is in sharp contrast to the static random potential, where R^{α}_{dis} is divergent, see Eq.(9.27). For the space- and time-dependent contribution we obtain from Eqs.(6.16) and (6.17)

$$\text{Re}S^{\alpha}_{\text{dis}}(r^{\alpha}_{\parallel} \hat{v}^{\alpha}, \tau) = -C_0 \frac{|\tau|}{2V} \sum_{q} \cos(\hat{v}^{\alpha} \cdot q r^{\alpha}_{\parallel}) e^{-|q|_1/q_c} e^{-|v^{\alpha} \cdot q||\tau|} \quad , \tag{9.39}$$

$$\text{Im}S^{\alpha}_{\text{dis}}(r^{\alpha}_{\parallel} \hat{v}^{\alpha}, \tau) = -C_0 \frac{\tau}{2V} \sum_{q} \sin(|\hat{v}^{\alpha} \cdot q| r^{\alpha}_{\parallel}) e^{-|q|_1/q_c} e^{-|v^{\alpha} \cdot q||\tau|} \quad . \tag{9.40}$$

Note that in this limit only the term $L^{\alpha}_{q}(\tau)$ in Eqs.(6.16) and (6.17) survives. With the above simple form of C_q the q-integration is trivial. We obtain for the total Debye-Waller factor

$$Q^{\alpha}_{\text{dis}}(r^{\alpha}_{\parallel} \hat{v}^{\alpha}, \tau) = -S^{\alpha}_{\text{dis}}(r^{\alpha}_{\parallel} \hat{v}^{\alpha}, \tau) = \frac{iW\tau}{r^{\alpha}_{\parallel} + i|v^{\alpha}|\tau + i\,\text{sgn}(\tau) q_c^{-1}} \quad , \tag{9.41}$$

where we have defined

$$W = \frac{C_0}{2\pi} \left(\frac{q_c}{\pi} \right)^{d-1} \quad . \tag{9.42}$$

Note that W has units of velocity. We conclude that the average sector Green's function is given by

$$G^{\alpha}(r, \tau) = G^{\alpha}_{0}(r, \tau) \exp \left[\frac{iW\tau}{\hat{v}^{\alpha} \cdot r + i|v^{\alpha}|\tau + i\,\text{sgn}(\tau) q_c^{-1}} \right] \quad . \tag{9.43}$$

Because the Debye-Waller factor vanishes at $\tau = 0$, we have $G^{\alpha}(r, 0) = G^{\alpha}_{0}(r, 0)$, so that the white noise dynamic random potential does not affect the momentum distribution. Hence, the average momentum distribution exhibits the same jump discontinuity as in the absence of randomness. In contrast, a static random potential completely washes out any singularities in the average momentum distribution.

In Fourier space Eq.(9.43) looks rather peculiar. Let us first calculate the imaginary frequency Fourier transform,

$$G^{\alpha}(q, i\tilde{\omega}_n) = \int_0^{\beta} d\tau e^{i\tilde{\omega}_n \tau} G^{\alpha}(q, \tau) \quad , \tag{9.44}$$

where

$$G^{\alpha}(\boldsymbol{q},\tau) = \int \mathrm{d}\boldsymbol{r} \mathrm{e}^{-\mathrm{i}\boldsymbol{q}\cdot\boldsymbol{r}} G^{\alpha}(\boldsymbol{r},\tau)$$

$$= \frac{-\mathrm{i}}{2\pi} \int_{-\infty}^{\infty} \mathrm{d}x \frac{\mathrm{e}^{-\mathrm{i}q_{\parallel}^{\alpha} x}}{x + \mathrm{i}|\boldsymbol{v}^{\alpha}|\tau} \exp\left[\frac{\mathrm{i}W\tau}{x + \mathrm{i}|\boldsymbol{v}^{\alpha}|\tau + \mathrm{i}\,\mathrm{sgn}(\tau) q_{\mathrm{c}}^{-1}}\right] , \tag{9.45}$$

with $q_{\parallel}^{\alpha} = \hat{\boldsymbol{v}}^{\alpha} \cdot \boldsymbol{q}$. Because the argument of the exponential in the last factor of Eq.(9.45) is always finite, we may expand the exponential in an infinite series and exchange the order of integration and summation. For $\beta \to \infty$ the resulting integrals can then be done by means of contour integration. Assuming for simplicity $q_{\parallel}^{\alpha} \geq 0$ and $\tau > 0$, the relevant residue is

$$\mathrm{Res}\left[\frac{\mathrm{e}^{-\mathrm{i}q_{\parallel}^{\alpha} z}}{[z + \mathrm{i}|\boldsymbol{v}^{\alpha}|\tau][z + \mathrm{i}|\boldsymbol{v}^{\alpha}|\tau + \mathrm{i}q_{\mathrm{c}}^{-1}]^{n}}\right]_{z=-\mathrm{i}|\boldsymbol{v}^{\alpha}|\tau - \mathrm{i}q_{\mathrm{c}}^{-1}}$$

$$= -\mathrm{e}^{-|\boldsymbol{v}^{\alpha}| q_{\parallel}^{\alpha}\tau} \mathrm{e}^{-q_{\parallel}^{\alpha}/q_{\mathrm{c}}} \sum_{k=0}^{\infty}\sum_{m=0}^{\infty} \delta_{n,k+m+1} \frac{(-\mathrm{i}q_{\parallel}^{\alpha})^{k}(-\mathrm{i}q_{\mathrm{c}})^{m+1}}{k!} . \tag{9.46}$$

After some straightforward algebra we obtain

$$G^{\alpha}(\boldsymbol{q},\tau) = -\mathrm{e}^{-|\boldsymbol{v}^{\alpha}| q_{\parallel}^{\alpha}\tau} \mathrm{e}^{-q_{\parallel}^{\alpha}/q_{\mathrm{c}}} \sum_{n=0}^{\infty} \frac{(\frac{q_{\parallel}^{\alpha}}{q_{\mathrm{c}}})^{n}}{n!} \sum_{m=0}^{n} \frac{(Wq_{\mathrm{c}}\tau)^{m}}{m!} . \tag{9.47}$$

Substituting this expression into Eq.(9.44), the τ-integration is trivial, so that we obtain in the limit $\beta \to \infty$

$$G^{\alpha}(\boldsymbol{q},\mathrm{i}\omega) = -\frac{\mathrm{e}^{-q_{\parallel}^{\alpha}/q_{\mathrm{c}}}}{\boldsymbol{v}^{\alpha}\cdot\boldsymbol{q} - \mathrm{i}\omega} \sum_{n=0}^{\infty} \frac{(\frac{q_{\parallel}^{\alpha}}{q_{\mathrm{c}}})^{n}}{n!} \sum_{m=0}^{n} \left[\frac{Wq_{\mathrm{c}}}{\boldsymbol{v}^{\alpha}\cdot\boldsymbol{q} - \mathrm{i}\omega}\right]^{m} . \tag{9.48}$$

The summations are now elementary, and we finally obtain

$$G^{\alpha}(\boldsymbol{q},i\omega) = \frac{1}{Wq_{\mathrm{c}} + \mathrm{i}\omega - \boldsymbol{v}^{\alpha}\cdot\boldsymbol{q}} \times \left\{1 + \frac{Wq_{\mathrm{c}}\mathrm{e}^{-\hat{\boldsymbol{v}}^{\alpha}\cdot\boldsymbol{q}/q_{\mathrm{c}}}}{\mathrm{i}\omega - \boldsymbol{v}^{\alpha}\cdot\boldsymbol{q}} \exp\left[-W\frac{\hat{\boldsymbol{v}}^{\alpha}\cdot\boldsymbol{q}}{\mathrm{i}\omega - \boldsymbol{v}^{\alpha}\cdot\boldsymbol{q}}\right]\right\} . \tag{9.49}$$

Recall that we have assumed $\hat{\boldsymbol{v}}^{\alpha}\cdot\boldsymbol{q} \geq 0$. For $|\omega| \ll Wq_{\mathrm{c}}$ and $q_{\parallel}^{\alpha} \ll \min\{q_{\mathrm{c}}, Wq_{\mathrm{c}}/|\boldsymbol{v}^{\alpha}|\}$ this reduces to

$$G^{\alpha}(\boldsymbol{q},\mathrm{i}\omega) \sim \frac{1}{\mathrm{i}\omega - \boldsymbol{v}^{\alpha}\cdot\boldsymbol{q}} \exp\left[-W\frac{\hat{\boldsymbol{v}}^{\alpha}\cdot\boldsymbol{q}}{\mathrm{i}\omega - \boldsymbol{v}^{\alpha}\cdot\boldsymbol{q}}\right] . \tag{9.50}$$

If we now analytically continue this expression to real frequencies by replacing $\mathrm{i}\omega \to \omega + \mathrm{i}0^{+}$, we encounter an essential singularity at $\omega = \boldsymbol{v}^{\alpha}\cdot\boldsymbol{q}$. As will be explained in Sect. 9.4, we believe that this singularity is an artefact of the linearization of the energy dispersion.

9.3.2 Finite correlation time

A dynamic random potential with a finite correlation time can be modeled by the covariance function

$$C_q = Z_q \frac{2\Omega_q}{\omega_m^2 + \Omega_q^2} \quad , \tag{9.51}$$

with some residue Z_q and frequency Ω_q. In the time domain this implies for $\beta \to \infty$

$$C(\boldsymbol{q}, \tau) \equiv \frac{1}{\beta} \sum_m C_q \mathrm{e}^{-\mathrm{i}\omega_m \tau} = Z_q \mathrm{e}^{-\Omega_q |\tau|} \quad . \tag{9.52}$$

Note that we can rewrite Eq.(9.51) as

$$C_q = \int_0^\infty \mathrm{d}\omega S_{\mathrm{col}}(\boldsymbol{q}, \omega) \frac{2\omega}{\omega_m^2 + \omega^2} \quad , \tag{9.53}$$

with

$$S_{\mathrm{col}}(\boldsymbol{q}, \omega) = Z_q \delta(\omega - \Omega_q) \quad . \tag{9.54}$$

Comparison with Eq.(9.36) shows that the exponentially decaying imaginary time correlator in Eq.(9.52) corresponds to an undamped collective mode of an underlying many-body system. To calculate the Green's function, we simply compare Eqs.(9.36) and (9.53), and note that both expressions agree if we set $f_q \to 0$ and $f_q^2 S_{\mathrm{RPA}}(\boldsymbol{q}, \omega) \to S_{\mathrm{col}}(\boldsymbol{q}, \omega)$. Hence we can obtain the spectral representation of the Debye-Waller factor by making these replacements in Eqs.(6.14), (6.16) and (6.17). The constant part is given by

$$R^\alpha_{\mathrm{dis}} = - \int \frac{\mathrm{d}\boldsymbol{q}}{(2\pi)^d} \frac{Z_q}{(\Omega_q + |\boldsymbol{v}^\alpha \cdot \boldsymbol{q}|)^2} \quad . \tag{9.55}$$

In contrast to the Gaussian white noise random potential, the finite correlation time leads to a renormalization of the quasi-particle residue. Similarly, $S^\alpha_{\mathrm{dis}}(r^\alpha_\parallel \hat{\boldsymbol{v}}^\alpha, \tau)$ can be obtained by substituting Eq.(9.54) into Eqs.(6.16) and (6.17). For simplicity we shall assume here that the frequency $\Omega_q = \Omega_0$ is dispersionless and larger than all other energy scales in the problem, and choose $Z_q = Z_0 \mathrm{e}^{-|\boldsymbol{q}|_1 / q_c}$. Keeping the next-to-leading order in Ω_0^{-1} we obtain

$$Q^\alpha_{\mathrm{dis}}(r^\alpha_\parallel \hat{\boldsymbol{v}}^\alpha, \tau) \approx \frac{Z_0}{\pi |\boldsymbol{v}^\alpha| \Omega_0} \left(\frac{q_c}{\pi} \right)^{d-1} \left[\frac{\mathrm{i} |\boldsymbol{v}^\alpha| \tau}{r^\alpha_\parallel + \mathrm{i} |\boldsymbol{v}^\alpha| \tau + \mathrm{i} \,\mathrm{sgn}(\tau) q_c^{-1}} \right.$$
$$\left. + \frac{|\boldsymbol{v}^\alpha| q_c}{\Omega_0} \left(1 - \frac{\mathrm{e}^{-\Omega_0 |\tau|}}{1 + (r^\alpha_\parallel q_c)^2} \right) \right] \quad . \tag{9.56}$$

If we take the limit $\Omega_0 \to \infty$ while keeping Z_0 / Ω_0 constant, we recover Eq.(9.41), with $W = (Z_0 / (\pi \Omega_0)) (q_c / \pi)^{d-1}$. Because the leading term in Eq.(9.56) has the same structure as Eq.(9.41), the spectral function exhibits again an essential singularity at $\omega = \boldsymbol{v}^\alpha \cdot \boldsymbol{q}$. Therefore the essential singularity in Eq.(9.50) is *not* a special feature of the Gaussian white noise limit.

9.4 Summary and outlook

In this chapter we have used our background field method developed in Chap. 5 to calculate the average Green's function of electrons subject to a long-range random potential. For simplicity, we have worked with linearized energy dispersion. Although for static disorder we have correctly reproduced the usual perturbative result of the Born approximation, for time-dependent disorder we have obtained the rather peculiar expression (9.50) for the Fourier transform of the Green's function, which involves an essential singularity on resonance (i.e. for $\omega = \boldsymbol{v}^{\alpha} \cdot \boldsymbol{q}$). We believe that this singularity is an artefact of the linearization of the energy dispersion. This is based on the observation that in the white noise limit considered in Sect. 9.3.1 the long-distance behavior of the Debye-Waller factor is completely determined by the term $L_{\boldsymbol{q}}^{\alpha}(\tau)$ of Eqs.(6.16) and (6.17). As discussed in detail in Chap. 6.1.3, this term is generated by the *double pole* in the Debye-Waller factor for linearized energy dispersion. On the other hand, for non-linear energy dispersion this double pole is split into two separate poles, so that a term similar to $L_{\boldsymbol{q}}^{\alpha}(\tau)$ does not appear. Thus, an interesting open problem is the **evaluation of the Debye-Waller factor due to dynamic disorder for non-linear energy dispersion**. In this context we would also like mention that a numerical analysis [9.16] of the higher-dimensional bosonization result for the Green's function *with linearized energy dispersion* (see Eqs.(5.31)–(5.33) and (5.37)–(5.39)) indicates that also for generic density-density interactions there exists some kind of unphysical singularity in the spectral function close to the resonance $\omega = \boldsymbol{v}^{\alpha} \cdot \boldsymbol{q}$. We believe that this singularity has precisely the same origin as the singularity in Eq.(9.50), namely the double pole in the Debye-Waller factor for linearized energy dispersion.

Another interesting unsolved problem is the correct description of the **diffusive motion** of the electrons within the framework of higher-dimensional bosonization[5]. The signature of diffusion is known to manifest itself also in the low-energy behavior of the single-particle Green's function of an *interacting* disordered Fermi system. Evidently our result (9.22) for the average Green's function in the presence of electron-electron interactions does not contain interference terms describing the interplay between disorder and electron-electron interactions. Note that the perturbative calculation of the average Green's function for disordered electrons in the presence of electron-electron interactions leads to singular terms due to multiple impurity scattering. These appear even at the first order in the effective electron-electron interaction and involve the so-called *Diffuson* and *Cooperon* propagators [9.10–9.12]. While the Cooperon involves momentum transfers of the order of $2k_{\mathrm{F}}$, the Diffuson is most singular for small momentum transfers. Because our approach attempts

[5] The case of one dimension [9.17] is special, because, at least in the absence of interactions, the localization length of one-dimensional disordered fermions has the same order of magnitude as the elastic mean free path [9.11, 9.12]. Therefore the diffusive regime does not exist in $d = 1$.

to treat the complete forward scattering problem non-perturbatively, the Diffuson should not be neglected. In fact, it is well known that the Diffuson qualitatively modifies the effective screened interaction at long wavelengths [9.11]. Furthermore, the so-called g_1-contribution to the self-energy [9.12], which to lowest order in the electron-electron interaction involves two Diffuson propagators, can be viewed as an effective long-range interaction between the electrons. This interaction is generated by many successive impurity scatterings and is a consequence of the diffusive motion of the electrons in a disordered metal. Obviously, such a motion cannot be correctly described within the approximations inherent in higher-dimensional bosonization at the level of the Gaussian approximation. However, in Chap. 5.2 we have developed a general method for calculating the Green's function beyond the Gaussian approximation, which might lead to a new non-perturbative approach to the problem of electron-electron interactions in disordered Fermi systems.

10. Transverse gauge fields

We generalize our functional bosonization approach to the case of fermions that are coupled to transverse abelian gauge fields. This is perhaps the physically most important application of higher-dimensional bosonization, because transverse gauge fields appear in effective low-energy theories for strongly correlated electrons and quantum Hall systems. In this chapter we shall restrict ourselves to the formal development of the methods. An important physical application to the quantum Hall effect is given in the Letter [10.1]. For linearized energy dispersion we have discussed the gauge field problem in the work [10.2]. It turns out, however, that in physically relevant cases quantitatively correct results for the single-particle Green's function can only be obtained if one retains the quadratic terms in the expansion of the energy dispersion close to the Fermi surface.

As shown in the classic textbook by Feynman and Hibbs [1.45], the static Coulomb interaction $4\pi e^2/\boldsymbol{q}^2$ between electrons can be obtained by coupling the electronic density to the scalar potential ϕ of the Maxwell field and integrating in the functional integral over all complexions of ϕ. The transverse radiation field $\boldsymbol{A}$ is usually neglected in condensed matter, because the coupling between the current density and the transverse radiation field involves an extra factor of v_{F}/c. At metallic densities the Fermi velocity v_{F} is two orders of magnitude smaller than the velocity of light c, so that for all practical applications it is justified to ignore the radiation field. The leading correction to the static Coulomb interaction is an effective retarded interaction between paramagnetic current densities, mediated by the transverse radiation field. Within the RPA the propagator of the transverse radiation field is in Coulomb gauge and for frequencies $|\omega_m| \ll v_{\mathrm{F}}|\boldsymbol{q}|$ given by (see Eq.(10.106) below)

$$h_q^{\mathrm{RPA},\alpha} = -\frac{1}{\nu}\left(\frac{v_{\mathrm{F}}}{c}\right)^2 \frac{1-(\hat{\boldsymbol{k}}^{\alpha}\cdot\hat{\boldsymbol{q}})^2}{\left(\frac{\boldsymbol{q}}{\kappa}\right)^2 + \frac{\pi}{4}\left(\frac{v_{\mathrm{F}}}{c}\right)^2 \frac{|\omega_m|}{v_{\mathrm{F}}|\boldsymbol{q}|}} \quad . \tag{10.1}$$

Here ν is the density of states at the Fermi surface, and κ is the usual Thomas-Fermi wave-vector in three dimensions. In 1973 Holstein, Norton and Pincus [10.3] showed that the associated effective current-current interaction gives rise to logarithmic singularities in the perturbative expansion of the electronic self-energy, and concluded that the low-energy behavior of the single-particle

Green's function is not of the Fermi liquid type. However, they also showed that due to the smallness of the parameter v_F/c the deviations from conventional Fermi liquid behavior are beyond experimental resolution, so that they have little practical consequences. Later the behavior of the electrodynamic field in metals was studied in more detail by Reizer [10.4]. A nice pedagogical discussion of this problem can be found in the textbook by Tsvelik [10.5].

The recent excitement about the unusual normal-state properties of the high-temperature superconductors [1.24,10.6–10.9] as well as half-filled quantum Hall systems [10.10–10.12] has revived the interest in the problem of electrons coupled to gauge fields from a more general point of view. In theoretical models for these systems the transverse gauge field is not necessarily the Maxwell field, so that in principle the magnitude of the velocity associated with the gauge field can be comparable with the Fermi velocity, and the effective coupling constant can be of the order of unity. Moreover, the effective dimensionality is not necessarily $d = 3$. Thus, we are led to the general problem of fermions in d dimensions that are coupled to transverse gauge fields with RPA propagator given by

$$h_q^{\mathrm{RPA},\alpha} = -\frac{1}{\nu} \frac{1 - (\hat{\boldsymbol{k}}^{\alpha} \cdot \hat{\boldsymbol{q}})^2}{\left(\frac{|\boldsymbol{q}|}{q_c}\right)^{\eta} + \lambda_d \frac{|\omega_m|}{v_F |\boldsymbol{q}|}} \quad . \tag{10.2}$$

Here $\eta > 0$ is some exponent, λ_d is a numerical constant, and q_c is some characteristic momentum scale. We recover Eq.(10.1) by setting $\eta = 2$, $\lambda_3 = \pi/4$, and $q_c = v_F \kappa / c$. On the other hand, the gauge field propagator in the two-dimensional Maxwell-Chern-Simons action (which is believed to describe the low-energy physics of composite Fermions in the half-filled Landau level [10.12]) corresponds to the choice $\eta = 1$, $\lambda_2 = 1$, and $q_c = (2k_F)^2/\kappa$, where κ is in this case the Thomas-Fermi wave-vector in $d = 2$, see Eq.(A.50).

The low-energy behavior of the Green's function of fermions that are coupled to transverse gauge fields with propagator (10.2) has recently been studied with the help of a variety of non-perturbative techniques, such as renormalization group and scaling methods [10.13–10.16], a $1/N$-expansion [10.17], higher-dimensional bosonization [1.33, 10.1, 10.2, 10.18], a quantum Boltzmann equation [10.19], and other non-perturbative resummation schemes [1.51, 5.11, 10.20, 10.21]. According to Ioffe *et al.* [10.17] as well as Castellani and Di Castro [1.51], in the case of transverse gauge fields it is not allowed to locally linearize the energy dispersion (thus approximating the Fermi surface by a collection of flat patches) because the effective interaction mediated by the gauge field is dominated by momentum transfers parallel to the Fermi surface. In fact, the method used by Ioffe *et al.* [10.17] produces results that are in disagreement with the predictions of higher-dimensional bosonization with linearized energy dispersion [10.2,10.18]. Because the linearization of the energy dispersion is one of the main (and a priori uncontrolled) approximations inherent in earlier formulations of higher-dimensional bosonization [1.31–1.37], one might suspect that the linearization is at least partially responsible for this disagreement.

Let us give a simple argument why for the effective interaction mediated by transverse gauge fields the curvature of the Fermi surface might indeed be more important than in the case of the conventional density-density interactions discussed in Chap. 6. Consider a fermion with momentum $\boldsymbol{k} = \boldsymbol{k}^{\alpha} + \boldsymbol{q}$ such that $|\boldsymbol{q}| \ll k_{\mathrm{F}}$. From Eq. (10.2) we see that the typical momentum q_{ω} transfered by the gauge field in a low-energy process with energy ω is determined by $(q_{\omega}/q_{\mathrm{c}})^{\eta} = \lambda_d \omega / (v_{\mathrm{F}} q_{\omega})$, so that

$$q_{\omega} = \left(\frac{\lambda_d q_{\mathrm{c}}^d}{v_{\mathrm{F}}} \right)^{\frac{1}{1+\eta}} \omega^{\frac{1}{1+\eta}} \quad . \tag{10.3}$$

Because the factor $1 - (\hat{\boldsymbol{k}}^{\alpha} \cdot \hat{\boldsymbol{q}})^2$ in Eq.(10.2) is maximal for wave-vectors $\boldsymbol{q}$ that are perpendicular to $\boldsymbol{k}^{\alpha}$ (see Fig. 10.1), we conclude that the typical momentum transfer $q_{\perp}$ parallel to the Fermi surface is of the order of q_{ω}. On

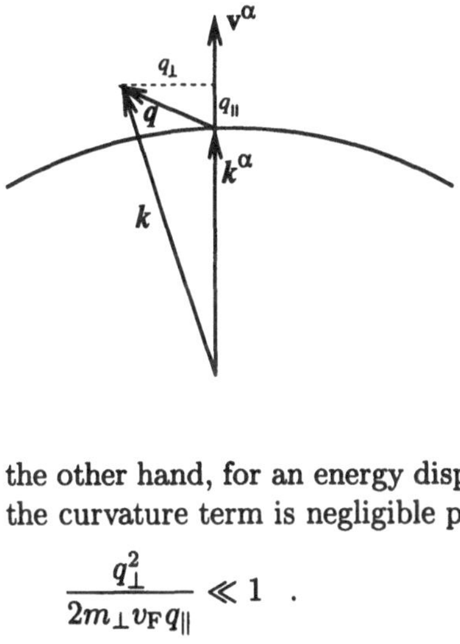

Fig. 10.1. Local coordinate system on the Fermi surface and definition of the components $q_{\parallel}$ and $q_{\perp}$ of $\boldsymbol{q} = \boldsymbol{k} - \boldsymbol{k}^{\alpha}$.

the other hand, for an energy dispersion of the form[1] $\xi_{\boldsymbol{q}}^{\alpha} = v_{\mathrm{F}} q_{\parallel} + q_{\perp}^2/(2m_{\perp})$, the curvature term is negligible provided

$$\frac{q_{\perp}^2}{2m_{\perp} v_{\mathrm{F}} q_{\parallel}} \ll 1 \quad . \tag{10.4}$$

Setting $q_{\perp} \approx q_{\omega}$ and using the fact that $v_{\mathrm{F}} q_{\parallel} \approx \omega$ close to the poles of the Green's function, we see that Eq.(10.4) reduces to

$$\frac{1}{2m_{\perp}} \left(\frac{\lambda_d q_{\mathrm{c}}^d}{v_{\mathrm{F}}} \right)^{\frac{2}{1+\eta}} \omega^{\frac{1-\eta}{1+\eta}} \ll 1 \quad . \tag{10.5}$$

For $\eta < 1$ this condition is always satisfied at sufficiently low energies, so that in this case curvature should be irrelevant. On the other hand, for $\eta > 1$

[1] To study curvature effects, we may omit the term quadratic in $q_{\parallel} = \boldsymbol{q} \cdot \hat{\boldsymbol{v}}^{\alpha}$, i.e. $q_{\parallel}^2/(2m_{\parallel})$. As discussed in Chaps. 5.2 and 7.2.2, this term does not describe the curvature of the Fermi surface and is irrelevant. For convenience we have also omitted the patch index on $q_{\parallel}^{\alpha}$ and $q_{\perp}^{\alpha}$.

the left-hand side becomes arbitrarily large for small ω, so that we expect that in the low-energy regime the curvature of the Fermi surface will become important. Of course, the above arguments are rather hand-waving, so that more rigorous methods are necessary to examine the role of curvature in the bosonization approach to the gauge field problem. Having developed a non-perturbative method to include curvature effects into higher-dimensional bosonization (see Chap. 5.2), we shall in this chapter examine the role of curvature by explicitly calculating the effect of the quadratic term in the energy dispersion on the gauge field contribution to the Debye-Waller factor.

10.1 Effective actions

We define a general field theory for non-relativistic electrons that are coupled to transverse abelian gauge fields. This theory contains the usual Maxwell action as a special case. We discuss in some detail the effective matter action that is obtained by integrating first over the gauge field, and the effective gauge field action that results from the integration over the matter degrees of freedom.

10.1.1 The coupled matter gauge field action

Transverse gauge fields can be viewed as Hubbard-Stratonovich fields that couple to the fermionic current density. We show how to obtain the propagators via functional integration and how to impose the Coulomb gauge constraint with the help of the Fadeev-Popov method.

Measuring wave-vectors with respect to local coordinate systems centered at the Fermi surface, the Euclidean Maxwell action [2.7] can be written as

$$S\{\psi, \phi^{\alpha}, \boldsymbol{A}^{\alpha}\} = S_0\{\psi\} + S_1\{\psi, \phi^{\alpha}, \boldsymbol{A}^{\alpha}\} + S_2\{\phi^{\alpha}, \boldsymbol{A}^{\alpha}\} \quad , \tag{10.6}$$

where the matter action $S_0\{\psi\}$ is defined in Eq.(3.3), and

$$S_1\{\psi, \phi^{\alpha}, \boldsymbol{A}^{\alpha}\} = \sum_{q} \sum_{\alpha} \left[\mathrm{i} \rho_q^{\alpha} \phi_{-q}^{\alpha} - \boldsymbol{j}_q^{\alpha} \cdot \boldsymbol{A}_{-q}^{\alpha} \right] \quad , \tag{10.7}$$

$$S_2\{\phi^{\alpha}, \boldsymbol{A}^{\alpha}\} = \frac{1}{2} \sum_{q} \sum_{\alpha\alpha'} \left[[\underline{\tilde{f}}_q^{-1}]^{\alpha\alpha'} \phi_{-q}^{\alpha} \phi_q^{\alpha'} + [\underline{\tilde{h}}_q^{-1}]^{\alpha\alpha'} \boldsymbol{A}_{-q}^{\alpha} \cdot \boldsymbol{A}_q^{\alpha'} \right] . \tag{10.8}$$

For convenience we have used the *Coulomb gauge*,

$$\boldsymbol{q} \cdot \boldsymbol{A}_q^{\alpha} = 0 \quad , \tag{10.9}$$

because then the longitudinal and transverse components of the gauge field $A_{\mu}^{\alpha} = [\phi^{\alpha}, \boldsymbol{A}^{\alpha}]$ in Eq.(10.8) are decoupled. The sector density ρ_q^{α} is defined in Eq.(3.5), and the gauge invariant sector current density $\boldsymbol{j}_q^{\alpha}$ has a para- and a diamagnetic contribution,

$$j_q^\alpha = j_q^{\mathrm{para},\alpha} + j_q^{\mathrm{dia},\alpha} \quad , \tag{10.10}$$

with

$$j_q^{\mathrm{para},\alpha} = \sum_k \Theta^\alpha(k) \frac{(\boldsymbol{k} + \boldsymbol{q}/2)}{mc} \psi_k^\dagger \psi_{k+q} \quad , \tag{10.11}$$

$$j_q^{\mathrm{dia},\alpha} = -\frac{1}{2mc^2\beta} \sum_{q'} \boldsymbol{A}_{q-q'}^\alpha \rho_{q'}^\alpha \quad . \tag{10.12}$$

For arbitrary matrices $\underline{\tilde{f}}_q$ and $\underline{\tilde{h}}_q$ in Eq.(10.8) the above action is more general than the usual Maxwell action. The latter can be obtained by choosing the matrix elements of $\underline{\tilde{f}}_q$ and $\underline{\tilde{h}}_q$ to be independently of the sector indices given by[2]

$$[\underline{\tilde{f}}_q]^{\alpha\alpha'} = \tilde{f}_q = \frac{\beta}{V} \frac{4\pi e^2}{q^2} \quad , \quad [\underline{\tilde{h}}_q]^{\alpha\alpha'} = \tilde{h}_q = \frac{\beta}{V} \frac{4\pi e^2}{q^2 + (\frac{\omega_m}{c})^2} \quad . \tag{10.13}$$

If we set $\boldsymbol{A}^\alpha = 0$ in Eq.(10.6), then $S\{\psi, \phi^\alpha, 0\}$ agrees precisely with the action given $S\{\psi, \phi^\alpha\}$ defined in Eq.(3.25), which has been obtained from the original density-density interaction by means of the Hubbard-Stratonovich transformation discussed in Chap. 3.2. Evidently the gauge field A_μ^α can be viewed as a generalized Hubbard-Stratonovich field which couples to the (gauge invariant) current density. The Maxwell-Chern-Simons action, which plays an important role in the theory of the half-filled Landau level [10.12], contains an additional term involving the coupling between the ϕ^α- and the $\boldsymbol{A}^\alpha$-field [10.18]. This coupling is due to the fact that in these theories density fluctuations are effectively mapped onto fluctuations of the gauge field. By a proper choice of the propagator of the ϕ^α field, one can therefore control the value of the exponent η that characterizes the dispersion of the gauge field propagator in Eq.(10.2). Because in this section we would like to discuss the basic concepts, we shall ignore at this point the Chern-Simons coupling between ϕ^α- and $\boldsymbol{A}^\alpha$-field. From the general structure of the final result for the single-particle Green's function the modifications arising from the Chern-Simons coupling will become obvious.

In complete analogy with Eq.(3.24), the exact single-particle Green's function is now given by

$$G(k) = -\beta \frac{\int \mathcal{D}\{\psi\}\mathcal{D}\{\phi^\alpha\}\mathcal{D}\{\boldsymbol{A}^\alpha\} \mathrm{e}^{-S\{\psi,\phi^\alpha,\boldsymbol{A}^\alpha\}} \psi_k \psi_k^\dagger}{\int \mathcal{D}\{\psi\}\mathcal{D}\{\phi^\alpha\}\mathcal{D}\{\boldsymbol{A}^\alpha\} \mathrm{e}^{-S\{\psi,\phi^\alpha,\boldsymbol{A}^\alpha\}}} \quad , \tag{10.14}$$

[2] Of course, matrices with all equal elements are not invertible, so that we should regularize $\underline{\tilde{f}}_q^{-1}$ and $\underline{\tilde{h}}_q^{-1}$ in some convenient way. As already mentioned in Chap. 3.2.1, our final results for physical quantities can be entirely expressed in terms of the original matrices $\underline{\tilde{f}}_q$ and $\underline{\tilde{h}}_q$, so that for our purpose it is sufficient to assume at intermediate stages that $\underline{\tilde{f}}_q^{-1}$ and $\underline{\tilde{h}}_q^{-1}$ have been properly regularized.

where the functional integration over the $\boldsymbol{A}^\alpha$-field is subject to the Coulomb gauge condition (10.9). Although in a gauge theory the single-particle propagator is in general gauge dependent [10.5], we expect physical quantities derived from it to be gauge invariant[3]. Moreover, as recently shown by Syljuåsen [10.23], for a particular class of gauge choices, the most singular part of the fermionic self-energy in non-relativistic quantum electrodynamics is independent of the gauge.

The matrices $\underline{\tilde{f}}_q$ and $\underline{\tilde{h}}_q$ determine the free propagator of the gauge field,

$$D^{\alpha\alpha'}_{0,\mu\nu}(q) = \frac{\int \mathcal{D}\{\phi^\alpha\}\mathcal{D}\{\boldsymbol{A}^\alpha\}\mathrm{e}^{-S_2\{\phi^\alpha,\boldsymbol{A}^\alpha\}} A^\alpha_{q,\mu} A^{\alpha'}_{-q,\nu}}{\int \mathcal{D}\{\phi^\alpha\}\mathcal{D}\{\boldsymbol{A}^\alpha\}\mathrm{e}^{-S_2\{\phi^\alpha,\boldsymbol{A}^\alpha\}}} \quad , \tag{10.15}$$

where we use the convention that $A^\alpha_{q,0} = \phi^\alpha_q$. *In Coulomb gauge* the longitudinal and transverse components do not mix, so that

$$D^{\alpha\alpha'}_{0,00}(q) = [\underline{\tilde{f}}_q]^{\alpha\alpha'} \quad , \tag{10.16}$$

$$D^{\alpha\alpha'}_{0,0i}(q) = D^{\alpha\alpha'}_{0,i0}(q) = 0 \quad , \quad i = 1,\ldots,d \quad , \tag{10.17}$$

$$D^{\alpha\alpha'}_{0,ij}(q) = [\underline{\tilde{h}}_q]^{\alpha\alpha'} \left[\delta_{ij} - (\boldsymbol{e}_i \cdot \hat{\boldsymbol{q}})(\boldsymbol{e}_j \cdot \hat{\boldsymbol{q}})\right] \quad , \quad i,j = 1,\ldots,d \quad , \tag{10.18}$$

where $\hat{\boldsymbol{q}} = \boldsymbol{q}/|\boldsymbol{q}|$. A simple way to derive Eq.(10.18) is to impose the Coulomb gauge condition (10.9) in the functional integral by means of the Fadeev-Popov method [2.4]. A very nice pedagogical discussion of this method can be found in the recent textbook by Sterman [2.18, page 190]. In the problem at hand, the Fadeev-Popov method amounts to inserting the following integral representation of the functional δ-function into the integrand of the denominator and the numerator of Eq.(10.15),

$$\prod_\alpha \delta\{\nabla \cdot \boldsymbol{A}^\alpha(\boldsymbol{r},\tau)\} = \int \mathcal{D}\{\lambda^\alpha\}\, \mathrm{e}^{-\sum_{q\alpha} \boldsymbol{A}^\alpha_{-q}\cdot \boldsymbol{q}\lambda^\alpha_q} \quad , \tag{10.19}$$

and then treating the $\boldsymbol{A}^\alpha$-integrations as unrestricted. The integration over the auxiliary fields λ^α_q, $\alpha = 1,\ldots,M$ enforces the Coulomb gauge condition (10.9) for each sector. Shifting

$$\boldsymbol{A}^\alpha_q \to \boldsymbol{A}^\alpha_q - \sum_{\alpha'} [\underline{\tilde{h}}_q]^{\alpha\alpha'} \boldsymbol{q}\lambda^{\alpha'}_q \quad , \tag{10.20}$$

and using $[\underline{\tilde{h}}_q]^{\alpha\alpha'} = [\underline{\tilde{h}}_{-q}]^{\alpha'\alpha}$ (see also Eq.(3.19)), we replace

$$\frac{1}{2}\sum_q \sum_{\alpha\alpha'} [\underline{\tilde{h}}_q^{-1}]^{\alpha\alpha'} \boldsymbol{A}^\alpha_{-q} \cdot \boldsymbol{A}^{\alpha'}_q + \sum_{q\alpha} \boldsymbol{A}^\alpha_{-q} \cdot \boldsymbol{q}\lambda^\alpha_q \to$$

$$\frac{1}{2}\sum_q \sum_{\alpha\alpha'} [\underline{\tilde{h}}_q^{-1}]^{\alpha\alpha'} \boldsymbol{A}^\alpha_{-q} \cdot \boldsymbol{A}^{\alpha'}_q + \frac{1}{2}\sum_q \sum_{\alpha\alpha'} \lambda^\alpha_{-q} \boldsymbol{q}^2 [\underline{\tilde{h}}_q]^{\alpha\alpha'} \lambda^{\alpha'}_q \quad . \tag{10.21}$$

---

[3] In particular, the imaginary part of the retarded Green's function can be directly related to the photoemission spectrum as long as certain standard approximations (which are discussed in detail in [10.22]) are assumed to be correct. Thus, we expect that $\mathrm{Im}G(\boldsymbol{k}, \omega + \mathrm{i}0^+)$ is to a large extent gauge invariant. I would like to thank C. Kübert and A. Muramatsu for pointing this out to me.

Using the fact that the integration measure is invariant with respect to shift transformations [2.5], we obtain

$$D^{\alpha\alpha'}_{0,ij}(q) = \left\{ \int \mathcal{D}\{\lambda^\alpha\} \mathcal{D}\{\boldsymbol{A}^\alpha\} \mathrm{e}^{-\tilde{S}_2\{\boldsymbol{A}^\alpha,\lambda^\alpha\}} \right\}^{-1}$$
$$\times \int \mathcal{D}\{\lambda^\alpha\} \mathcal{D}\{\boldsymbol{A}^\alpha\} \mathrm{e}^{-\tilde{S}_2\{\boldsymbol{A}^\alpha,\lambda^\alpha\}}$$
$$\times \left[A^\alpha_{q,i} A^{\alpha'}_{-q,j} - q_i q_j \sum_{\alpha_1\alpha_2} [\underline{\tilde{h}}_q]^{\alpha\alpha_1} \lambda^{\alpha_1}_q \lambda^{\alpha_2}_{-q} [\underline{\tilde{h}}_q]^{\alpha_2\alpha'} \right] , \qquad (10.22)$$

where

$$\tilde{S}_2\{\boldsymbol{A}^\alpha, \lambda^\alpha\} = \frac{1}{2} \sum_q \sum_{\alpha\alpha'} \left[[\underline{\tilde{h}}_q^{-1}]^{\alpha\alpha'} \boldsymbol{A}^\alpha_{-q} \cdot \boldsymbol{A}^{\alpha'}_q + \boldsymbol{q}^2 [\underline{\tilde{h}}_q]^{\alpha\alpha'} \lambda^\alpha_{-q} \lambda^{\alpha'}_q \right] . \qquad (10.23)$$

The unrestricted Gaussian integrations are now easily done, and we finally arrive at Eq.(10.18).

10.1.2 The effective matter action

... can be obtained by integrating first over the gauge field.

To see the connection with the conventional many-body approach more clearly, it is instructive to calculate the effective interaction between the matter degrees of freedom mediated by the gauge field. Performing in Eq.(10.14) the integration over the gauge field first, we obtain an expression of the same form as Eq.(3.6), with matter action $S_{\mathrm{mat}}\{\psi\} = S_0\{\psi\} + S_{\mathrm{int}}\{\psi\}$, where now

$$S_{\mathrm{int}}\{\psi\} = -\ln \left(\int \mathcal{D}\{\phi^\alpha\} \int \mathcal{D}\{\boldsymbol{A}^\alpha\} \mathrm{e}^{-S_1\{\psi,\phi^\alpha,\boldsymbol{A}^\alpha\} - S_2\{\phi^\alpha,\boldsymbol{A}^\alpha\}} \right) . \qquad (10.24)$$

In Coulomb gauge the integration over the longitudinal component ϕ^α is trivial, and corresponds just to undoing the Hubbard-Stratonovich transformation of Chap. 3.2. Hence,

$$S_{\mathrm{int}}\{\psi\} = \frac{1}{2} \sum_q \sum_{\alpha\alpha'} \tilde{f}^{\alpha\alpha'}_q \rho^\alpha_{-q} \rho^{\alpha'}_q + S^{\mathrm{rad}}_{\mathrm{int}}\{\psi\} , \qquad (10.25)$$

with

$$S^{\mathrm{rad}}_{\mathrm{int}}\{\psi\} = -\ln \left(\int \mathcal{D}\{\boldsymbol{A}^\alpha\} \mathrm{e}^{-S_3\{\psi,\boldsymbol{A}^\alpha\}} \right) , \qquad (10.26)$$

where the integration is subject to the constraint $\boldsymbol{q} \cdot \boldsymbol{A}^\alpha_q = 0$, and

$$S_3\{\psi, \boldsymbol{A}^\alpha\} = \frac{1}{2} \sum_q \sum_{\alpha\alpha'} [\underline{\tilde{h}}_q^{-1}]^{\alpha\alpha'} \boldsymbol{A}^\alpha_{-q} \cdot \boldsymbol{A}^{\alpha'}_q - \sum_q \sum_\alpha \boldsymbol{j}^{\mathrm{para},\alpha}_q \cdot \boldsymbol{A}^\alpha_{-q}$$
$$+ \frac{1}{2mc^2\beta} \sum_{qq'} \sum_\alpha \rho^\alpha_{q-q'} \boldsymbol{A}^\alpha_{-q} \cdot \boldsymbol{A}^\alpha_{q'} . \qquad (10.27)$$

Because the diamagnetic part of the current density gives rise to a term in Eq.(10.27) which is not diagonal in momentum space, the functional integration in Eq.(10.26) cannot be carried out exactly. The higher order diamagnetic contributions generate also ***retarded density-density interactions***, which should be combined with the static Coulomb interaction due to the longitudinal component of the gauge field. For the Maxwell field these corrections are of higher order in $e^2/c \approx \frac{1}{137}$, so that it is allowed to ignore them. Calculating the action $S_{\text{int}}^{\text{rad}}\{\psi\}$ perturbatively, we find to leading order that the transverse gauge field generates the following effective action for the matter degrees of freedom,

$$
\begin{aligned}
S_{\text{int}}^{\text{rad}}\{\psi\} &\approx -\frac{1}{2}\sum_{q}\sum_{\alpha\alpha'}\sum_{ij} j_{-q,i}^{\text{para},\alpha} D_{0,ij}^{\alpha\alpha'}(q) j_{q,j}^{\text{para},\alpha'} \\
&= -\frac{1}{2}\sum_{q}\sum_{\alpha\alpha'}\sum_{ij} j_{-q,i}^{\text{para},\alpha} [\underline{\tilde{h}}_q]^{\alpha\alpha'} \left[\delta_{ij} - (\hat{e}_i \cdot \hat{q})(\hat{e}_j \cdot \hat{q})\right] j_{q,j}^{\text{para},\alpha'} \ . \quad (10.28)
\end{aligned}
$$

Hence, the coupling between radiation field and matter gives rise to an effective interaction between the transverse parts of the paramagnetic current densities [10.24, 10.25]. For $d = 3$ we may use the vector product to rewrite the second line in Eq.(10.28) as

$$
S_{\text{int}}^{\text{rad}}\{\psi\} \approx -\frac{1}{2}\sum_{q}\sum_{\alpha\alpha'} [\underline{\tilde{h}}_q]^{\alpha\alpha'} (\hat{q} \times j_{-q}^{\text{para},\alpha}) \cdot (\hat{q} \times j_{q}^{\text{para},\alpha'}) \ . \quad (10.29)
$$

In a conventional many-body approach, one would now treat the effective two-body interactions in $S_{\text{int}}\{\psi\}$ perturbatively [10.4]. However, a priori such an expansion cannot be justified, because the interaction becomes arbitrary large for small wave-vectors and frequencies. In the case of the longitudinal component of the gauge field the physics of screening comes as a rescue. By performing an infinite resummation of a formally divergent series (which is of course nothing but the RPA for the effective density-density interaction [1.6, 2.9]), it is possible to formulate the perturbative expansion such that at high densities the effective expansion parameter is small. Unfortunately, this strategy fails in the case of the effective current-current interaction mediated by the transverse radiation field, because in the static limit[4] transverse gauge fields are not screened as long as the gauge invariance is not spontaneously broken. Therefore the conventional many-body approach fails as far as the perturbative calculation of the effect of $S_{\text{int}}^{\text{rad}}\{\psi\}$ on the single-particle Green's function is concerned. This has first been noticed by Holstein, Norton and Pincus [10.3], and has been discussed later in more detail by Reizer [10.4].

---

[4] Note that at *finite frequencies* transverse gauge fields are dynamically screened, see Eq.(10.103) below. For the Maxwell field this is called the *skin effect*.

10.1.3 The effective gauge field action

... can be obtained by integrating first over the matter field. This is what we need for our functional bosonization approach.

If we integrate in Eq.(10.14) first over the Grassmann fields, we obtain, in complete analogy with Eqs.(3.34)–(3.37),

$$G(k) = \int \mathcal{D}\{\phi^\alpha\}\mathcal{D}\{\boldsymbol{A}^\alpha\}\mathcal{P}\{\phi^\alpha, \boldsymbol{A}^\alpha\}[\hat{G}]_{kk} \equiv \left\langle [\hat{G}]_{kk} \right\rangle_{S_{\mathrm{eff}}} , \tag{10.30}$$

where the probability distribution is now

$$\mathcal{P}\{\phi^\alpha, \boldsymbol{A}^\alpha\} = \frac{\mathrm{e}^{-S_{\mathrm{eff}}\{\phi^\alpha, \boldsymbol{A}^\alpha\}}}{\int \mathcal{D}\{\phi^\alpha\}\mathcal{D}\{\boldsymbol{A}^\alpha\}\mathrm{e}^{-S_{\mathrm{eff}}\{\phi^\alpha, \boldsymbol{A}^\alpha\}}} , \tag{10.31}$$

with

$$S_{\mathrm{eff}}\{\phi^\alpha, \boldsymbol{A}^\alpha\} = S_2\{\phi^\alpha, \boldsymbol{A}^\alpha\} + S_{\mathrm{kin}}\{\phi^\alpha, \boldsymbol{A}^\alpha\} . \tag{10.32}$$

The potential energy part $S_2\{\phi^\alpha, \boldsymbol{A}^\alpha\}$ of the effective action is given in Eq.(10.8), and the kinetic energy contribution is

$$S_{\mathrm{kin}}\{\phi^\alpha, \boldsymbol{A}^\alpha\} = -\mathrm{Tr}\ln[1 - \hat{G}_0\hat{V}] . \tag{10.33}$$

The matrix elements of $\hat{V}$ are

$$[\hat{V}]_{kk'} = \sum_\alpha \Theta^\alpha(\boldsymbol{k}) V^\alpha_{k-k'} , \tag{10.34}$$

$$V^\alpha_q = \frac{1}{\beta}\left[\mathrm{i}\phi^\alpha_q - \boldsymbol{u}^\alpha \cdot \boldsymbol{A}^\alpha_q + \frac{1}{2mc^2\beta}\sum_{q''} \boldsymbol{A}^\alpha_{-q''} \cdot \boldsymbol{A}^\alpha_{q+q''} \right] . \tag{10.35}$$

Here $\boldsymbol{u}^\alpha = \boldsymbol{k}^\alpha/(mc)$ is a dimensionless vector with magnitude of the order of v_{F}/c. The infinite matrix $\hat{G}$ is defined as in Eq.(5.7), with V^α_q now given in Eq.(10.35). For the calculation of the kinetic energy contribution to the effective gauge field action we shall use the Gaussian approximation. Note that the generalized closed loop theorem discussed in Chap. 4.1 implies that, at least in certain parameter regimes (at high densities and at long wavelengths), the corrections to the Gaussian approximation are small. In this case the expansion of the logarithm in Eq.(10.33) can be truncated at the second order, so that we may approximate (see Eq.(4.2))

$$\begin{aligned} S_{\mathrm{kin}}\{\phi^\alpha, \boldsymbol{A}^\alpha\} &\approx \mathrm{Tr}\left[\hat{G}_0\hat{V}\right] + \frac{1}{2}\mathrm{Tr}\left[\hat{G}_0\hat{V}\right]^2 \\ &\equiv S_{\mathrm{kin},1}\{\phi^\alpha, \boldsymbol{A}^\alpha\} + S_{\mathrm{kin},2}\{\phi^\alpha, \boldsymbol{A}^\alpha\} . \end{aligned} \tag{10.36}$$

The first term yields

$$S_{\mathrm{kin},1}\{\phi^\alpha, \boldsymbol{A}^\alpha\} = \sum_\alpha N_0^\alpha V_0^\alpha$$

$$= \sum_\alpha N_0^\alpha \left[\mathrm{i}\phi_0^\alpha - \boldsymbol{u}^\alpha \cdot \boldsymbol{A}_0^\alpha + \frac{1}{2mc^2\beta} \sum_q \boldsymbol{A}_{-q}^\alpha \cdot \boldsymbol{A}_q^\alpha \right] \quad , \qquad (10.37)$$

where N_0^α is the number of occupied states in sector $K_{\Lambda,\lambda}^\alpha$, see Eq.(4.19). If we neglect the terms with the transverse gauge field, Eq.(10.37) reduces to $S_{\mathrm{kin},1}\{\phi^\alpha\}$, see Eq.(4.20). As already mentioned in the first footnote of Chap. 4, the terms involving ϕ_0^α and $\boldsymbol{A}_0^\alpha$ do not contribute to fermionic correlation functions at zero temperature, and can be ignored for our purpose. Note, however, that the last term in Eq.(10.37) is quadratic and has to be retained within the Gaussian approximation. This *diamagnetic* contribution to the effective gauge field action can be written as

$$S_{\mathrm{kin},1}^{\mathrm{dia}}\{\boldsymbol{A}^\alpha\} = \frac{1}{2} \sum_q \sum_\alpha \tilde{\Delta}^\alpha \boldsymbol{A}_{-q}^\alpha \cdot \boldsymbol{A}_q^\alpha \quad , \quad \tilde{\Delta}^\alpha = \frac{N_0^\alpha}{\beta mc^2} \quad . \qquad (10.38)$$

The second order term in Eq.(10.36) is

$$S_{\mathrm{kin},2}\{\phi^\alpha, \boldsymbol{A}^\alpha\} = -\frac{\beta^2}{2} \sum_q \sum_\alpha \tilde{\Pi}_0^\alpha(q) V_{-q}^\alpha V_q^\alpha \quad , \qquad (10.39)$$

where $\tilde{\Pi}_0^\alpha(q) = \frac{V}{\beta} \Pi_0^\alpha(q)$ is the dimensionless sector polarization[5]. From Eq.(10.35) it is clear that Eq.(10.39) contains also terms that are cubic and quartic in the fields. The origin for these non-Gaussian terms are the diamagnetic fluctuations described by the last term in Eq.(10.35). Within the Gaussian approximation we shall simply ignore these terms. It is important to stress, however, that the closed loop theorem does not imply the cancellation of these terms, because it applies to the total field V_q^α. Thus, within the Gaussian approximation we have

$$S_{\mathrm{kin},2}\{\phi^\alpha, \boldsymbol{A}^\alpha\} \approx S_{\mathrm{kin},2}\{\phi^\alpha\} + S_{\mathrm{kin},2}^{\mathrm{para}}\{\boldsymbol{A}^\alpha\} + S_{\mathrm{kin},2}^{\mathrm{mix}}\{\phi^\alpha, \boldsymbol{A}^\alpha\} \quad , \qquad (10.40)$$

where $S_{\mathrm{kin},2}\{\phi^\alpha\}$ is given in Eq.(4.23) (with $\tilde{\Pi}_0^{\alpha\alpha'}(q) \approx \delta^{\alpha\alpha'} \tilde{\Pi}_0^\alpha(q)$) and

$$S_{\mathrm{kin},2}^{\mathrm{para}}\{\boldsymbol{A}^\alpha\} = -\frac{1}{2} \sum_q \sum_\alpha \tilde{\Pi}_0^\alpha(q) (\boldsymbol{u}^\alpha \cdot \boldsymbol{A}_{-q}^\alpha)(\boldsymbol{u}^\alpha \cdot \boldsymbol{A}_q^\alpha) \quad , \qquad (10.41)$$

$$S_{\mathrm{kin},2}^{\mathrm{mix}}\{\phi^\alpha, \boldsymbol{A}^\alpha\} = \mathrm{i} \sum_q \sum_\alpha \tilde{\Pi}_0^\alpha(q) (\boldsymbol{u}^\alpha \cdot \boldsymbol{A}_{-q}^\alpha) \phi_q^\alpha \quad . \qquad (10.42)$$

Collecting all quadratic terms, we obtain for the effective gauge field action defined in Eq.(10.32) within the Gaussian approximation

$$S_{\mathrm{eff},2}\{\phi^\alpha, \boldsymbol{A}^\alpha\} = S_{\mathrm{eff},2}\{\phi^\alpha\} + S_{\mathrm{eff},2}\{\boldsymbol{A}^\alpha\} + S_{\mathrm{kin},2}^{\mathrm{mix}}\{\phi^\alpha, \boldsymbol{A}^\alpha\} \quad , \qquad (10.43)$$

[5] See Eqs.(4.22) and (4.24); for simplicity we have assumed sufficiently small $|\boldsymbol{q}|/k_{\mathrm{F}}$ and large sectors $K_{\Lambda,\lambda}^\alpha$, so that only the diagonal element of Eq.(4.22) has to be retained.

with

$$S_{\text{eff},2}\{\phi^\alpha\} = \frac{1}{2}\sum_q \sum_{\alpha\alpha'} \phi^\alpha_{-q}[(\underline{\tilde{f}}_q^{\text{RPA}})^{-1}]^{\alpha\alpha'} \phi^{\alpha'}_q \quad , \tag{10.44}$$

$$S_{\text{eff},2}\{\boldsymbol{A}^\alpha\} = \frac{1}{2}\sum_q \sum_{\alpha\alpha'} \sum_{ij} A^\alpha_{-q,i}[(\underline{\tilde{h}}_q^{\text{RPA}})^{-1}]^{\alpha\alpha'}_{ij} A^{\alpha'}_{q,j} \quad . \tag{10.45}$$

Here the matrix $\underline{\tilde{f}}_q^{\text{RPA}} = \frac{\beta}{V}\underline{f}_q^{\text{RPA}}$ is the rescaled RPA interaction matrix (see also Eqs.(4.31) and (4.33)),

$$[(\underline{\tilde{f}}_q^{\text{RPA}})^{-1}]^{\alpha\alpha'} = [\underline{\tilde{f}}_q^{-1}]^{\alpha\alpha'} + \delta^{\alpha\alpha'}\tilde{\Pi}_0^\alpha(q) \quad , \tag{10.46}$$

and $(\underline{\tilde{h}}_q^{\text{RPA}})^{-1}$ is the following matrix in the sector and coordinate labels,

$$[(\underline{\tilde{h}}_q^{\text{RPA}})^{-1}]^{\alpha\alpha'}_{ij} = \delta_{ij}[\underline{\tilde{h}}_q^{-1}]^{\alpha\alpha'} + \delta^{\alpha\alpha'}\left[\delta_{ij}\tilde{\Delta}^\alpha - u_i^\alpha u_j^\alpha \tilde{\Pi}_0^\alpha(q)\right] \quad , \tag{10.47}$$

where

$$u_i^\alpha = \boldsymbol{e}_i \cdot \boldsymbol{u}^\alpha \quad , \quad \boldsymbol{u}^\alpha = \frac{\boldsymbol{k}^\alpha}{mc} \quad . \tag{10.48}$$

The diamagnetic term $\tilde{\Delta}^\alpha$ in Eq.(10.47) represents the increase in energy due to diamagnetic fluctuations of the transverse gauge field, while the last term represents the lowering of the energy due to paramagnetism. In Sect. 10.3 we shall show that in the static limit there exists an exact cancellation between these two terms, so that the transverse gauge field is not screened. The action $S^{\text{mix}}_{\text{kin},2}\{\phi^\alpha, \boldsymbol{A}^\alpha\}$ describes the mixing between longitudinal and transverse components of the gauge field, which arises due to the presence of the matter degrees of freedom. Note that *in Coulomb gauge* the isolated gauge field action $S_2\{\phi^\alpha, \boldsymbol{A}^\alpha\}$ does not contain such a mixing term. In Sect. 10.3 we shall show that in the special case when the elements of the interaction matrices $\underline{\tilde{f}}_q$ and $\underline{\tilde{h}}_q$ are constants independent of the patch indices, this mixing term does not contribute to the final expression for the Green's function.

10.2 The Green's function in Gaussian approximation

Using our background field method described in Chap. 5, we derive a non-perturbative expression for the single-particle Green's function in Coulomb gauge. Gauge fixing is again imposed with the help of the Fadeev-Popov method. We use the Gaussian approximation, but work with non-linear energy dispersion.

10.2.1 The Green's function for fixed gauge field

For simplicity let us first consider the case of linearized energy dispersion, and then discuss the modifications due to the quadratic term in the energy dispersion.

For linearized energy dispersion we may copy the results of Chap. 5.1. To obtain the Green's function from Eq.(10.30), we first need to calculate the diagonal matrix elements $[\hat{G}]_{kk}$ for a fixed configuration of the gauge fields. Obviously this can be done in precisely the same way as described in Chap. 5.1.1; we simply should substitute the modified form (10.35) of the potential V_q^α into the expression for $\hat{G}^{-1}$ given in Eq.(5.7). Using Eqs.(5.10), (5.17) and (5.25), we obtain for the interacting Matsubara Green's function within the Gaussian approximation

$$G(k) = \sum_\alpha \Theta^\alpha(\boldsymbol{k}) \int \mathrm{d}\boldsymbol{r} \int_0^\beta \mathrm{d}\tau \mathrm{e}^{-\mathrm{i}[(\boldsymbol{k}-\boldsymbol{k}^\alpha)\cdot\boldsymbol{r}-\tilde{\omega}_n\tau]} \times G_0^\alpha(\boldsymbol{r},\tau) \left\langle \mathrm{e}^{\Phi^\alpha(\boldsymbol{r},\tau)-\Phi^\alpha(0,0)} \right\rangle_{S_{\mathrm{eff},2}} , \qquad (10.49)$$

where now, in complete analogy with Eq.(5.26),

$$\Phi^\alpha(\boldsymbol{r},\tau) - \Phi^\alpha(0,0) = \sum_q \mathcal{J}_{-q}^\alpha(\boldsymbol{r},\tau) \left[\phi_q^\alpha + \mathrm{i}\boldsymbol{u}^\alpha \cdot \boldsymbol{A}_q^\alpha - \frac{\mathrm{i}}{2mc^2\beta} \sum_{q''} \boldsymbol{A}_{-q''}^\alpha \cdot \boldsymbol{A}_{q+q''}^\alpha \right] , \qquad (10.50)$$

with $\mathcal{J}_q^\alpha(\boldsymbol{r},\tau)$ given in Eq.(5.27). The last term in Eq.(10.50) it is not diagonal in momentum space, and represents higher order diamagnetic fluctuations beyond the RPA. Because in our derivation of the effective gauge field action we have already ignored these higher order fluctuations, it is consistent to drop this term here as well.

From Chap. 5.2.1 we know that for fermions with *quadratic energy dispersion* Eq.(10.49) should be replaced by

$$G(k) = \sum_\alpha \Theta^\alpha(\boldsymbol{k}) \int \mathrm{d}\boldsymbol{r} \int_0^\beta \mathrm{d}\tau \mathrm{e}^{-\mathrm{i}[(\boldsymbol{k}-\boldsymbol{k}^\alpha)\cdot\boldsymbol{r}-\tilde{\omega}_n\tau]} \times \left\langle \mathcal{G}_1^\alpha(\boldsymbol{r},0,\tau,0) \mathrm{e}^{\Phi^\alpha(\boldsymbol{r},\tau)-\Phi^\alpha(0,0)} \right\rangle_{S_{\mathrm{eff},2}} , \qquad (10.51)$$

where the functional $\Phi^\alpha(\boldsymbol{r},\tau)$ satisfies the eikonal equation (5.108), and the Green's function $\mathcal{G}_1^\alpha(\boldsymbol{r},\boldsymbol{r}',\tau,\tau')$ is the solution of the differential equation (5.109). The potential $V^\alpha(\boldsymbol{r},\tau)$ in these expressions should now be identified with the Fourier transform of Eq.(10.35). Because in this chapter we would like to restrict ourselves to the Gaussian approximation, it is consistent to truncate the eikonal expansion (5.110) at the first order. In this case

$\Phi^{\alpha}(\boldsymbol{r},\tau)-\Phi^{\alpha}(0,0)$ is formally identical with Eq.(10.50), except that $\mathcal{J}^{\alpha}_{q}(\boldsymbol{r},\tau)$ is now defined in Eq.(5.123).

10.2.2 Gaussian averaging

We would like to emphasize again that we do not linearize the energy dispersion, because later we shall show that in the case of transverse gauge fields the curvature of the Fermi surface qualitatively changes the long-distance behavior of the Debye-Waller factor.

Let us begin with the calculation of the average eikonal $Q^{\alpha}(\boldsymbol{r},\tau)$. Within the Gaussian approximation $Q^{\alpha}(\boldsymbol{r},\tau)$ is given by (see Eq.(5.133))

$$e^{Q^{\alpha}(\boldsymbol{r},\tau)} = \left\langle e^{\Phi^{\alpha}(\boldsymbol{r},\tau)-\Phi^{\alpha}(0,0)}\right\rangle_{S_{\mathrm{eff},2}} \quad . \tag{10.52}$$

It is convenient to integrate first over the longitudinal field ϕ^{α} before averaging over the transverse components $\boldsymbol{A}^{\alpha}$ of the gauge field. Because of the coupling between the longitudinal and transverse fields in $S^{\mathrm{mix}}_{\mathrm{kin},2}\{\phi^{\alpha},\boldsymbol{A}^{\alpha}\}$, the integration over the ϕ^{α}-field generates also a contribution to the effective action for the transverse gauge fields. From Eq.(10.42) we have

$$\begin{aligned}
&\sum_{q}\mathcal{J}^{\alpha}_{-q}(\boldsymbol{r},\tau)\phi^{\alpha}_{q} - S^{\mathrm{mix}}_{\mathrm{kin},2}\{\phi^{\alpha},\boldsymbol{A}^{\alpha}\}\\
&=\sum_{q}\sum_{\alpha'}\left[\delta^{\alpha'\alpha}\mathcal{J}^{\alpha}_{-q}(\boldsymbol{r},\tau)-\mathrm{i}\tilde{\Pi}^{\alpha'}_{0}(q)\boldsymbol{u}^{\alpha'}\cdot\boldsymbol{A}^{\alpha'}_{-q}\right]\phi^{\alpha'}_{q} \quad ,
\end{aligned} \tag{10.53}$$

so that the ϕ^{α}-integration in Eq.(10.52) yields

$$\begin{aligned}
&\int\mathcal{D}\{\phi^{\alpha}\}\exp\left[-S_{\mathrm{eff},2}\{\phi^{\alpha}\}-S^{\mathrm{mix}}_{\mathrm{kin},2}\{\phi^{\alpha},\boldsymbol{A}^{\alpha}\}+\sum_{q}\mathcal{J}^{\alpha}_{-q}(\boldsymbol{r},\tau)\phi^{\alpha}_{q}\right] = \mathrm{const}\times\\
&\exp\left\{\frac{1}{2}\sum_{q}\sum_{\alpha'\alpha''}\left\langle\phi^{\alpha'}_{q}\phi^{\alpha''}_{-q}\right\rangle_{S_{\mathrm{eff},2}}\left[\delta^{\alpha'\alpha}\mathcal{J}^{\alpha}_{-q}(\boldsymbol{r},\tau)-\mathrm{i}\tilde{\Pi}^{\alpha'}_{0}(q)\boldsymbol{u}^{\alpha'}\cdot\boldsymbol{A}^{\alpha'}_{-q}\right]\right.\\
&\qquad\left.\times\left[\delta^{\alpha''\alpha}\mathcal{J}^{\alpha}_{q}(\boldsymbol{r},\tau)-\mathrm{i}\tilde{\Pi}^{\alpha''}_{0}(q)\boldsymbol{u}^{\alpha''}\cdot\boldsymbol{A}^{\alpha''}_{q}\right]\right\} \quad .
\end{aligned} \tag{10.54}$$

From Eq.(4.32) we know that the Gaussian propagator of the ϕ^{α}-field is simply given by the rescaled RPA interaction (see also Eq.(10.46)), so that

$$\begin{aligned}
&\left\langle e^{\Phi^{\alpha}(\boldsymbol{r},\tau)-\Phi^{\alpha}(0,0)}\right\rangle_{S_{\mathrm{eff},2}} = e^{Q^{\alpha}_{1}(\boldsymbol{r},\tau)}\\
&\times\frac{\int\mathcal{D}\{\boldsymbol{A}^{\alpha}\}\exp\left[-S'_{\mathrm{eff},2}\{\boldsymbol{A}^{\alpha}\}+\mathrm{i}\sum_{q\alpha' i}\mathcal{K}^{\alpha\alpha'}_{-q,i}(\boldsymbol{r},\tau)A^{\alpha'}_{q,i}\right]}{\int\mathcal{D}\{\boldsymbol{A}^{\alpha}\}\exp\left[-S'_{\mathrm{eff},2}\{\boldsymbol{A}^{\alpha}\}\right]} \quad ,
\end{aligned} \tag{10.55}$$

where the Debye-Waller factor $Q^{\alpha}_{1}(\boldsymbol{r},\tau)$ due to the longitudinal component of the gauge field is given in Eqs.(5.151)–(5.153), and

$$\mathcal{K}_{q,i}^{\alpha\alpha'}(\boldsymbol{r},\tau) = \mathcal{J}_q^{\alpha}(\boldsymbol{r},\tau) U_{q,i}^{\alpha\alpha'} \quad , \tag{10.56}$$

with

$$U_{q,i}^{\alpha\alpha'} = u_i^{\alpha}\delta^{\alpha\alpha'} - u_i^{\alpha'} \tilde{\Pi}_0^{\alpha'}(q)[\underline{\tilde{f}}_q^{\mathrm{RPA}}]^{\alpha\alpha'} \quad . \tag{10.57}$$

Note that the label α of $\mathcal{K}_{-q,i}^{\alpha\alpha'}(\boldsymbol{r},\tau)$ in Eq.(10.55) is an external label, and *not* a summation label. The renormalized Gaussian action $S'_{\mathrm{eff},2}\{\boldsymbol{A}^{\alpha}\}$ differs from the action $S_{\mathrm{eff},2}\{\boldsymbol{A}^{\alpha}\}$ given in Eq.(10.45) by an additional term that is generated because of the coupling between the ϕ^{α}- and $\boldsymbol{A}^{\alpha}$-fields in $S_{\mathrm{kin},2}^{\mathrm{mix}}\{\phi^{\alpha}, \boldsymbol{A}^{\alpha}\}$,

$$\begin{aligned} S'_{\mathrm{eff},2}\{\boldsymbol{A}^{\alpha}\} &= S_{\mathrm{eff},2}\{\boldsymbol{A}^{\alpha}\} \\ &\quad + \frac{1}{2}\sum_{q}\sum_{\alpha\alpha'} [\underline{\tilde{f}}_q^{\mathrm{RPA}}]^{\alpha\alpha'} \tilde{\Pi}_0^{\alpha}(q)\tilde{\Pi}_0^{\alpha'}(q)(\boldsymbol{u}^{\alpha}\cdot\boldsymbol{A}_{-q}^{\alpha})(\boldsymbol{u}^{\alpha'}\cdot\boldsymbol{A}_q^{\alpha'}) \\ &\equiv \frac{1}{2}\sum_{q}\sum_{\alpha\alpha'}\sum_{ij} [\underline{\mathrm{H}}_q^{-1}]_{ij}^{\alpha\alpha'} A_{-q,i}^{\alpha} A_{q,j}^{\alpha'} \quad , \end{aligned} \tag{10.58}$$

where we have defined

$$[\underline{\mathrm{H}}_q^{-1}]_{ij}^{\alpha\alpha'} = [\underline{\tilde{h}}_q^{\mathrm{RPA}-1}]_{ij}^{\alpha\alpha'} + u_i^{\alpha}u_j^{\alpha'}\tilde{\Pi}_0^{\alpha}(q)\tilde{\Pi}_0^{\alpha'}(q)[\underline{\tilde{f}}_q^{\mathrm{RPA}}]^{\alpha\alpha'} \quad . \tag{10.59}$$

Next, let us integrate over the transverse gauge field in Eq.(10.55). The Gaussian integration generates another Debye-Waller factor, so that

$$\left\langle \mathrm{e}^{\Phi^{\alpha}(\boldsymbol{r},\tau)-\Phi^{\alpha}(0,0)} \right\rangle_{S_{\mathrm{eff},2}} = \mathrm{e}^{Q_1^{\alpha}(\boldsymbol{r},\tau)}\mathrm{e}^{Q_{\mathrm{tr}}^{\alpha}(\boldsymbol{r},\tau)} \quad , \tag{10.60}$$

where

$$Q_{\mathrm{tr}}^{\alpha}(\boldsymbol{r},\tau) = -\frac{1}{2}\sum_{q}\sum_{\alpha'\alpha''}\sum_{ij} \left\langle A_{q,i}^{\alpha'} A_{-q,j}^{\alpha''} \right\rangle_{S'_{\mathrm{eff},2}} \mathcal{K}_{-q,i}^{\alpha\alpha'}(\boldsymbol{r},\tau)\mathcal{K}_{q,j}^{\alpha\alpha''}(\boldsymbol{r},\tau) \, , \tag{10.61}$$

with

$$\begin{aligned} \left\langle A_{q,i}^{\alpha} A_{-q,j}^{\alpha'} \right\rangle_{S'_{\mathrm{eff},2}} &\equiv [D^{\mathrm{RPA}}(q)]_{ij}^{\alpha\alpha'} \\ &= \frac{\int \mathcal{D}\{\boldsymbol{A}^{\alpha}\} \mathrm{e}^{-S'_{\mathrm{eff},2}\{\boldsymbol{A}^{\alpha}\}} A_{q,i}^{\alpha} A_{-q,j}^{\alpha'}}{\int \mathcal{D}\{\boldsymbol{A}^{\alpha}\} \mathrm{e}^{-S'_{\mathrm{eff},2}\{\boldsymbol{A}^{\alpha}\}}} \quad . \end{aligned} \tag{10.62}$$

To calculate this propagator, we impose again the Coulomb gauge condition by inserting functional δ-functions in the form (10.19) and then shifting

$$A_{q,i}^{\alpha} \to A_{q,i}^{\alpha} - \sum_{\alpha'}\sum_{j} [\underline{\mathrm{H}}_q]_{ij}^{\alpha\alpha'} q_j \lambda_q^{\alpha'} \quad . \tag{10.63}$$

This leads to the replacement

$$\begin{aligned} & S'_{\mathrm{eff},2}\{\boldsymbol{A}^{\alpha}\} + \sum_{q}\sum_{\alpha} \boldsymbol{A}_{-q}^{\alpha}\cdot\boldsymbol{q}\lambda_q^{\alpha} \to \\ & \qquad S'_{\mathrm{eff},2}\{\boldsymbol{A}^{\alpha}\} + \frac{1}{2}\sum_{q}\sum_{\alpha\alpha'} \lambda_{-q}^{\alpha}[(\boldsymbol{q}\underline{\mathrm{H}}_q\boldsymbol{q})]^{\alpha\alpha'}\lambda_q^{\alpha'} \quad , \end{aligned} \tag{10.64}$$

where $(\boldsymbol{q}\underline{\mathsf{H}}_q\boldsymbol{q})$ is a matrix in the patch labels, with elements given by

$$[(\boldsymbol{q}\underline{\mathsf{H}}_q\boldsymbol{q})]^{\alpha\alpha'} = \sum_{ij} q_i [\underline{\mathsf{H}}_q]_{ij}^{\alpha\alpha'} q_j \quad . \tag{10.65}$$

Performing the independent Gaussian integrations we finally obtain

$$[D^{\mathrm{RPA}}(q)]_{ij}^{\alpha\alpha'} = [\underline{\mathsf{H}}_q]_{ij}^{\alpha\alpha'} - \left[(\boldsymbol{e}_i\underline{\mathsf{H}}_{-q}\boldsymbol{q})(\boldsymbol{q}\underline{\mathsf{H}}_q\boldsymbol{q})^{-1}(\boldsymbol{q}\underline{\mathsf{H}}_{-q}\boldsymbol{e}_j)\right]^{\alpha\alpha'} \ , \tag{10.66}$$

where the product in the last term should be understood as a product of matrices in the patch indices. Using Eq.(5.123), the transverse part of the average eikonal can also be written as

$$Q_{\mathrm{tr}}^{\alpha}(\boldsymbol{r},\tau) = R_{\mathrm{tr}}^{\alpha} - S_{\mathrm{tr}}^{\alpha}(\boldsymbol{r},\tau) \ , \tag{10.67}$$

with

$$R_{\mathrm{tr}}^{\alpha} = \frac{1}{\beta V}\sum_q \frac{h_q^{\mathrm{RPA},\alpha}}{[\mathrm{i}\omega_m - \xi_{\boldsymbol{q}}^{\alpha}][\mathrm{i}\omega_m + \xi_{-\boldsymbol{q}}^{\alpha}]} = S_{\mathrm{tr}}^{\alpha}(0,0) \ , \tag{10.68}$$

$$S_{\mathrm{tr}}^{\alpha}(\boldsymbol{r},\tau) = \frac{1}{\beta V}\sum_q \frac{h_q^{\mathrm{RPA},\alpha}\cos(\boldsymbol{q}\cdot\boldsymbol{r} - \omega_m\tau)}{[\mathrm{i}\omega_m - \xi_{\boldsymbol{q}}^{\alpha}][\mathrm{i}\omega_m + \xi_{-\boldsymbol{q}}^{\alpha}]} \quad . \tag{10.69}$$

The effective interaction is

$$h_q^{\mathrm{RPA},\alpha} = -\frac{V}{\beta}\sum_{\alpha'\alpha''}\sum_{ij} U_{-q,i}^{\alpha\alpha'}[D^{\mathrm{RPA}}(q)]_{ij}^{\alpha'\alpha''}U_{q,j}^{\alpha\alpha''} \quad . \tag{10.70}$$

Note that these equations are valid for arbitrary patch geometry and arbitrary patch-dependent bare interactions $[\underline{\tilde{f}}_q]^{\alpha\alpha'}$ and $[\underline{h}_q]^{\alpha\alpha'}$. In deriving Eq.(10.70) we have assumed that the effective mass tensor is proportional to the unit matrix. To discuss quasi-one-dimensional anisotropic systems, it is necessary to allow for different effective masses m_i^{α}, $i = 1,\ldots,d$. In this case Eq.(10.70) is still correct, provided we take the different effective masses in the definition of $\boldsymbol{v}^{\alpha}$ into account. Then we have $\boldsymbol{v}^{\alpha} = (\mathsf{M}^{\alpha})^{-1}\boldsymbol{k}^{\alpha}$, where the effective mass tensor M^{α} is defined in Eq.(5.106). Hence we should replace in Eq.(10.47)

$$\delta_{ij}\tilde{\Delta}^{\alpha} \to \delta_{ij}\tilde{\Delta}_i^{\alpha} \ , \ \tilde{\Delta}_i^{\alpha} = \frac{N_0^{\alpha}}{\beta m_i^{\alpha} c^2} \ , \ \boldsymbol{u}^{\alpha} \to (\mathsf{M}^{\alpha})^{-1}\frac{\boldsymbol{k}^{\alpha}}{c} \ . \tag{10.71}$$

For the special case of patch-independent bare interactions and *linear energy dispersion* (i.e. $\xi_{\boldsymbol{q}}^{\alpha} \to \boldsymbol{v}^{\alpha}\cdot\boldsymbol{q}$) Eqs.(10.67)–(10.70) are equivalent with the expression given by Kwon, Houghton and Marston [10.18]. However, as will be shown in Sect. 10.4, for physically relevant forms of the gauge field propagator the linearization of the energy dispersion is not allowed. In fact, in [10.1] we have shown that in the case of the two-dimensional Chern-Simons action the low-energy behavior of the spectral function is completely dominated by the prefactor self-energy $\Sigma_1^{\alpha}(\tilde{q})$ and vertex function $Y^{\alpha}(\tilde{q})$ discussed in Chap. 5.3.2, which are ignored for linearized energy dispersion.

Comparing Eqs.(10.67)–(10.69) with (5.151)–(5.153), it is obvious that at the level of the Gaussian approximation the contributions from the longitudinal and transverse components to the total Debye-Waller factor are additive and formally identical; we simply have to use the corresponding RPA screened propagators. Of course, this is only true in Gaussian approximation, which produces the first order term in an expansion in powers of the RPA interaction. Clearly, the leading contributions to the prefactor Green's function are also additive so that we may simply copy the relevant equations from Chap. 5.3.2. Thus, in complete analogy with Eqs.(5.132) and (5.163) we obtain

$$\left\langle \mathcal{G}_1^\alpha(\boldsymbol{r},0,\tau,0) e^{\Phi^\alpha(\boldsymbol{r},\tau)-\Phi^\alpha(0,0)} \right\rangle_{S_{\text{eff},2}} = \tilde{G}^\alpha(\boldsymbol{r},\tau) e^{Q_1^\alpha(\boldsymbol{r},\tau)+Q_{\text{tr}}^\alpha(\boldsymbol{r},\tau)} \; , \tag{10.72}$$

with

$$\begin{aligned} \tilde{G}^\alpha(\boldsymbol{r},\tau) &\equiv G_1^\alpha(\boldsymbol{r},\tau) + G_2^\alpha(\boldsymbol{r},\tau) \\ &= \frac{1}{\beta V} \sum_{\tilde{q}} e^{i(\boldsymbol{q}\cdot\boldsymbol{r}-\tilde{\omega}_n\tau)} \frac{1 + Y^\alpha(\tilde{q}) + Y_{\text{tr}}^\alpha(\tilde{q})}{i\tilde{\omega}_n - \epsilon_{\boldsymbol{k}^\alpha+\boldsymbol{q}} + \mu - \Sigma_1^\alpha(\tilde{q}) - \Sigma_{1,\text{tr}}^\alpha(\tilde{q})} \; . \end{aligned} \tag{10.73}$$

Here $\Sigma_1^\alpha(\tilde{q})$ and $Y^\alpha(\tilde{q})$ are given in Eqs.(5.160) and (5.162), while $\Sigma_{1,\text{tr}}^\alpha(\tilde{q})$ and $Y_{\text{tr}}^\alpha(\tilde{q})$ can be obtained by replacing in these equations $f_q^{\text{RPA},\alpha} \to h_q^{\text{RPA},\alpha}$. In the special case of a spherical Fermi surface with radius $k_{\text{F}} = m v_{\text{F}}$ the prefactor self-energy due to the transverse gauge field is explicitly given by (see Eq.(5.186))

$$\begin{aligned} \Sigma_{1,\text{tr}}^\alpha(\tilde{q}) = &-\frac{1}{\beta V} \sum_{q'} h_{q'}^{\text{RPA},\alpha} G_1^\alpha(\tilde{q}+q') \\ &\times \frac{(\boldsymbol{q}\cdot\boldsymbol{q}')\boldsymbol{q}'^2 + (\boldsymbol{q}\cdot\boldsymbol{q}')^2}{m^2[i\omega_{m'} - \xi_{\boldsymbol{q}'}^\alpha][i\omega_{m'} + \xi_{-\boldsymbol{q}'}^\alpha]} \; , \end{aligned} \tag{10.74}$$

with $h_q^{\text{RPA},\alpha}$ given in Eq.(10.70). The corresponding vertex function is (see Eq.(5.187))

$$\begin{aligned} Y_{\text{tr}}^\alpha(\tilde{q}) = &\frac{1}{\beta V} \sum_{q'} h_{q'}^{\text{RPA},\alpha} G_1^\alpha(\tilde{q}+q') \\ &\times \frac{\boldsymbol{q}'^2 + 2\boldsymbol{q}\cdot\boldsymbol{q}'}{m[i\omega_{m'} - \xi_{\boldsymbol{q}'}^\alpha][i\omega_{m'} + \xi_{-\boldsymbol{q}'}^\alpha]} \; . \end{aligned} \tag{10.75}$$

As discussed in Chaps. 2.5 and 5.4, for spherical Fermi surfaces there is no need to introduce several patches. Then the index α simply indicates that all wave-vectors are measured with respect to a point $\boldsymbol{k}^\alpha$ on the Fermi surface, as shown in Fig. 2.8. *In this case there are no uncontrolled corrections due to around-the-corner processes to the above expressions.* In the following section we shall simplify $h_q^{\text{RPA},\alpha}$ such that we see more clearly that it contains the physics of transverse screening.

10.3 Transverse screening

Assuming patch-independent bare interactions, we derive from Eq.(10.70) the transverse dielectric tensor and show that in the static limit the transverse gauge field is not screened. We then discuss in some detail the form of $h_q^{\mathrm{RPA},\alpha}$ for spherical d-dimensional Fermi surfaces, where the effective mass tensor is isotropic.

10.3.1 The transverse dielectric tensor

Here and in the following section we assume that the effective masses m_i are independent of the patch index, i.e. $[\mathsf{M}^\alpha]_{ij} = m_i \delta_{ij}$. The expression for $h_q^{\mathrm{RPA},\alpha}$ in Eq.(10.70) can be simplified if we assume that all elements of the bare matrix $\underline{\tilde{h}}_q$ are identical, $[\underline{\tilde{h}}_q]^{\alpha\alpha'} = \tilde{h}_q \equiv \frac{\beta}{V} h_q$. Using the same method as in Eq.(4.34), we find that in this case also the matrix $\underline{\mathsf{H}}_q$ is independent of the patch indices,

$$[\underline{\mathsf{H}}_q]_{ij}^{\alpha\alpha'} = \tilde{h}_q [\mathsf{E}_q^{-1}]_{ij} \quad , \tag{10.76}$$

where E_q is a matrix in the spatial indices, with matrix elements given by

$$[\mathsf{E}_q]_{ij} = \delta_{ij} + h_q \left[\delta_{ij} \Delta_i - P_{ij}(q) + (\boldsymbol{e}_i \cdot \boldsymbol{\Pi}_q)(\boldsymbol{e}_j \cdot \boldsymbol{\Pi}_q) f_q^{\mathrm{RPA}} \right] \quad , \tag{10.77}$$

and

$$\Delta_i = \frac{\beta}{V} \sum_\alpha \tilde{\Delta}_i^\alpha = \frac{N}{V m_i c^2} \quad , \tag{10.78}$$

$$P_{ij}(q) = \sum_\alpha (\boldsymbol{e}_i \cdot \boldsymbol{u}^\alpha)(\boldsymbol{e}_j \cdot \boldsymbol{u}^\alpha) \Pi_0^\alpha(q) \quad , \tag{10.79}$$

$$\boldsymbol{\Pi}_q = \sum_\alpha \boldsymbol{u}^\alpha \Pi_0^\alpha(q) \quad . \tag{10.80}$$

The first term in Eq.(10.77) is the diamagnetic transverse polarization tensor, the second term is the paramagnetic one, and the last term describes the coupling between the longitudinal and the transverse fluctuations. Substituting Eq.(10.76) into the general expression for the gauge field propagator in Eq.(10.66), we obtain

$$[D^{\mathrm{RPA}}(q)]_{ij}^{\alpha\alpha'} = \tilde{h}_q \left[[\mathsf{E}_q^{-1}]_{ij} - \frac{\sum_{kl=1}^d [\mathsf{E}_q^{-1}]_{ik} q_k q_l [\mathsf{E}_q^{-1}]_{lj}}{\boldsymbol{q} \mathsf{E}_q^{-1} \boldsymbol{q}} \right] \quad , \tag{10.81}$$

where we have used the same notation as in Eq.(5.105). From Eqs.(10.70) and Eq.(10.81) we finally obtain

$$h_q^{\mathrm{RPA},\alpha} = -h_q \left[\boldsymbol{u}_q^\alpha \mathsf{E}_q^{-1} \boldsymbol{u}_q^\alpha - \frac{(\boldsymbol{u}_q^\alpha \mathsf{E}_q^{-1} \boldsymbol{q})(\boldsymbol{q} \mathsf{E}_q^{-1} \boldsymbol{u}_q^\alpha)}{\boldsymbol{q} \mathsf{E}_q^{-1} \boldsymbol{q}} \right] \quad , \tag{10.82}$$

where $\boldsymbol{u}_q^\alpha = \boldsymbol{u}^\alpha - \boldsymbol{\Pi}_q f_q^{\mathrm{RPA}}$. Eq.(10.82) can be further simplified by choosing an appropriate coordinate system. Because the scalar products are independent of the choice of the coordinate system and $q \equiv [\boldsymbol{q}, \mathrm{i}\omega_m]$ appears as an external parameter, we may choose the orientation of the coordinate system such that one of its axis ($\boldsymbol{e}_d$, for example) matches the direction of $\hat{\boldsymbol{q}}$, as shown in Fig. 10.2. Because for small $\boldsymbol{q}$ the function $\Pi_0^\alpha(q)$ in Eq.(10.80) depends on $\boldsymbol{q}$

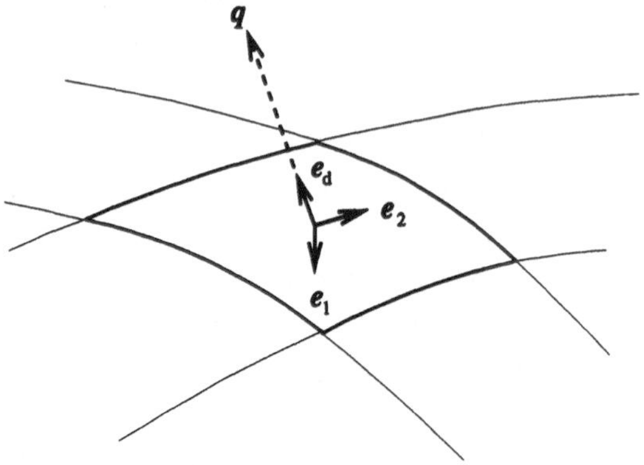

Fig. 10.2. Local coordinate system associated with a patch on the Fermi surface in which the transverse dielectric tensor E_q is diagonal.

only via $\boldsymbol{v}^\alpha \cdot \boldsymbol{q}$, it is easy to see that in the long wavelength limit $\hat{\boldsymbol{q}}_\perp \cdot \boldsymbol{\Pi}_q = 0$ for any direction $\hat{\boldsymbol{q}}_\perp$ that is orthogonal to $\boldsymbol{q}$. Note that by construction the $d-1$ directions $\boldsymbol{e}_i$, $i = 1, \ldots, d-1$ are all perpendicular to $\boldsymbol{q}$, so that the last term in Eq.(10.77) does not contribute to Eq.(10.82). For the same reason $P_{ij}(q) = \delta_{ij} P_{ii}(q)$. It follows that in this basis the matrix E_q is diagonal. The eigenvalues corresponding to the $d-1$ directions orthogonal to $\hat{\boldsymbol{q}}$ are simply given by

$$\epsilon_i(q) = [\mathsf{E}_q]_{ii} = 1 + h_q \Pi_i(q) \quad , \quad i = 1, \ldots, d-1 \quad , \tag{10.83}$$

where the transverse polarization in direction $\boldsymbol{e}_i$ is

$$\Pi_i(q) = \Delta_i - \sum_\alpha (\boldsymbol{e}_i \cdot \boldsymbol{u}^\alpha)^2 \Pi_0^\alpha(q) \quad , \quad i = 1, \ldots, d-1 \quad . \tag{10.84}$$

We finally obtain for the effective screened interaction

$$h_q^{\mathrm{RPA},\alpha} = -h_q \sum_{i=1}^{d-1} \frac{(\boldsymbol{e}_i \cdot \boldsymbol{u}^\alpha)^2}{\epsilon_i(q)} \quad . \tag{10.85}$$

Note that Eq.(10.85) involves only the transverse eigenvalues of E_q, because $h_q^{\mathrm{RPA},\alpha}$ is by construction the propagator of the *transverse* components of the gauge field. The dimensionless functions $\epsilon_i(q)$ are called the *transverse dielectric functions.*

10.3.2 Screening and gauge invariance

According to Eqs.(4.24), (4.36), (10.78) and (10.84) the longitudinal and transverse polarizations are for small $\boldsymbol{q}$ and arbitrary frequencies given by

$$\Pi_0(q) = \sum_\alpha \nu^\alpha \frac{\boldsymbol{v}^\alpha \cdot \boldsymbol{q}}{\boldsymbol{v}^\alpha \cdot \boldsymbol{q} - \mathrm{i}\omega_m} \tag{10.86}$$

$$\Pi_i(q) = \frac{1}{m_i c^2} \sum_\alpha \left[\frac{N_0^\alpha}{V} - m_i (\boldsymbol{v}^\alpha \cdot \boldsymbol{e}_i)^2 \nu^\alpha \frac{\boldsymbol{v}^\alpha \cdot \boldsymbol{q}}{\boldsymbol{v}^\alpha \cdot \boldsymbol{q} - \mathrm{i}\omega_m} \right] \quad , \tag{10.87}$$

where $\boldsymbol{e}_i \cdot \boldsymbol{q} = 0$, and we have used the fact that $m_i \boldsymbol{v}^\alpha \cdot \boldsymbol{e}_i = \boldsymbol{k}^\alpha \cdot \boldsymbol{e}_i$ (see Eq.(10.71)). In the static limit we have $\Pi_0(\boldsymbol{q}, 0) = \sum_\alpha \nu^\alpha = \nu$, where the total density of states is given in Eq.(A.2). A finite value of the longitudinal polarization implies that long-range interactions are screened. For example, for the three-dimensional Coulomb interaction the static longitudinal dielectric function is within the RPA given by $\epsilon_{\mathrm{RPA}}(\boldsymbol{q}, 0) = 1 + \kappa^2 / \boldsymbol{q}^2$, see Eqs.(2.52) and (A.55). For wave-vectors $|\boldsymbol{q}| \ll \kappa$ the longitudinal dielectric function is large compared with unity, so that the interaction is screened at length scales larger than the Thomas-Fermi length κ^{-1}. On the other hand, as long as the gauge symmetry is not spontaneously broken, the transverse gauge field is not screened in the static limit. Using Eq.(2.57), it is easy to see that the transverse polarization in a direction $\boldsymbol{e}_i$ orthogonal to $\boldsymbol{q}$ can in the static limit and for small $|\boldsymbol{q}|$ be written as

$$\Pi_i(\boldsymbol{q}, 0) = \frac{1}{m_i c^2} \int \frac{\mathrm{d}\boldsymbol{k}}{(2\pi)^d} \left[\Theta(\mu - \epsilon_{\boldsymbol{k}}) - \frac{k_i^2}{m_i} \delta(\mu - \epsilon_{\boldsymbol{k}}) \right] \quad . \tag{10.88}$$

But

$$\delta(\mu - \epsilon_{\boldsymbol{k}}) = - \frac{m_i}{k_i} \frac{\partial}{\partial k_i} \Theta(\mu - \epsilon_{\boldsymbol{k}}) \quad , \tag{10.89}$$

so that we obtain after an integration by parts

$$\Pi_i(\boldsymbol{q}, 0) = \frac{1}{m_i c^2} \int \frac{\mathrm{d}\boldsymbol{k}}{(2\pi)^d} \left[\Theta(\mu - \epsilon_{\boldsymbol{k}}) + k_i \frac{\partial}{\partial k_i} \Theta(\mu - \epsilon_{\boldsymbol{k}}) \right] = 0 \quad . \tag{10.90}$$

The vanishing of the transverse polarization tensor in the static limit is due to a perfect cancellation between the dia- and paramagnetic contributions. The fundamental symmetry which is responsible for this cancellation is gauge invariance, which insures that the transverse gauge field remains massless in the presence of matter. Hence, as long as the gauge symmetry is not spontaneously broken, the transverse gauge field is not screened in the static limit. However, as shown by Kohn and Luttinger [10.26], any interacting Fermi system shows at very low temperatures a superconducting instability (Kohn-Luttinger effect), so that gauge invariance is in fact broken at very low temperatures, and the transverse gauge field is eventually screened. This instability is not included in our calculation.

10.3.3 The transverse dielectric function for spherical Fermi surfaces

For spherical Fermi surfaces the effective mass tensor is proportional to the unit matrix. The $d-1$ transverse eigenvalues of E_q are then degenerate, and are called the *transverse dielectric function* [1.7],

$$\epsilon_\perp(q) = 1 + h_q \Pi_\perp(q) \quad . \tag{10.91}$$

From Eq.(10.84) we see that the transverse polarization $\Pi_\perp(q)$ is within the RPA given by

$$\Pi_\perp(q) = \Delta - \sum_\alpha (\boldsymbol{e}_i \cdot \boldsymbol{u}^\alpha)^2 \Pi_0^\alpha(q) \quad , \tag{10.92}$$

with $\Delta = N/(Vmc^2)$ (see Eq.(10.78)). Here $\boldsymbol{e}_i$ is any of the $d-1$ unit vectors perpendicular to $\hat{\boldsymbol{q}} = \boldsymbol{e}_d$. From Eq.(10.87) it is easy to show that for a spherical Fermi surface

$$\Pi_\perp(q) = \left(\frac{v_{\mathrm{F}}}{c} \right)^2 \nu \Lambda_d \left(\frac{\mathrm{i}\omega_m}{v_{\mathrm{F}} |\boldsymbol{q}|} \right) \quad , \tag{10.93}$$

where the dimensionless function $\Lambda_d(z)$ is

$$\Lambda_d(z) = \frac{1}{d} - \left\langle (\boldsymbol{e}_i \cdot \hat{\boldsymbol{k}})^2 \frac{\hat{\boldsymbol{q}} \cdot \hat{\boldsymbol{k}}}{\hat{\boldsymbol{q}} \cdot \hat{\boldsymbol{k}} - z} \right\rangle_{\hat{\boldsymbol{k}}} \quad . \tag{10.94}$$

Here the angular average is defined as in Eq.(A.4). By symmetry, the average is independent of the choice of $\boldsymbol{e}_i$. Using the fact that $\left\langle (\boldsymbol{e}_i \cdot \hat{\boldsymbol{k}})^2 \right\rangle_{\hat{\boldsymbol{k}}} = 1/d$ it is easy to see that Eq.(10.94) can also be written as

$$\Lambda_d(z) = -z \left\langle \frac{(\boldsymbol{e}_i \cdot \hat{\boldsymbol{k}})^2}{\hat{\boldsymbol{q}} \cdot \hat{\boldsymbol{k}} - z} \right\rangle_{\hat{\boldsymbol{k}}} \quad . \tag{10.95}$$

Because all transverse directions are equivalent, we may replace in the average

$$(\boldsymbol{e}_i \cdot \hat{\boldsymbol{k}})^2 \rightarrow \frac{\sum_{i=1}^{d-1} (\boldsymbol{e}_i \cdot \hat{\boldsymbol{k}})^2}{d-1} = \frac{1 - (\hat{\boldsymbol{q}} \cdot \hat{\boldsymbol{k}})^2}{d-1} \quad , \tag{10.96}$$

so that

$$\Lambda_d(z) = -\frac{z}{d-1} \left\langle \frac{1 - (\hat{\boldsymbol{q}} \cdot \hat{\boldsymbol{k}})^2}{\hat{\boldsymbol{q}} \cdot \hat{\boldsymbol{k}} - z} \right\rangle_{\hat{\boldsymbol{k}}} \quad . \tag{10.97}$$

From this expression we find

$$\begin{aligned} \mathrm{Im}\Lambda_d(x + \mathrm{i}0^+) &= -\frac{\pi x (1 - x^2)}{d-1} \left\langle \delta(\hat{\boldsymbol{q}} \cdot \hat{\boldsymbol{k}} - x) \right\rangle_{\hat{\boldsymbol{k}}} \\ &\sim -\frac{\pi \gamma_d}{d-1} x \quad , \quad \text{for } |x| \ll 1 \quad , \end{aligned} \tag{10.98}$$

with the numerical constant γ_d given in Eq.(A.9). Note that, in contrast to γ_d, the quantity

$$\tilde{\gamma}_d \equiv \frac{\gamma_d}{d-1} = \frac{\Gamma(\frac{d}{2})}{(d-1)\sqrt{\pi}\Gamma(\frac{d-1}{2})} \tag{10.99}$$

has a finite limit as $d \to 1$. In particular,

$$\tilde{\gamma}_1 = \frac{1}{2} \quad , \quad \tilde{\gamma}_2 = \frac{1}{\pi} \quad , \quad \tilde{\gamma}_3 = \frac{1}{4} \quad . \tag{10.100}$$

On the imaginary axis Eq.(10.98) implies

$$\Lambda_d\,(\mathrm{i}y) \sim \lambda_d |y| \quad , \quad \text{for } |y| \ll 1 \quad , \tag{10.101}$$

where

$$\lambda_d = \pi\tilde{\gamma}_d = \frac{\pi\gamma_d}{d-1} \quad . \tag{10.102}$$

For the Maxwell action discussed in Sect. 10.1.1 the bare interaction h_q is given in Eq.(10.13). Then we obtain from Eqs.(10.91) and (10.93) for the transverse dielectric function

$$\epsilon_{\perp}(q) = 1 + \left(\frac{v_{\mathrm{F}}}{c}\right)^2 \frac{\Lambda_d(\frac{\mathrm{i}\omega_m}{v_{\mathrm{F}}|\boldsymbol{q}|})}{(\frac{q}{\kappa})^2 + (\frac{\omega_m}{c\kappa})^2} \quad , \quad \kappa^2 = 4\pi e^2 \nu \quad . \tag{10.103}$$

After some simple rescalings we obtain for the effective interaction (10.69)

$$h_q^{\mathrm{RPA},\alpha} = -\frac{1}{\nu}\left(\frac{v_{\mathrm{F}}}{c}\right)^2 \frac{1 - (\hat{\boldsymbol{k}}^{\alpha} \cdot \hat{\boldsymbol{q}})^2}{\left(\frac{q}{\kappa}\right)^2 + \left(\frac{v_{\mathrm{F}}}{c}\right)^2 \left[(\frac{\omega_m}{v_{\mathrm{F}}\kappa})^2 + \Lambda_d\left(\frac{\mathrm{i}\omega_m}{v_{\mathrm{F}}|\boldsymbol{q}|}\right)\right]} \quad . \tag{10.104}$$

In the regime $|\omega_m| \ll v_{\mathrm{F}}|\boldsymbol{q}|$ we obtain from Eq.(10.101)

$$\Lambda_d\left(\frac{\mathrm{i}\omega_m}{v_{\mathrm{F}}|\boldsymbol{q}|}\right) \sim \lambda_d \frac{|\omega_m|}{v_{\mathrm{F}}|\boldsymbol{q}|} \quad . \tag{10.105}$$

To this order in ω_m the term proportional to ω_m^2 in the denominator of Eq.(10.104) is negligible, so that the effective interaction can be approximated by

$$h_q^{\mathrm{RPA},\alpha} \approx -\frac{1}{\nu}\left(\frac{v_{\mathrm{F}}}{c}\right)^2 \frac{1 - (\hat{\boldsymbol{k}}^{\alpha} \cdot \hat{\boldsymbol{q}})^2}{\left(\frac{q}{\kappa}\right)^2 + \lambda_d(\frac{v_{\mathrm{F}}}{c})^2 \frac{|\omega_m|}{v_{\mathrm{F}}|\boldsymbol{q}|}} \quad , \quad \text{for } |\omega_m| \ll v_{\mathrm{F}}|\boldsymbol{q}| \quad . \tag{10.106}$$

For $d = 3$ we recover Eq.(10.1). The term proportional to $|\omega_m|/(v_{\mathrm{F}}|\boldsymbol{q}|)$ in the denominator describes the dynamical screening of the fluctuations of the gauge field due to Landau damping. This term is responsible for the dynamical screening of the magnetic field in a clean metal, i.e. the *anomalous skin effect* [10.5].

10.4 The transverse Debye-Waller factor

We now analyze the transverse Debye-Waller factor $Q^{\alpha}_{\mathrm{tr}}(\boldsymbol{r},\tau)$ in Eq.(10.67) in more detail. We determine the parameter regime where $Q^{\alpha}_{\mathrm{tr}}(\boldsymbol{r},\tau)$ is bounded for large distances or times, and where the non-linear terms in the energy dispersion must be retained in order to obtain qualitatively correct results.

In Sect. 10.2 we have derived a non-perturbative expression for the single-particle Green's function $G(k)$ in Coulomb gauge (see Eqs.(10.51) and (10.72)). The effect of the Gaussian fluctuations of the gauge field is parameterized in terms of three distinct contributions to the Green's function: the Debye-Waller factor $Q^{\alpha}_{\mathrm{tr}}(\boldsymbol{r},\tau)$ in Eqs.(10.67)–(10.69), the prefactor self-energy $\Sigma^{\alpha}_{1,\mathrm{tr}}(\tilde{q})$ in Eq.(10.74), and the prefactor vertex $Y^{\alpha}_{\mathrm{tr}}(\tilde{q})$ in Eq.(10.75). Thus, the calculation of $G(k)$ and the resulting spectral function has been reduced to the purely mathematical problem of doing the relevant integrations. Unfortunately, it is impossible to perform these integrations analytically, so that a complete analysis of our non-perturbative result for $G(k)$ requires extensive numerical work, which is beyond the scope of this book. In the recent Letter [10.1] we have made some progress in this problem. In particular, we have shown that in physically relevant cases the low-energy behavior of the spectral function is essentially determined by the functions $\Sigma^{\alpha}_{1,\mathrm{tr}}(\tilde{q})$ and $Y^{\alpha}_{\mathrm{tr}}(\tilde{q})$, and *not* by the Debye-Waller factor $Q^{\alpha}_{\mathrm{tr}}(\boldsymbol{r},\tau)$. Because for linearized energy dispersion both functions $\Sigma^{\alpha}_{1,\mathrm{tr}}$ and Y^{α}_{tr} vanish, the problem of fermions that are coupled to gauge fields can only be studied via higher-dimensional bosonization if the quadratic terms in the expansion of the energy dispersion close to the Fermi surface are retained[6]. To see more clearly why the curvature of the Fermi surface is so important in the present problem, we shall in this section study the Debye-Waller factor $Q^{\alpha}_{\mathrm{tr}}(\boldsymbol{r},\tau)$ in some detail.

10.4.1 Exact rescalings

Let us consider a spherical Fermi surface in d dimensions and a general gauge field propagator of the form

$$h^{\mathrm{RPA},\alpha}_{q} = -\frac{1}{\nu}\frac{1-(\hat{\boldsymbol{k}}^{\alpha}\cdot\hat{\boldsymbol{q}})^2}{\left(\frac{|\boldsymbol{q}|}{q_c}\right)^{\eta} + \Lambda_d\left(\frac{\mathrm{i}\omega_m}{v_F|\boldsymbol{q}|}\right)} \quad , \tag{10.107}$$

where $\Lambda_d(\mathrm{i}y) \sim \lambda_d|y|$ for small $|y|$, see Eq.(10.97). Substituting Eq.(10.107) into Eq.(10.67), we obtain

$$Q^{\alpha}_{\mathrm{tr}}(\boldsymbol{r},\tau) = -\frac{1}{\beta V\nu}\sum_{q}\frac{1-(\hat{\boldsymbol{k}}^{\alpha}\cdot\hat{\boldsymbol{q}})^2}{\left(\frac{|\boldsymbol{q}|}{q_c}\right)^{\eta} + \Lambda_d\left(\frac{\mathrm{i}\omega_m}{v_F|\boldsymbol{q}|}\right)}\frac{1-\cos(\boldsymbol{q}\cdot\boldsymbol{r}-\omega_m\tau)}{(\mathrm{i}\omega_m-\xi^{\alpha}_{\boldsymbol{q}})(\mathrm{i}\omega_m+\xi^{\alpha}_{-\boldsymbol{q}})} \; . \tag{10.108}$$

[6] Because of the formal similarity between higher-dimensional bosonization and the leading term in the conventional eikonal expansion [5.12], it seems that this is true for any eikonal type of approach to this problem [5.11, 10.20].

As discussed in Chap. 5.1.3, for *linearized energy dispersion* we may replace $\boldsymbol{r} \to r_{\parallel}^{\alpha} \hat{\boldsymbol{v}}^{\alpha}$ in Eq.(10.108), because the sector Green's function $G_0^{\alpha}(\boldsymbol{r}, \tau)$ is proportional to $\delta^{(d-1)}(\boldsymbol{r}_{\perp}^{\alpha})$, see Eq.(5.48). Although for non-linear energy dispersion we should consider $Q_{\mathrm{tr}}^{\alpha}(\boldsymbol{r}, \tau)$ for all $\boldsymbol{r}$, we shall restrict ourselves here to the direction $\boldsymbol{r} = r_{\parallel}^{\alpha} \hat{\boldsymbol{v}}^{\alpha}$. This is sufficient for investigating whether the non-linear terms in the energy dispersion qualitatively modify the result obtained for linearized energy dispersion. Obviously, for $\boldsymbol{r} = r_{\parallel}^{\alpha} \hat{\boldsymbol{v}}^{\alpha}$ the $\boldsymbol{q}$-dependence of the right-hand side of Eq.(10.108) involves only the absolute value of $\boldsymbol{q}$ and the component[7] $q_{\parallel}^{\alpha} = \hat{\boldsymbol{v}}^{\alpha} \cdot \boldsymbol{q}$. Then the $d+1$-dimensional integration in Eq.(10.108) can be reduced to a three-dimensional one with the help of d-dimensional spherical coordinates: for $V \to \infty$ and $\beta \to \infty$ we have for any function $f(|\boldsymbol{q}|, \hat{\boldsymbol{q}} \cdot \hat{\boldsymbol{v}}^{\alpha}, \mathrm{i}\omega_m)$

$$\frac{1}{\beta V \nu} \sum_q f(|\boldsymbol{q}|, \hat{\boldsymbol{q}} \cdot \hat{\boldsymbol{v}}^{\alpha}, \mathrm{i}\omega_m) \to$$

$$\frac{v_{\mathrm{F}}}{k_{\mathrm{F}}^{d-1}} \gamma_d \int_0^{\infty} \mathrm{d}q q^{d-1} \int_0^{\pi} \mathrm{d}\vartheta (\sin\vartheta)^{d-2} \int_{-\infty}^{\infty} \frac{\mathrm{d}\omega}{2\pi} f(q, \cos\vartheta, \mathrm{i}\omega) \quad , \tag{10.109}$$

where the numerical constant γ_d is given in Eq.(A.10), and we have used Eq.(A.5). Introducing the dimensionless integration variables

$$p = \frac{q}{q_{\mathrm{c}}} \quad , \quad y = \frac{\omega}{v_{\mathrm{F}} q} = \frac{\omega}{v_{\mathrm{F}} q_{\mathrm{c}} p} \quad , \tag{10.110}$$

and noting that $Q_{\mathrm{tr}}^{\alpha}(r_{\parallel}^{\alpha} \hat{\boldsymbol{v}}^{\alpha}, \tau)$ depends on the sector index α only via $r_{\parallel}^{\alpha} = \hat{\boldsymbol{v}}^{\alpha} \cdot \boldsymbol{r}$, we may write

$$Q_{\mathrm{tr}}^{\alpha}(r_{\parallel}^{\alpha} \hat{\boldsymbol{v}}^{\alpha}, \tau) = Q_{\mathrm{tr}}(q_{\mathrm{c}} r_{\parallel}^{\alpha}, v_{\mathrm{F}} q_{\mathrm{c}} \tau) \quad , \tag{10.111}$$

where the function $Q_{\mathrm{tr}}(\tilde{x}, \tilde{\tau})$ is given by

$$Q_{\mathrm{tr}}(\tilde{x}, \tilde{\tau}) = -\frac{\gamma_d g^{d-1}}{2\pi} \int_0^{\infty} \mathrm{d}p p^{d-2} \int_0^{\pi} \mathrm{d}\vartheta (\sin\vartheta)^d \int_{-\infty}^{\infty} \mathrm{d}y$$
$$\times \frac{1 - \cos[p(\tilde{x}\cos\vartheta - \tilde{\tau} y)]}{[p^{\eta} + \Lambda_d(\mathrm{i}y)]\,[\mathrm{i}y - \cos\vartheta - \frac{g}{2}p]\,[\mathrm{i}y - \cos\vartheta + \frac{g}{2}p]} \quad , \tag{10.112}$$

and the dimensionless coupling constant g is simply

$$g = \frac{q_{\mathrm{c}}}{k_{\mathrm{F}}} \quad . \tag{10.113}$$

Note that the linearization of the energy dispersion corresponds to setting $g = 0$ in the integrand of Eq.(10.112). The evaluation of Eq.(10.108) for $\boldsymbol{r} = r_{\parallel}^{\alpha} \hat{\boldsymbol{v}}^{\alpha}$ is now reduced to the three-dimensional integration. Possible non-Fermi liquid behavior due to the coupling between fermions and the gauge field should be due to the regime $|y| = |\omega_m|/(v_{\mathrm{F}}|\boldsymbol{q}|) \lesssim 1$, because here the gauge field propagator is most singular. To further investigate this point, we

[7] Note that for a spherical Fermi surface $\hat{\boldsymbol{v}}^{\alpha} = \hat{\boldsymbol{k}}^{\alpha}$.

may approximate $\Lambda_d(iy) \approx \lambda_d|y|$ (see Eqs.(10.101) and (10.102)). Of course, when substituting this expression into Eq.(10.112), we should restrict the y-integration to the regime $|y| \leq y_c = O(1)$. Moreover, physically it is clear that the power-law $(|\boldsymbol{q}|/q_c)^\eta$ of the gauge field propagator in Eq.(10.107) can only be valid up to some finite cutoff Q_c, because at short wavelengths non-universal short-range interactions will dominate[8]. Assuming that the form (10.107) remains valid up to $|\boldsymbol{q}| \leq Q_c$, we should impose a cutoff $p_c = Q_c/q_c$ on the p-integration in Eq.(10.112). With these cutoffs the integration volume is finite, so that possible non-Fermi liquid behavior must be due to infrared singularities.

To exhibit the infrared behavior of the integrand in Eq.(10.112) more clearly, it is advantageous to perform another rescaling of the integration variables, substituting

$$y = |\cos\vartheta| u \quad , \quad p = |\cos\vartheta|^{\frac{1}{\eta}} k \quad . \tag{10.114}$$

Then we obtain (taking the above ultraviolet cutoffs into account)

$$Q_{\mathrm{tr}}(\tilde{x}, \tilde{\tau}) =$$

$$-\frac{\gamma_d g^{d-1}}{2\pi} \int_0^\pi d\vartheta (\sin\vartheta)^d |\cos\vartheta|^{\frac{d-1}{\eta}-2} \int_0^{p_c|\cos\vartheta|^{-\frac{1}{\eta}}} dk k^{d-2} \int_{-y_c|\cos\vartheta|^{-1}}^{y_c|\cos\vartheta|^{-1}} du$$

$$\times \frac{1 - \cos\left[k|\cos\vartheta|^{\frac{1}{\eta}+1}(\tilde{x}s_\vartheta - \tilde{\tau}u)\right]}{\left[k^\eta + \lambda_d|u|\right]\left[iu - s_\vartheta - \frac{g}{2}k|\cos\vartheta|^{\frac{1}{\eta}-1}\right]\left[iu - s_\vartheta + \frac{g}{2}k|\cos\vartheta|^{\frac{1}{\eta}-1}\right]} , \tag{10.115}$$

where we have defined $s_\vartheta = \mathrm{sgn}(\cos\vartheta)$. From Eq.(10.115) it is now evident that the regime $\vartheta \approx \pi/2$ can give rise to singular behavior (in the sense that the ϑ-integral diverges if we retain only the space- and time-independent contribution R_{tr}), because the integral over the factor $|\cos\vartheta|^{\frac{d-1}{\eta}-2}$ does not exist for $\frac{d-1}{\eta} - 2 < -1$. In fact, let us *assume* for the moment that the rest of the integrand does not modify the small-ϑ behavior of the integral. Evidently, this assumption will be correct provided it is allowed to set $g = 0$ in the rest of the integral, corresponding to the linearization of the energy dispersion. In this case the angular integration is free of singularities as long as the integral

$$A_{d,\eta} = \int_0^\pi d\vartheta (\sin\vartheta)^d |\cos\vartheta|^{\frac{d-1}{\eta}-2} = \frac{\Gamma(\frac{d+1}{2})\Gamma(\frac{d-1-\eta}{2\eta})}{\Gamma(\frac{(1+\eta)d-1}{2\eta})} \tag{10.116}$$

is finite. This is the case for $\frac{d-1}{\eta} - 2 > -1$, or

$$\eta < d - 1 \quad . \tag{10.117}$$

[8] In the case of the three-dimensional Coulomb interaction we should choose $Q_c = q_c \approx \kappa$ (the Thomas-Fermi wave-vector), so that $p_c \approx 1$. In general, however, Q_c and q_c need not be equal. For example, in Sect. 10.4.3 we shall show that in the two-dimensional Chern-Simons theory for the half-filled Landau level Q_c can be much larger than q_c.

This is precisely the criterion for the existence of the quasi-particle residue that is obtained for linearized energy dispersion [10.2], where one sets $g = 0$ in the integrand of Eq.(10.115). In particular, for linearized energy dispersion higher-dimensional bosonization predicts for the three-dimensional Maxwell action ($\eta = 2$) and the two-dimensional Maxwell-Chern-Simons action ($\eta = 1$) non-Fermi liquid behavior due to a logarithmic divergence of $A_{d,\eta}$. As a consequence, the momentum distribution exhibits an algebraic singularity at the Fermi surface [10.2], just like in the one-dimensional Tomonaga-Luttinger model. The crucial point is, however, that for $\eta > 1$ the assumption that the rest of the ϑ-dependence of the second line in Eq.(10.115) does not modify the infrared behavior of the integrand is *not* correct, because for $\eta > 1$ and any finite g the curvature terms in the denominator of Eq.(10.115) become arbitrarily large for $\cos\vartheta \to 0$. *Hence, for $\eta > 1$ the non-linear terms in the expansion of the energy dispersion close to the Fermi surface cannot be ignored!* On the other hand, for $\eta < 1$ these terms vanish for $\cos\vartheta \to 0$, so that in this case the criterion (10.117) is valid. Only then the finiteness of $A_{d,\eta}$ implies the existence of R^{α}_{tr}, so that the system shows Fermi liquid behavior. But in the physically interesting cases of the Maxwell action ($d = 3$, $\eta = 2$) and the Maxwell-Chern-Simons action ($d = 2$, $\eta = 1$) the curvature term in the denominator of Eq.(10.115) cannot be neglected. Interestingly, for $\eta = 1$ there terms are independent of ϑ, so that their relevance cannot be determined by simple power counting. We shall come back to this point in Sect. 10.4.3, where we show by explicit evaluation of the relevant integral that even then the curvature terms are essential. Note that the above analysis confirms our intuitive arguments based on the simple estimate (10.5). In the following section we shall study the effect of the curvature terms more carefully.

10.4.2 The relevance of curvature

We study the effect of the quadratic term in the energy dispersion on the constant part R_{tr} of the Debye-Waller factor. This is sufficient to see whether the curvature of the Fermi surface is relevant or not.

According to Eq.(10.115) the constant part R_{tr} of the Debye-Waller factor can be written as

$$R_{\mathrm{tr}} = -\frac{(d-1)g^{d-1}}{\pi^2} \int_0^{\pi/2} d\vartheta (\sin\vartheta)^d |\cos\vartheta|^{\frac{d-1}{\eta}-2}$$

$$\times \int_0^{p_{\mathrm{c}}|\cos\vartheta|^{-\frac{1}{\eta}}} dk k^{d-2} F(\lambda_d^{-1} k^{\eta}, gk(\cos\vartheta)^{\frac{1}{\eta}-1}) \quad , \tag{10.118}$$

with

$$F(E,\gamma) = \int_{-\infty}^{\infty} du \frac{1}{[E + |u|] \left[iu - 1 - \frac{\gamma}{2}\right] \left[iu - 1 + \frac{\gamma}{2}\right]}$$

$$= \frac{2}{\gamma} \left\{ (1 - \frac{\gamma}{2}) \int_0^{\infty} du \frac{1}{[u + E] \left[u^2 + (1 - \frac{\gamma}{2})^2\right]} \right.$$

$$\left. - (1 + \frac{\gamma}{2}) \int_0^{\infty} du \frac{1}{[u + E] \left[u^2 + (1 + \frac{\gamma}{2})^2\right]} \right\} \quad . \tag{10.119}$$

In deriving the prefactor in Eq.(10.118) we have used $\gamma_d / \lambda_d = (d-1)/\pi$, see Eq.(10.102). Because we are interested in the singularities of the integrand for small $\cos\vartheta$, we have replaced upper cutoff $\pm y_c / |\cos\vartheta|$ for the u-integration by $\pm\infty$. We shall verify a posteriori that the integral without cutoff is convergent, so that this procedure is justified. Note, however, that we retain the cutoff p_c for the k-integration in Eq.(10.118), because for the two-dimensional Maxwell-Chern-Simons theory the value of the integral will crucially depend on this cutoff (see Sect. 10.4.3 below). Using

$$\int_0^{\infty} du \frac{1}{[u + E][u^2 + a^2]} = \frac{1}{a^2 + E^2} \left[\frac{\pi E}{2a} - \ln(\frac{E}{a}) \right] \quad , \tag{10.120}$$

the integrations in Eq.(10.119) are easily done, and we obtain

$$F(E,\gamma) = \frac{\pi E}{\gamma} \left\{ \frac{\mathrm{sgn}(1 - \frac{\gamma}{2})}{E^2 + (1 - \frac{\gamma}{2})^2} - \frac{1}{E^2 + (1 + \frac{\gamma}{2})^2} \right\}$$

$$+ \frac{2}{\gamma} \left\{ \frac{(1 - \frac{\gamma}{2}) \ln \left[\frac{|1 - \frac{\gamma}{2}|}{E} \right]}{E^2 + (1 - \frac{\gamma}{2})^2} - \frac{(1 + \frac{\gamma}{2}) \ln \left[\frac{|1 + \frac{\gamma}{2}|}{E} \right]}{E^2 + (1 + \frac{\gamma}{2})^2} \right\} \quad . \tag{10.121}$$

It is easy to show that the function $F(E,\gamma)$ has a finite limit as $\gamma \to 0$, which is given by [10.2]

$$F(E,0) = 2 \frac{\pi E - E^2 - 1 + (E^2 - 1) \ln E}{(1 + E^2)^2} \quad . \tag{10.122}$$

The cancellation of the singular prefactor $1/\gamma$ in Eq.(10.121) can be traced back to the factor $\mathrm{sgn}(\xi_{\boldsymbol{q}}^{\alpha})$ in our general spectral representation given in Eq.(6.4). In fact, it is instructive to re-derive Eqs.(10.118) and (10.121) from Eq.(6.4). Therefore we simply note that the gauge field propagator in Eq.(10.107) can also be written as

$$h_{\boldsymbol{q}}^{\mathrm{RPA},\alpha} = -(h_{\boldsymbol{q}}^{\alpha})^2 \int_0^{\infty} d\omega S_{\mathrm{RPA}}(\boldsymbol{q}, \omega) \frac{2\omega}{\omega^2 + \omega_m^2} \quad , \tag{10.123}$$

with

$$(h_{\boldsymbol{q}}^{\alpha})^2 \equiv \frac{1 - (\hat{\boldsymbol{k}}^{\alpha} \cdot \hat{\boldsymbol{q}})^2}{\nu^2} \left(\frac{q_c}{|\boldsymbol{q}|} \right)^{\eta} \quad , \tag{10.124}$$

and

$$S_{\mathrm{RPA}}(\boldsymbol{q},\omega) = \frac{\nu}{\pi}\mathrm{Im}\left\{\frac{1}{1+(\frac{q_c}{|\boldsymbol{q}|})^\eta \Lambda_d(\frac{\omega}{v_{\mathrm{F}}|\boldsymbol{q}|}+\mathrm{i}0^+)}\right\} \quad . \tag{10.125}$$

For $|\omega_m| \ll v_{\mathrm{F}}|\boldsymbol{q}|$ we may approximate (see Eq.(10.98))

$$\Lambda_d(\frac{\omega}{v_{\mathrm{F}}|\boldsymbol{q}|}+\mathrm{i}0^+) \approx -\mathrm{i}\lambda_d\frac{\omega}{v_{\mathrm{F}}|\boldsymbol{q}|} \quad , \tag{10.126}$$

so that in this regime Eq.(10.125) reduces to the usual dynamic structure factor due to an overdamped mode,

$$S_{\mathrm{RPA}}(\boldsymbol{q},\omega) = \frac{\nu}{\pi}\frac{\omega\Gamma_{\boldsymbol{q}}}{\omega^2+\Gamma_{\boldsymbol{q}}^2} \quad , \quad \Gamma_{\boldsymbol{q}} = \frac{v_{\mathrm{F}}|\boldsymbol{q}|^{1+\eta}}{\lambda_d q_c^\eta} \quad . \tag{10.127}$$

Substituting Eq.(10.123) for the gauge field propagator into Eq.(10.108), and taking the limit $\beta \to \infty$, the ω_m-integral is easily done. The result is formally identical with Eqs.(6.4)–(6.6), except that we should replace $f_{\boldsymbol{q}}^2 \to (h_{\boldsymbol{q}}^\alpha)^2$ and omit the terms proportional to $f_{\boldsymbol{q}}$. After rescaling the integration variables as above, we obtain the following alternative expression for R_{tr},

$$\begin{aligned} R_{\mathrm{tr}} &= -\frac{2(d-1)g^{d-2}}{\pi^2}\int_0^\pi \mathrm{d}\vartheta(\sin\vartheta)^d \\ &\times \int_0^{p_c}\mathrm{d}p p^{d-3}\left[\Theta(-\cos\vartheta-\frac{gp}{2})-\Theta(\cos\vartheta+\frac{gp}{2})\right] \\ &\times \int_0^\infty \mathrm{d}x\frac{x}{[x^2+\lambda_d^{-2}p^{2\eta}][x+|\cos\vartheta+\frac{gp}{2}|]} \quad . \end{aligned} \tag{10.128}$$

The x-integration is easily performed using

$$\int_0^\infty \mathrm{d}x\frac{x}{[x^2+E^2][x+a]} = \frac{E}{E^2+a^2}\left[\frac{\pi}{2}+\frac{a}{E}\ln(\frac{a}{E})\right] \quad , \quad a>0\,, \tag{10.129}$$

and after rescaling $p = |\cos\vartheta|^{\frac{1}{\eta}}k$ we recover Eq.(10.118).

From Eqs.(10.118) and (10.121) it is easy to determine whether R_{tr} is finite or not. First of all, if we linearize the energy dispersion, we effectively replace the function $F(\lambda_d^{-1}k^\eta, gk(\cos\vartheta)^{\frac{1}{\eta}-1})$ in Eq.(10.118) by $F(\lambda_d^{-1}k^\eta,0)$. Because according to Eq.(10.122) this function is non-singular for small k, the existence of R_{tr} is determined by the singularity in the remaining ϑ-integration. In this way we recover the criterion (10.117). On the other hand, for finite g and $\eta > 1$ it is clear that the singularity of the integrand of Eq.(10.118) for small $\cos\vartheta$ is determined by the large-γ behavior of the function $F(E,\gamma)$, which is given by

$$F(E,\gamma) \sim \frac{4\ln\gamma}{\gamma^2} \quad , \quad \gamma \to \infty \quad . \tag{10.130}$$

Obviously the curvature term in the energy dispersion gives rise to an additional factor of $(\cos\vartheta)^{2-\frac{2}{\eta}}/g^2$, so that the most singular part of the integral in Eq.(10.118) becomes

$$R_{\rm tr}^{\rm sing} = -B_{d,\eta} g^{d-3} \int_0^{\pi/2} d\vartheta (\sin\vartheta)^d (\cos\vartheta)^{\frac{d-3}{\eta}} \quad , \quad \eta > 1 \quad , \tag{10.131}$$

where $B_{d,\eta}$ is some numerical constant which depends on d and η in a complicated way, but remains finite as long as $\eta > 1$. By simple power counting, we see that this integral exists for $\frac{d-3}{\eta} > -1$, i.e.

$$\eta > 3 - d \quad . \tag{10.132}$$

Combining this result with the criterion (10.117) for $\eta < 1$, and using the fact that a finite value of $R_{\rm tr}$ implies the boundedness of the total Debye-Waller factor $Q_{\rm tr}(\tilde{x}, \tilde{\tau})$ for all $\tilde{x}$ and $\tilde{\tau}$, we conclude that outside the shaded region shown in Fig. 10.3 the contribution of the transverse gauge fields to the Debye-Waller factor remains bounded. Hence, non-Fermi liquid behavior due to $Q^{\alpha}_{\rm tr}(r^{\alpha}_{\parallel}\hat{v}^{\alpha}, \tau)$ is only possible in the shaded regime shown in Fig. 10.3. In particular, for the three-dimensional Maxwell action ($\eta = 2$, $d = 3$) higher-dimensional bosonization with linearized energy dispersion predicts that the static Debye-Waller factor grows logarithmically with distance (as in the one-dimensional Tomonaga-Luttinger model), while the inclusion of curvature leads to a bounded Debye-Waller factor!

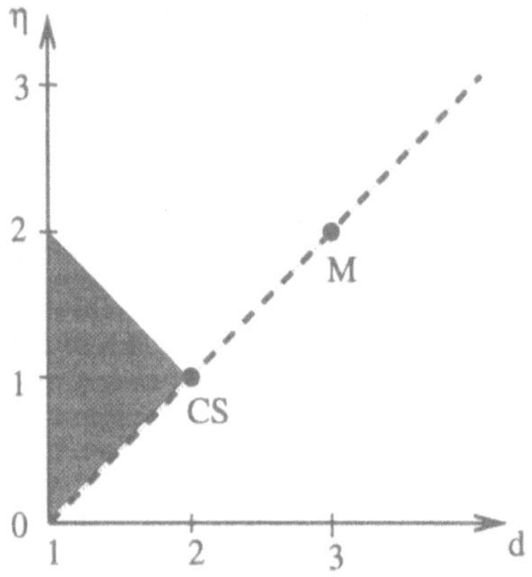

Fig. 10.3. The shaded triangle is the parameter regime in the $d - \eta$-plane where the long-distance and large-time behavior of the Debye-Waller $Q^{\alpha}_{\rm tr}(r^{\alpha}_{\parallel}\hat{v}^{\alpha}, \tau)$ due to transverse gauge fields gives rise to non-Fermi liquid behavior. Note that for linearized energy dispersion one incorrectly obtains non-Fermi liquid behavior for all points above the dashed line $\eta = d - 1$. The points M and CS correspond to the three-dimensional Maxwell theory and the two-dimensional Maxwell-Chern-Simons theory, respectively.

In the marginal case $\eta = 1$ our simple power-counting analysis is not sufficient, and we cannot avoid explicitly performing the relevant integrations. Because in two dimensions the case $\eta = 1$ is of particular physical interest in connection with the half-filled Landau level, we shall analyze this case in some detail in the following section.

10.4.3 Two-dimensional Maxwell-Chern-Simons theory

We evaluate the constant part $R_{\rm tr}$ of the Debye-Waller factor given in Eq.(10.118) for the special case of the two-dimensional Maxwell-Chern-Simons theory ($\eta = 1, d = 2$).

Two-dimensional electron systems in strong external magnetic fields are difficult to handle within the framework of conventional many-body theory. In fact, in this problem the most successful theories are based on variational wave-functions, and do not make use of the standard methods of second quantization [10.27]. But also functional methods have been very fruitful [10.10, 10.28, 10.29]. Of particular recent interest has been the case when the areal density of the electron gas and the strength of the external magnetic field are such that the lowest Landau level is exactly half-filled[9]. Then the two-dimensional electron system is mathematically equivalent to a system of fermions interacting with a Chern-Simons gauge field such that the average gauge field acting on the fermions is zero [10.10–10.12]. The Chern-Simons field effectively attaches two flux quanta to each electron. The resulting spinless fermions are called *composite fermions* [10.30]. At the mean-field level, where fluctuations of the Chern-Simons gauge field are ignored, the magnetic field generated by the Chern-Simons field exactly cancels the external magnetic field, so that mean-field theory predicts that composite fermions in the half-filled Landau level should behave like free spinless fermions without magnetic field, with Fermi wave-vector $k_{\mathrm{F}} = (4\pi n_{\mathrm{e}})^{1/2}$. Here n_{e} is the areal density of the two-dimensional electron gas. Although the existence of a well-defined Fermi surface in the half-filled Landau level has been confirmed by several experiments [10.31–10.35] and there exists general agreement that experimentally composite fermions manifest themselves as well-defined quasi-particles, theoretically the situation is less clear. For a summary of the current status of the fermionic Chern-Simons description of quantum Hall systems see the recent review by Halperin [10.36].

In the usual perturbative approach the leading correction to the Green's function of composite fermions due to the fluctuations of the Chern-Simons field is obtained from the GW self-energy (see Eq.(5.74))

$$\Sigma^{\alpha}_{\mathrm{GW,tr}}(\tilde{q}) = -\frac{1}{\beta V}\sum_{q'} h^{\mathrm{RPA},\alpha}_{q'} G^{\alpha}_{0}(\tilde{q}+q') \quad . \tag{10.133}$$

In $d=2$ this expression is easily evaluated if one introduces circular coordinates centered at $\boldsymbol{k}^{\alpha}$ and first performs the angular integration exactly [10.1]. In the regime where $v_{\mathrm{F}}|\boldsymbol{q}|$ is not much larger than $|\tilde{\omega}_n|$, one finds to leading order for small frequencies [4.7, 10.12]

$$\Sigma^{\alpha}_{\mathrm{GW,tr}}(\tilde{q}) \propto -\mathrm{i}\tilde{\omega}_n \ln\left(\frac{v_{\mathrm{F}} q_{\mathrm{c}}}{|\tilde{\omega}_n|}\right) \quad . \tag{10.134}$$

After analytic continuation to real frequencies, the real part of the self-energy vanishes as $\omega \ln(v_{\mathrm{F}} q_{\mathrm{c}}/|\omega|)$ for $\omega \to 0$, implying a logarithmically vanishing

[9] This means that $N\phi_0/\phi = 1/2$, where N is the number of electrons in the system, $\phi_0 = hc/e$ is the flux quantum, and ϕ is the total magnetic flux through the system.

quasi-particle residue[10]. Because Z^{α} vanishes, the momentum distribution does not exhibit a step-discontinuity at the Fermi surface (see Eq.(2.26)). Thus, lowest order perturbation theory suggests that composite fermions are *not* well defined quasi-particles. This seems to disagree with the experimental evidence [10.31–10.35] that composite fermions behave like well-defined quasi-particles.

Because the leading perturbative correction completely changes the mean-field picture, the single-particle Green's function can only be calculated by means of non-perturbative methods, which sum infinite orders in perturbation theory. However, controlled non-perturbative methods in $d > 1$ are rare, and it is at least controversial whether the methods applied so far to the problem of composite fermions in the half-filled Landau level are really valid. These include the so-called eikonal approximation [5.11], a $1/N$-expansion [10.17], and higher-dimensional bosonization with linearized energy dispersion [1.33, 10.18]. All of these methods predict some kind of non-Fermi liquid behavior, but there is no general agreement on the detailed form of the Green's function. Although this might be related to the gauge-dependence of the single-particle Green's function (see, however, the work [10.23] and the footnote after Eq.(10.14)), the discrepancies between the various approaches could also be related to uncontrolled approximations inherent in each of the different resummation schemes[11]. Moreover, as we shall show in this section, even in the marginal case of $\eta = 1$ higher-dimensional bosonization *with linearized energy dispersion* does not correctly resum the leading singularities in the perturbation series. Note that in two dimensions the case $\eta = 1$ corresponds to the right corner CS of the shaded triangle in Fig. 10.3. Naively, one might expect that precisely on the boundary of the triangle one obtains logarithmic singularities in the transverse Debye-Waller factor, which are correctly predicted by bosonization with linearized energy dispersion [10.18]. We now show that at the special point $d = 2$ and $\eta = 1$ this is not the case.

Using the fact that in two dimensions $\lambda_2 = 1$ (see Eqs.(10.100) and (10.102)), we obtain from Eq.(10.118) for $\eta = 1$

$$R_{\mathrm{tr}} = -\frac{g}{\pi^2} \int_0^{\pi/2} \mathrm{d}\vartheta \frac{(\sin\vartheta)^2}{\cos\vartheta} \int_0^{p_c (\cos\vartheta)^{-1}} \mathrm{d}k F(k, gk) \quad . \tag{10.135}$$

[10] In this section we shall consider only the case $\eta = 1$ (the Maxwell-Chern-Simons theory), corresponding to the unscreened Coulomb interaction. In [10.1] we have studied the general Chern-Simons theory with $\eta > 1$. Then $\Sigma^{\alpha}_{\mathrm{GW,tr}}(\tilde{q}) \propto -\mathrm{i}\,\mathrm{sgn}(\tilde{\omega}_n)|\tilde{\omega}_n|^{\frac{2}{1+\eta}}$, so that the quasi-particle residue vanishes like a power law for $\omega \to 0$. Note that $\eta = 2$ describes experiments where the long-range part of the Coulomb interaction is screened by metal plates.

[11] For example, in the work [5.12] we have shown that in one dimension the usual eikonal approximation [5.10] does not correctly reproduce the exact solution of the Tomonaga-Luttinger model.

To determine the proper values of our dimensionless constant g, we note that in case of the Chern-Simons propagator the wave-vector q_c in Eq.(10.2) is given by $q_c = (2k_F)^2/\kappa$, where $\kappa = 2\pi e^2 \nu = e^2 m$ is the Thomas-Fermi wave-vector in two dimensions [10.12]. Hence we obtain from Eq.(10.113)

$$g = \frac{4k_F}{\kappa} = \frac{4v_F}{e^2} \quad . \tag{10.136}$$

To obtain the relevant ultraviolet cutoff p_c, we note that the dimensionless Coulomb interaction becomes larger than unity for $|\boldsymbol{q}| \lesssim \kappa$, see Appendix A.3.1. Hence,

$$p_c = \frac{\kappa}{q_c} = \left(\frac{\kappa}{2k_F} \right)^2 = \frac{4}{g^2} \quad . \tag{10.137}$$

Note that the charge e in Eq.(10.136) should be understood as effective screened charge, which takes the dielectric screening due to the background into account. Halperin, Lee and Read [10.12] have estimated $v_F/e^2 \approx 0.3$ in the experimentally relevant regime, so that $g \approx 1.2$. Let us also emphasize that the prefactor of $g = 4k_F/\kappa$ in Eq.(10.135) is the ***inverse*** of the prefactor of κ/k_F that appears in the constant part R^α of the Debye-Waller factor for conventional density-density interactions in $d = 2$, see Eq.(6.49). Hence the wave-vector scale $q_c = (2k_F)^2/\kappa$ now plays the same role as the Thomas-Fermi wave-vector κ in the case of density-density interactions. Note, however, that $g = q_c/k_F$ is the only small parameter which formally justifies the truncation of the eikonal expansion (5.142) at the first order[12]. We therefore conclude that Eq.(10.135) can only be qualitatively correct for $g \ll 1$. If this condition is satisfied, the higher-order corrections $Q_n^\alpha(\boldsymbol{r}, \tau)$, $n \geq 2$, to the average eikonal (see Eq.(5.142)) are controlled by higher powers of g, which are generated by additional loop integrations. Obviously, in the experimentally relevant regime [10.31] the condition $g \ll 1$ is not satisfied, so that for an accurate quantitative comparison with experiments it is not sufficient to retain only the leading term in the eikonal expansion.

In order to perform a controlled calculation, we shall restrict ourselves from now on to the regime $g \ll 1$, with the hope that the qualitative behavior of the Green's function does not change for larger g. Naively one might be tempted to replace the upper limit for k-integration in Eq.(10.135) by infinity, because $p_c = 4/g^2 \gg 1$ for small g, and the ϑ-integration seems to be dominated by the regime $\cos\vartheta \ll 1$. If we ***linearize*** the energy dispersion, such

[12] Recall that in Chap. 4.3.4 we have shown by explicit calculation of corrections to the density-density correlation function beyond the RPA that the loop integrations give rise to additional powers of q_c/k_F, see Eq.(4.115). Because the non-Gaussian corrections $Q_n^\alpha(\boldsymbol{r}, \tau)$, $n \geq 2$, to the average eikonal involve additional powers of the interaction, these corrections are controlled by higher orders in q_c/k_F. Note that for a spherical Fermi surface the curvature parameter C^α in Eq.(4.115) is of the order of unity. Furthermore, for the Chern-Simons action the value of the relevant dimensionless effective interaction is not small, which leaves us with q_c/k_F as the only small parameter in the problem.

a procedure is indeed correct, because in this case the k-integration yields a finite number, which is according to Eq.(10.122) given by

$$\int_0^\infty \mathrm{d}k F(k,0) = 2\int_0^\infty \mathrm{d}k \frac{\pi k - k^2 - 1 + (k^2-1)\ln k}{(1+k^2)^2} \quad . \tag{10.138}$$

To perform the integration, we need [6.3, 6.4]

$$I_\mu = \int_0^\infty \mathrm{d}x \frac{x^{2\mu-1}}{[1+x^2]^2} = \frac{1}{2}\Gamma(\mu)\Gamma(2-\mu) \quad , \quad 0<\mu<2 \quad , \tag{10.139}$$

$$\begin{aligned} \tilde{I}_\mu &= \int_0^\infty \mathrm{d}x \frac{x^{2\mu-1}\ln x}{[1+x^2]^2} \\ &= \begin{cases} \frac{(\mu-1)\pi}{4\sin(\pi\mu)}\left[\cot(\pi\mu) - \frac{1}{\mu-1}\right] & , \ 0<\mu<2\,,\ \mu\neq 1 \\ 0 & , \ \mu = 1 \end{cases} \quad . \end{aligned} \tag{10.140}$$

Hence,

$$\int_0^\infty \mathrm{d}k F(k,0) = 2\left[\pi I_{\frac{3}{4}} - I_{\frac{3}{2}} - I_{\frac{1}{2}} + \tilde{I}_{\frac{3}{2}} - \tilde{I}_{\frac{1}{2}}\right] \quad . \tag{10.141}$$

With $I_1 = 1/2$, $I_{\frac{1}{2}} = I_{\frac{3}{2}} = \pi/4$, $\tilde{I}_{\frac{1}{2}} = -\pi/4$, and $\tilde{I}_{\frac{3}{2}} = \pi/4$ we finally obtain

$$\int_0^\infty \mathrm{d}k F(k,0) = \pi \quad . \tag{10.142}$$

It is now easy to see that R_{tr} is logarithmically divergent. Of course, in this case we should consider the total Debye-Waller factor $Q_{\mathrm{tr}}(\tilde{x},\tilde{\tau})$ in Eq.(10.115), which grows logarithmically for large $\tilde{x}$ or $\tilde{\tau}$. Setting for simplicity $\tilde{\tau}=0$, it is easy to show from Eqs.(10.115) and (10.135) that, to leading logarithmic order for large $\tilde{x}$, one obtains with linearized energy dispersion

$$Q_{\mathrm{tr}}(\tilde{x},0) \sim -\frac{g}{\pi}\int_0^{\pi/2} \mathrm{d}\vartheta \frac{(\sin\vartheta)^2}{\cos\vartheta}\left[1-\cos[(\cos\vartheta)^2\tilde{x}]\right] \sim -\frac{g}{2\pi}\ln\tilde{x} \,. \tag{10.143}$$

This implies anomalous scaling characteristic for Luttinger liquids, with anomalous dimension given by $\gamma_{\mathrm{CS}} = g/(2\pi)$.

The crucial point is now that the above result is completely changed by the quadratic term in the energy dispersion, because even for small g it is *not* allowed to set $g=0$ in the integrand $F(k,gk)$ of Eq.(10.135). To see this, consider the function

$$J(g,h) = \int_0^h \mathrm{d}k F(k,gk) \quad , \quad g>0 \quad . \tag{10.144}$$

According to Eq.(10.135) the constant part R_{tr} of the Debye-Waller factor is determined by $J(g, p_{\mathrm{c}}(\cos\vartheta)^{-1})$, where $p_{\mathrm{c}} \propto g^{-2}$, see Eq.(10.137). For $g=0$

we have from Eq.(10.141) $\lim_{h\to\infty} J(0,h) = \pi$. To evaluate $J(g,h)$ for finite g, we use Eq.(10.121) to write $J(g,h) = \sum_{n=1}^{4} J_n(g,h)$, where[13]

$$J_1(g,h) = \frac{\pi}{g}\left[\int_0^{2/g} \mathrm{d}k \frac{1}{k^2+(1-\frac{gk}{2})^2} - \int_{2/g}^{h} \mathrm{d}k \frac{1}{k^2+(1-\frac{gk}{2})^2}\right] , \quad (10.145)$$

$$J_2(g,h) = -\frac{\pi}{g}\int_0^{h} \mathrm{d}k \frac{(1+\frac{gk}{2})}{k^2+(1+\frac{gk}{2})^2} , \quad (10.146)$$

$$J_3(g,h) = \frac{2}{g}\int_0^{h} \frac{\mathrm{d}k}{k} \frac{(1-\frac{gk}{2})\ln\left[\frac{|1-\frac{gk}{2}|}{k}\right]}{k^2+(1-\frac{gk}{2})^2} , \quad (10.147)$$

$$J_4(g,h) = -\frac{2}{g}\int_0^{h} \frac{\mathrm{d}k}{k} \frac{(1+\frac{gk}{2})\ln\left[\frac{|1+\frac{gk}{2}|}{k}\right]}{k^2+(1+\frac{gk}{2})^2} . \quad (10.148)$$

From Eq.(10.137) we see that the upper limit for the k-integration in Eq.(10.118) is large compared with $g/2$, so that we may assume $h > g/2$. In the first integral on the right-hand side of Eq.(10.145) and in Eq.(10.147) we substitute $x = k/(1-\frac{gk}{2})$ (so that $\frac{\mathrm{d}x}{x^2} = \frac{\mathrm{d}k}{k^2}$), and in the second integral of Eq.(10.145) we set $x = k/(\frac{gk}{2}-1)$ (so that $\frac{\mathrm{d}x}{x^2} = -\frac{\mathrm{d}k}{k^2}$). Similarly, in the above expressions for J_2 and J_4 we substitute $x = k/(1+\frac{gk}{2})$ (so that again $\frac{\mathrm{d}x}{x^2} = \frac{\mathrm{d}k}{k^2}$). With these substitutions it is easy to show that

$$\begin{aligned} J_1(g,h) + J_2(g,h) &= \frac{\pi}{g}\int_{\frac{2}{g+2/h}}^{\frac{2}{g-2/h}} \mathrm{d}x \frac{1}{1+x^2} \\ &= \frac{\pi}{g}\left\{\arctan\left[\frac{2}{g(1-\frac{2}{gh})}\right] - \arctan\left[\frac{2}{g(1+\frac{2}{gh})}\right]\right\} , \end{aligned} \quad (10.149)$$

and

$$J_3(g,h) + J_4(g,h) = -\frac{2}{g}\int_{\frac{2}{g+2/h}}^{\frac{2}{g-2/h}} \mathrm{d}x \frac{\ln x}{x(1+x^2)} . \quad (10.150)$$

In the limit of interest ($g \ll 1$, $gh \gg 1$) the width of the interval of integration is small,

$$\frac{2}{g(1-\frac{2}{gh})} - \frac{2}{g(1+\frac{2}{gh})} \approx \frac{8}{g^2 h} . \quad (10.151)$$

Hence, to leading order, the integrals can be approximated by the product of the value of the integrand at $x = 2/g$ and the width of the interval of integration. Then we obtain to leading order

[13] Although J_3 and J_4 are logarithmically divergent, the divergence cancels in the sum $J_3 + J_4$, which is the only relevant combination.

$$J_1(g,h) + J_2(g,h) \approx \frac{2\pi}{gh} \quad , \tag{10.152}$$

$$J_3(g,h) + J_4(g,h) \approx -\frac{4\ln g^{-1}}{h} \quad . \tag{10.153}$$

Note that for small g the contribution $J_1 + J_2$ is dominant. Taking the limit $h \to \infty$, we obtain $\lim_{h\to\infty} J(g,h) = 0$, so that we conclude that

$$\int_0^\infty \mathrm{d}k F(k, gk) = 0 \quad , \quad g > 0 \quad , \tag{10.154}$$

which should be compared with Eq.(10.142). Because Eq.(10.152) depends on the ***product*** of the small parameter g and the large parameter h, it is clear that for any finite g the limiting behavior of the integral $J(g,h)$ for large h is very different from $\lim_{h\to\infty} J(0,h) = \pi$. This is the mathematical reason why the linearization of the energy dispersion in the two-dimensional Chern-Simons theory is not allowed. Using Eq.(10.154), we see that Eq.(10.135) can be rewritten as

$$R_{\mathrm{tr}} = \frac{g}{\pi^2} \int_0^{\pi/2} \mathrm{d}\vartheta \frac{(\sin\vartheta)^2}{\cos\vartheta} \int_{p_{\mathrm{c}}(\cos\vartheta)^{-1}}^\infty \mathrm{d}k F(k, gk) \quad . \tag{10.155}$$

From this expression it is evident that the cutoff-dependence of the k-integral gives rise to an additional power of $\cos\vartheta$ in the numerator, which removes the logarithmic divergence that has been artificially generated by linearizing the energy dispersion. According to Eq.(10.137) we should choose $p_{\mathrm{c}} = 4/g^2$, so that we obtain to leading order for small g (see Eq.(10.152))

$$\int_{\frac{4}{g^2}(\cos\vartheta)^{-1}}^\infty \mathrm{d}k F(k, gk) \sim -\frac{\pi}{2} g \cos\vartheta \quad . \tag{10.156}$$

Hence Eq.(10.155) reduces to

$$R_{\mathrm{tr}} \sim -\frac{1}{8} g^2 \quad , \quad g \ll 1 \quad . \tag{10.157}$$

The precise numerical value of the prefactor 1/8 is the result of our special choice of the cutoff p_{c} in Eq.(10.137) and has no physical significance. However, Eqs.(10.143) and (10.157) imply that in the case of the two-dimensional Chern-Simons theory it is *not* allowed to linearize the energy dispersion [10.18]. Physically Eq.(10.157) represents a contribution from gauge field fluctuations with wavelengths large compared with the Thomas-Fermi screening length κ^{-1} to the reduction of the quasi-particle residue. While for linearized energy dispersion one finds that these fluctuations wash out any step-discontinuity at the Fermi surface, the quadratic term in the energy dispersion drastically changes this scenario: in the regime $g \ll 1$ under consideration the right-hand-side of Eq.(10.157) is very small, so that this term can be safely ignored and certainly does not modify the mean-field prediction of a step discontinuity at the Fermi surface.

10.5 Summary and outlook

In this chapter we have generalized our non-perturbative background field method for calculating the single-particle Green's function to the case of fermions that are coupled to transverse gauge fields. Let us summarize again our main result for the special case of a spherical Fermi surface in d dimensions. As discussed in Chaps. 2.5 and 5.4, in this case it is *not* necessary to partition the Fermi surface into several patches, so that uncontrolled corrections due to the around-the-corner processes discussed in Chap. 2.4.3 simply do not arise. The Matsubara Green's function can then be written as

$$G(\boldsymbol{k}^{\alpha} + \boldsymbol{q}, \mathrm{i}\tilde{\omega}_n) = \int \mathrm{d}\boldsymbol{r} \int_0^{\beta} \mathrm{d}\tau \mathrm{e}^{-\mathrm{i}(\boldsymbol{q}\cdot\boldsymbol{r} - \tilde{\omega}_n \tau)} \tilde{G}^{\alpha}(\boldsymbol{r}, \tau) \mathrm{e}^{Q^{\alpha}(\boldsymbol{r}, \tau)} \quad , \tag{10.158}$$

$$Q^{\alpha}(\boldsymbol{r}, \tau) = Q^{\alpha}_{1}(\boldsymbol{r}, \tau) + Q^{\alpha}_{\mathrm{tr}}(\boldsymbol{r}, \tau) \quad , \tag{10.159}$$

where the longitudinal Debye-Waller factor $Q^{\alpha}_{1}(\boldsymbol{r}, \tau)$ is given in Eqs.(5.151)–(5.153), and the contribution $Q^{\alpha}_{\mathrm{tr}}(\boldsymbol{r}, \tau)$ from the transverse gauge field to the Debye-Waller factor is given in Eqs.(10.67)–(10.69). The prefactor Green's function $\tilde{G}^{\alpha}(\boldsymbol{r}, \tau)$ has the following Fourier expansion,

$$\tilde{G}^{\alpha}(\boldsymbol{r}, \tau) = \frac{1}{\beta V} \sum_{\tilde{q}} \mathrm{e}^{\mathrm{i}(\boldsymbol{q}\cdot\boldsymbol{r} - \tilde{\omega}_n \tau)} \tilde{G}^{\alpha}(\tilde{q}) \quad , \tag{10.160}$$

$$\tilde{G}^{\alpha}(\tilde{q}) = \frac{1 + Y^{\alpha}(\tilde{q}) + Y^{\alpha}_{\mathrm{tr}}(\tilde{q})}{\mathrm{i}\tilde{\omega}_n - \epsilon_{\boldsymbol{k}^{\alpha}+\boldsymbol{q}} + \mu - \Sigma^{\alpha}_{1}(\tilde{q}) - \Sigma^{\alpha}_{1,\mathrm{tr}}(\tilde{q})} \quad , \tag{10.161}$$

where the self-energies and the vertex functions are given in Eqs.(5.186), (5.187), (10.74), and (10.75). Due to the spherical symmetry, it is sufficient to evaluate Eq.(10.158) for external wave-vectors of the form $\boldsymbol{q} = q^{\alpha}_{\parallel} \hat{\boldsymbol{k}}^{\alpha}$, and then replace $q^{\alpha}_{\parallel} \to |\boldsymbol{k}| - k_{\mathrm{F}}$ in the final result, see also Eqs.(5.181),(5.182) and the discussion in Chap. 2.5.

In Sect. 10.4 we have shown that for the calculation of the transverse Debye-Waller factor $Q^{\alpha}_{\mathrm{tr}}(\boldsymbol{r}, \tau)$ it is essential to retain the quadratic term in the expansion of the energy dispersion close to the Fermi surface. In physically relevant cases one obtains then a *bounded* Debye-Waller factor, which does not lead to a breakdown of the Fermi liquid state. This is in sharp contrast with the results of higher-dimensional bosonization with linearized energy dispersion [1.33, 10.2, 10.18]. We would like to emphasize that the quadratic term in the energy dispersion is *irrelevant* in the renormalization group sense. However, it is relevant in the sense that the *exponentiation of the perturbation series for the real-space Green's function*, which in arbitrary dimensions is the characteristic feature of bosonization with linearized energy dispersion, does not resum the dominant singularities.

One of the most interesting problems for further research is the **evaluation of the prefactor Green's function** $\tilde{G}^{\alpha}(\tilde{q})$ in the case of the Chern-Simons theory for the half-filled Landau level. Because by construction our approach exactly reproduces the leading term in a naive expansion of the Green's function in powers of the effective interaction (see Chap. 5.3.3), the perturbatively obtained signature (10.134) of non-Fermi liquid behavior is certainly contained in Eqs.(10.158)–(10.161). Very recently Castilla and the present author [10.1] have made considerable progress in evaluating the above expressions for the case $\eta > 1$. Because we know from Sect. 10.4 that the contribution from the Debye-Waller factor is finite and small, the low-energy behavior of the total spectral function is essentially determined by the imaginary part of the prefactor Green's function $\tilde{G}^{\alpha}(\boldsymbol{q}, \omega + i0^{+})$. Most importantly, we have shown in the Letter [10.1] that the Gaussian fluctuations of the Chern-Simons gauge field do not invalidate the quasi-particle picture for the composite fermions in the half-filled Landau level. On other words, our non-perturbative approach predicts a narrow peak in the spectral function, with a width that vanishes faster than the quasi-particle energy as $\boldsymbol{q} \rightarrow 0$. This clearly demonstrates that lowest order perturbation theory is not reliable, and explains the experimental fact that composite fermions in half-filled quantum Hall systems behave like well-defined quasi-particles [10.31–10.35].

In our opinion, the calculation of the Green's function of fermions that are coupled to gauge fields is the physically most interesting and important application of the non-perturbative method developed in this book. Gauge fields in non-relativistic condensed matter systems arise not only in connection with the quantum Hall effect, but also in effective low-energy theories for strongly correlated Fermi systems [1.24, 10.6–10.9]. Because the gauge field problem cannot be analyzed within perturbation theory, controlled non-perturbative methods are necessary. In the absence of any other small parameter, the Gaussian approximation employed in our background field approach is justified for $g = q_c/k_F \ll 1$ (see Eq.(10.113)). In this case the closed loop theorem discussed in Chap. 4.1 guarantees that the corrections to the Gaussian approximation involve higher powers of our small parameter g.

It seems that the potential of our approach is far from being exhausted. Let us point out two obvious directions for further research. First of all, the combination of the methods developed in Chap. 9 with the results of the present chapter might lead to a new non-perturbative approach to the **random gauge field problem**. Random gauge fields and the related problem of random magnetic fields have recently been analyzed with the help of many different methods [10.37–10.41]. Of course, this problem is interesting in connection with the quantum Hall effect, because experimental systems always have a finite amount of disorder. Another interesting and only partially solved problem is the **explicit calculation of the non-Gaussian corrections** to our non-perturbative result for the single-particle Green's function. Recall that in Chap. 4.3 we have performed such a calculation for the density-density correlation function. Although in Chap. 5.2 we have derived explicit expressions for the non-Gaussian corrections to the average eikonal (see Eqs.(5.142)–(5.147)), a detailed analysis of the leading correction to the Gaussian approximation still remains to be done. At this point we cannot exclude the possibility that, although the leading term in the expansion of the average eikonal (i.e. the Debye-Waller factor $Q^{\alpha}_{\mathrm{tr}}(\boldsymbol{r}, \tau)$ given in Eq.(10.108)) remains bounded for all $\boldsymbol{r}$ and τ, the higher order terms $Q^{\alpha}_{n}(\boldsymbol{r}, \tau)$, $n \geq 2$, in Eq.(5.143) exhibit singularities which lead to non-Fermi liquid behavior in the spectral function.

Appendix: Screening and collective modes

We summarize some useful expressions for the polarization, the dynamic structure factor and the long wavelength behavior of the collective plasmon mode within the RPA. The results presented in this chapter are not new, but a systematic discussion of the above quantities as function of dimensionality seems not to exist in the literature.

A.1 The non-interacting polarization for spherical Fermi surfaces

... which we need in order to calculate the dynamic structure factor within the RPA. Here and in the following two sections we assume spherical symmetry.

For a spherical Fermi surface in d dimensions it is easy to show from Eqs.(3.13) and (4.24) that the non-interacting polarization is in the limit $V, \beta \to \infty$ and $|\boldsymbol{q}| \ll k_{\mathrm{F}}$ given by

$$\Pi_0(q) = \nu g_d \left(\frac{\mathrm{i}\omega_m}{v_{\mathrm{F}} |\boldsymbol{q}|} \right) \quad , \tag{A.1}$$

where the density of states at the Fermi energy is (see Eq.(4.28))

$$\nu = \int \frac{\mathrm{d}\boldsymbol{k}}{(2\pi)^d} \delta(\epsilon_{\boldsymbol{k}} - \mu) \quad , \tag{A.2}$$

and the dimensionless function $g_d(z)$ is defined by

$$g_d(z) = \left\langle \frac{\hat{\boldsymbol{q}} \cdot \hat{\boldsymbol{k}}}{\hat{\boldsymbol{q}} \cdot \hat{\boldsymbol{k}} - z} \right\rangle_{\hat{\boldsymbol{k}}} \quad . \tag{A.3}$$

Here $\hat{\boldsymbol{k}} = \boldsymbol{k}/|\boldsymbol{k}|$, $\hat{\boldsymbol{q}} = \boldsymbol{q}/|\boldsymbol{q}|$, and $< \ldots >_{\hat{\boldsymbol{k}}}$ denotes angular average over the surface of the d-dimensional unit sphere in $\boldsymbol{k}$-space, i.e. for any function $f(\hat{\boldsymbol{k}})$

$$\left\langle f(\hat{\boldsymbol{k}}) \right\rangle_{\hat{\boldsymbol{k}}} = \frac{\int \mathrm{d}\Omega_{\hat{\boldsymbol{k}}} f(\hat{\boldsymbol{k}})}{\int \mathrm{d}\Omega_{\hat{\boldsymbol{k}}}} \quad , \tag{A.4}$$

where $d\Omega_{\hat{k}}$ is the differential solid angle at point $\hat{k}$ on the unit sphere. Note that by construction $g_d(0) = 1$. For a system of N spinless electrons with mass m in a d-dimensional volume V the density of states can be written as

$$\nu = \frac{d}{2\mu}\frac{N}{V} = \frac{\Omega_d}{(2\pi)^d}\frac{k_F^{d-1}}{v_F} = \frac{\Omega_d}{(2\pi)^d} m k_F^{d-2} \quad , \tag{A.5}$$

where Ω_d is the surface area of the unit sphere in d dimensions,

$$\Omega_d = \int d\Omega_{\hat{k}} = \frac{2\pi^{\frac{d}{2}}}{\Gamma(\frac{d}{2})} \quad . \tag{A.6}$$

The integrand in Eq.(A.3) depends only on $\cos\vartheta = \hat{q} \cdot \hat{k}$. For this type of functions it is convenient to use d-dimensional spherical coordinates,

$$\left\langle f(\hat{q}\cdot\hat{k})\right\rangle_{\hat{k}} = \gamma_d \int_0^\pi d\vartheta (\sin\vartheta)^{d-2} f(\cos\vartheta) \quad , \quad \text{for } d > 1 \quad , \tag{A.7}$$

$$\left\langle f(\hat{q}\cdot\hat{k})\right\rangle_{\hat{k}} = \frac{1}{2}\left[f(1) + f(-1)\right] \quad , \quad \text{for } d = 1 \quad . \tag{A.8}$$

Here the numerical constant γ_d is defined by

$$\gamma_d = \left\langle \delta(\hat{q}\cdot\hat{k})\right\rangle_{\hat{k}} = \left[\int_0^\pi d\vartheta (\sin\vartheta)^{d-2}\right]^{-1} \quad , \tag{A.9}$$

and can be identified with the ratio of the surfaces of the unit spheres in $d-1$ and d dimensions,

$$\gamma_d = \frac{\Omega_{d-1}}{\Omega_d} = \frac{\Gamma(\frac{d}{2})}{\sqrt{\pi}\,\Gamma(\frac{d-1}{2})} \quad . \tag{A.10}$$

In particular,

$$\gamma_1 = 0 \quad , \quad \gamma_2 = \frac{1}{\pi} \quad , \quad \gamma_3 = \frac{1}{2} \quad . \tag{A.11}$$

For $z = iy$ and real y the function $g_d(iy)$ is an even and positive function of y, and is in $d = 1, 2, 3$ explicitly given by

$$g_1(iy) = 1 - \frac{y^2}{1+y^2} = \frac{1}{1+y^2} \quad , \tag{A.12}$$

$$g_2(iy) = 1 - \frac{|y|}{\sqrt{1+y^2}} \quad , \tag{A.13}$$

$$g_3(iy) = 1 - |y| \arctan\left(\frac{1}{|y|}\right) \quad . \tag{A.14}$$

On the real axis we have

$$g_1(x+\mathrm{i}0^+) = \frac{1}{1-(x+\mathrm{i}0^+)^2} \quad , \tag{A.15}$$

$$g_2(x+\mathrm{i}0^+) = 1 - \frac{x}{\sqrt{(x+\mathrm{i}0^+)^2-1}} \quad , \tag{A.16}$$

$$g_3(x+\mathrm{i}0^+) = 1 - \frac{x}{2}\ln\left(\frac{x+\mathrm{i}0^+ + 1}{x+\mathrm{i}0^+ - 1}\right) \quad . \tag{A.17}$$

For $|x| < 1$ the function $g_d(x + \mathrm{i}0^+)$ has in $d > 1$ a finite imaginary part. In the expression for the RPA dynamic structure factor discussed below this imaginary part describes the decay of density fluctuations into particle-hole excitations, i.e. Landau damping [1.7]. From Eq.(A.3) it is easy to show that

$$\mathrm{Im}g_d(x+\mathrm{i}0^+) = \pi x \left\langle \delta(\hat{\boldsymbol{q}}\cdot\hat{\boldsymbol{k}} - x)\right\rangle_{\hat{\boldsymbol{k}}} \quad , \tag{A.18}$$

so that

$$\mathrm{Im}g_d(x+\mathrm{i}0^+) = \pi\gamma_d x \quad , \quad \text{for } |x| \ll 1 \quad . \tag{A.19}$$

Keeping in mind that $g_d(0) = 1$, this implies on the imaginary axis

$$g_d(\mathrm{i}y) = 1 - \pi\gamma_d|y| \quad , \quad \text{for } |y| \ll 1 \quad . \tag{A.20}$$

For large $|z|$ we have in any dimension

$$g_d(z) \sim -\frac{1}{dz^2} \quad , \quad \text{for } |z| \gg 1 \quad . \tag{A.21}$$

Finally, on the real axis we have in the vicinity of unity to leading order in $\delta = x - 1 > 0$

$$g_d(1+\delta) \sim \begin{cases} g_d(1) < 0 & \text{for } d > 3 \\ -\frac{1}{2}\ln(1/\delta) & \text{for } d = 3 \\ -c_d/\delta^{\frac{3-d}{2}} & \text{for } d < 3 \end{cases} \quad , \tag{A.22}$$

where c_d is a positive numerical constant. In particular, $c_1 = \frac{1}{2}$ and $c_2 = \frac{1}{\sqrt{2}}$.

A.2 The dynamic structure factor for spherical Fermi surfaces

Within the RPA the dynamic structure factor consists of two contributions: The first one is a featureless function and describes the decay of density fluctuations into particle-hole pairs, i.e. Landau damping; the second one is a δ-function peak due to the collective plasmon mode.

For simplicity we shall assume in the rest of this chapter that the bare interaction is frequency-independent, i.e. $f_q = f_{\boldsymbol{q}}$. Introducing the dimensionless interaction

$$F_q = \nu f_q \quad , \tag{A.23}$$

we obtain from Eqs.(2.47) and (A.1) for the RPA density-density correlation function in the long-wavelength limit

$$\Pi_{\mathrm{RPA}}(q) = \nu \frac{g_d(\frac{\mathrm{i}\omega_m}{v_{\mathrm{F}}|\boldsymbol{q}|})}{1 + F_q g_d(\frac{\mathrm{i}\omega_m}{v_{\mathrm{F}}|\boldsymbol{q}|})} \quad . \tag{A.24}$$

According to Eq.(2.45) the RPA dynamic structure factor is then given by

$$S_{\mathrm{RPA}}(\boldsymbol{q}, \omega) = \frac{\nu}{\pi} \mathrm{Im} \left\{ \frac{g_d(\frac{\omega}{v_{\mathrm{F}}|\boldsymbol{q}|} + \mathrm{i}0^+)}{1 + F_q g_d(\frac{\omega}{v_{\mathrm{F}}|\boldsymbol{q}|} + \mathrm{i}0^+)} \right\} \quad . \tag{A.25}$$

From the properties of the function $g_d(z)$ discussed above it is clear that there exist two separate contributions to the imaginary part in Eq.(A.25),

$$S_{\mathrm{RPA}}(\boldsymbol{q}, \omega) = S_{\mathrm{RPA}}^{\mathrm{sp}}(\boldsymbol{q}, \omega) + S_{\mathrm{RPA}}^{\mathrm{col}}(\boldsymbol{q}, \omega) \quad . \tag{A.26}$$

The first term $S_{\mathrm{RPA}}^{\mathrm{sp}}(\boldsymbol{q}, \omega)$ describes the creation and annihilation of a single particle-hole pair [1.7]. This process, which is called *Landau damping*, is only possible in $d > 1$ and for energies $0 < \omega \leq v_{\mathrm{F}}|\boldsymbol{q}|$. Mathematically Landau damping is due to the finite imaginary part of $g_d(x + \mathrm{i}0^+)$ for $x < 1$. Thus

$$S_{\mathrm{RPA}}^{\mathrm{sp}}(\boldsymbol{q}, \omega) = \Theta\left(1 - \frac{\omega}{v_{\mathrm{F}}|\boldsymbol{q}|}\right) \frac{\nu}{\pi} \mathrm{Im} \left\{ \frac{g_d(\frac{\omega}{v_{\mathrm{F}}|\boldsymbol{q}|} + \mathrm{i}0^+)}{1 + F_q g_d(\frac{\omega}{v_{\mathrm{F}}|\boldsymbol{q}|} + \mathrm{i}0^+)} \right\} \quad . \tag{A.27}$$

The second term $S_{\mathrm{RPA}}^{\mathrm{col}}(\boldsymbol{q}, \omega)$ arises from the poles of Eq.(A.24), which define the dispersion relation ω_q of the collective plasmon mode,

$$1 + F_q g_d\left(\frac{\omega_q}{v_{\mathrm{F}}|\boldsymbol{q}|}\right) = 0 \quad . \tag{A.28}$$

The formal solution of Eq.(A.28) is

$$\frac{\omega_q}{v_{\mathrm{F}}|\boldsymbol{q}|} = g_d^{-1}\left(-\frac{1}{F_q}\right) \quad , \tag{A.29}$$

where $g_d^{-1}(x)$ is the inverse of the function $g_d(x)$, i.e. $g_d^{-1}(g_d(x)) = x$. Because of the simple form of $g_1(x)$ and $g_2(x)$, the solution of Eq.(A.29) in $d = 1$ and $d = 2$ can be calculated analytically,

$$\frac{\omega_q}{v_{\mathrm{F}}|\boldsymbol{q}|} = \sqrt{1 + F_q} \quad , \quad \text{for } d = 1 \quad , \tag{A.30}$$

$$\frac{\omega_q}{v_{\mathrm{F}}|\boldsymbol{q}|} = \sqrt{1 + \frac{F_q^2}{1 + 2F_q}} = \frac{|1 + F_q|}{\sqrt{1 + 2F_q}} \quad , \quad \text{for } d = 2 \quad . \tag{A.31}$$

Note that ω_q is real, so that the plasmon mode is not damped. It is easy to see that for repulsive interactions in arbitrary dimensions the plasmon mode is not damped within the RPA [1.7], so that it gives rise to a δ-function contribution to the RPA dynamic structure factor,

$$S^{\rm col}_{\rm RPA}(\boldsymbol{q},\omega) = Z_{\boldsymbol{q}}\delta(\omega - \omega_{\boldsymbol{q}}) \quad , \tag{A.32}$$

with

$$Z_{\boldsymbol{q}} = \frac{1}{f^2_{\boldsymbol{q}} \frac{\partial}{\partial z} \Pi_0(\boldsymbol{q}, z)\big|_{z=\omega_{\boldsymbol{q}}}} = \frac{\nu}{F^2_{\boldsymbol{q}}} \frac{v_{\rm F}|\boldsymbol{q}|}{g'_d(\frac{\omega_{\boldsymbol{q}}}{v_{\rm F}|\boldsymbol{q}|})} \quad , \tag{A.33}$$

where $g'_d(z)$ is the derivative of the function $g_d(z)$. Because the dispersion relation of the collective mode satisfies Eq.(A.29), $g'_d(\frac{\omega_{\boldsymbol{q}}}{v_{\rm F}|\boldsymbol{q}|})$ can be considered as function of $F_{\boldsymbol{q}}$. We conclude that $Z_{\boldsymbol{q}}$ is of the form

$$Z_{\boldsymbol{q}} = \nu v_{\rm F}|\boldsymbol{q}| Z_d(F_{\boldsymbol{q}}) \quad , \tag{A.34}$$

where the function $Z_d(F)$ is given by

$$Z_d(F) = \frac{1}{F^2 g'_d(g_d^{-1}(-\frac{1}{F}))} \quad . \tag{A.35}$$

In $d = 1, 2, 3$ we have explicitly

$$Z_1(F) = \frac{1}{2\sqrt{1+F}} \quad , \tag{A.36}$$

$$Z_2(F) = \frac{F}{(1+2F)^{\frac{3}{2}}} \quad , \tag{A.37}$$

$$Z_3(F) = \frac{g_3^{-1}(-\frac{1}{F})}{F^2} \left[\frac{1}{[g_3^{-1}(-\frac{1}{F})]^2 - 1} - \frac{1}{F} \right]^{-1} \quad . \tag{A.38}$$

The strong and weak coupling behavior can be obtained analytically in any dimension. The collective mode for large $F_{\boldsymbol{q}}$ is determined by the asymptotic behavior of $g_d(x)$ for large x. From Eq.(A.21) it is easy to show that to leading order

$$g_d^{-1}(-\frac{1}{F}) \sim \sqrt{\frac{F}{d}} \quad , \quad F \gg 1 \quad , \tag{A.39}$$

and that

$$g'_d(x) \sim \frac{2}{dx^3} \quad , \quad x \gg 1 \quad . \tag{A.40}$$

Then it is easy to see that for $F \gg 1$

$$g'_d(g_d^{-1}(-\frac{1}{F})) \sim \frac{2\sqrt{d}}{F^{3/2}} \quad . \tag{A.41}$$

It follows that the leading behavior at strong coupling is

$$\omega_{\boldsymbol{q}} \sim \frac{v_{\rm F}|\boldsymbol{q}|}{\sqrt{d}} \sqrt{F_{\boldsymbol{q}}} \quad , \quad F_{\boldsymbol{q}} \gg 1 \tag{A.42}$$

$$Z_d(F_{\boldsymbol{q}}) = \frac{Z_{\boldsymbol{q}}}{\nu v_{\rm F}|\boldsymbol{q}|} \sim \frac{1}{2\sqrt{d}\sqrt{F_{\boldsymbol{q}}}} \quad , \quad F_{\boldsymbol{q}} \gg 1 \quad . \tag{A.43}$$

The dispersion of the plasmon mode at weak coupling is determined by the behavior of the function $g_d(1+\delta)$ for small positive δ, which is given in Eq.(A.22). Because $g_d(1)$ is finite for $d > 3$, the collective mode equation (A.28) does not have any solution for $F_q < 1/|g_d(1)|$ in $d > 3$. In this case there is at weak coupling no collective mode contribution to the dynamic structure factor. For $d \leq 3$ we find to leading order for small F

$$g_d^{-1}(-\frac{1}{F}) \sim \begin{cases} 1+e^{-2/F} & \text{for } d=3 \\ 1+(c_d F)^{\frac{2}{3-d}} & \text{for } d<3 \end{cases} , \tag{A.44}$$

$$g_d'(g_d^{-1}(-\frac{1}{F})) \sim \begin{cases} \frac{1}{2}e^{2/F} & \text{for } d=3 \\ \frac{3-d}{2}c_d(c_d F)^{-\frac{5-d}{3-d}} & \text{for } d<3 \end{cases} , \tag{A.45}$$

so that at weak coupling the collective mode and the residue are

$$\frac{\omega_q}{v_F|q|} \sim \begin{cases} 1+e^{-2/F_q} & \text{for } d=3 \\ 1+(c_d F_q)^{\frac{2}{3-d}} & \text{for } d<3 \end{cases} , \quad F_q \ll 1 , \tag{A.46}$$

$$Z_d(F_q) = \frac{Z_q}{\nu v_F|q|} \sim \begin{cases} \frac{2}{F_q^2}e^{-2/F_q} & \text{for } d=3 \\ \frac{2}{3-d}c_d(c_d F_q)^{\frac{d-1}{3-d}} & \text{for } d<3 \end{cases} , \quad F_q \ll 1 . \tag{A.47}$$

A.3 Collective modes for singular interactions

Here we explicitly calculate the dispersion relation of the plasmon mode and the associated residue for singular interactions that diverge in d dimensions as $|q|^{-\eta}$ for $q \to 0$ (see Chap. 6). We start with the physically most important Coulomb interaction and then discuss the general case.

A.3.1 The Coulomb interaction in $1 \leq d \leq 3$

The bare Coulomb potential between two charges separated by a distance r is $e^2/|r|$ in any dimension. For $1 < d \leq 3$ the Fourier transformation to momentum space is easily calculated using d-dimensional spherical coordinates (see Eq.(A.7)), with the result

$$f_q = \int dr e^{ik\cdot r}\frac{e^2}{|r|} = \frac{\Gamma(d-1)\Omega_d e^2}{|q|^{d-1}} , \quad d>1 . \tag{A.48}$$

In $d=1$ the integral in Eq.(A.48) is logarithmically divergent, and must be regularized. Introducing a short-distance cutoff a, one obtains

$$f_q = 2e^2 \ln\left(\frac{1}{|q|a}\right) , \quad d=1 . \tag{A.49}$$

In dimensions $d>1$ the *Thomas-Fermi screening wave-vector* κ is defined by

$$\kappa^{d-1} = \nu\Gamma(d-1)\Omega_d e^2 , \tag{A.50}$$

to that with the help of Eq.(A.5) we obtain

$$\left(\frac{\kappa}{k_{\mathrm{F}}}\right)^{d-1}=\frac{\Gamma(d-1)\Omega_d^2}{(2\pi)^d}\frac{e^2}{v_{\mathrm{F}}} \quad . \tag{A.51}$$

Thus, the requirement that the Thomas-Fermi screening wave-vector should be small compared with k_{F} is equivalent with $e^2/v_{\mathrm{F}} \ll 1$. Note that $e^2/v_{\mathrm{F}} = (k_{\mathrm{F}}a_{\mathrm{B}})^{-1} = \alpha c/v_{\mathrm{F}}$ where $a_{\mathrm{B}} = 1/(me^2)$ is the Bohr radius, and $\alpha = e^2/c \approx \frac{1}{137}$ is the fine structure constant [1]. Up to a numerical factor of the order of unity, the parameter $(\kappa/k_{\mathrm{F}})^{d-1}$ can be identified with the usual dimensionless Wigner-Seitz radius r_{s}, which is a measure for the density of the electron gas. In d dimensions r_{s} is defined by $V/N = V_d(a_{\mathrm{B}}r_{\mathrm{s}})^d$ where

$$V_d=\frac{\Omega_d}{d}=\frac{2\pi^{\frac{d}{2}}}{d\Gamma(\frac{d}{2})} \tag{A.52}$$

is the volume of the d-dimensional unit sphere. Using the fact that the density of spinless fermions in d dimensions can be written as $N/V = V_d k_{\mathrm{F}}^d/(2\pi)^d$, we obtain in d dimensions

$$r_{\mathrm{s}}=\left(\frac{1}{V_d}\right)^{\frac{2}{d}}\frac{2\pi e^2}{v_{\mathrm{F}}} \quad . \tag{A.53}$$

Combining this with Eq.(A.51), we conclude that

$$\left(\frac{\kappa}{k_{\mathrm{F}}}\right)^{d-1}=\frac{\Gamma(d-1)\Omega_d^2}{(2\pi)^{d+1}}V_d^{\frac{2}{d}}r_{\mathrm{s}} \quad . \tag{A.54}$$

In particular, in $d=2$ we have $\kappa/k_{\mathrm{F}} = e^2/v_{\mathrm{F}} = r_{\mathrm{s}}/2$, and in three dimensions $(\kappa/k_{\mathrm{F}})^2 = 2e^2/(\pi v_{\mathrm{F}}) \approx 0.263 r_{\mathrm{s}}$.

With the above definitions, the dimensionless Coulomb interaction $F_{\boldsymbol{q}} = \nu f_{\boldsymbol{q}}$ can be written as

$$F_{\boldsymbol{q}}=\left(\frac{\kappa}{|\boldsymbol{q}|}\right)^{d-1} \quad . \tag{A.55}$$

Because $F_{\boldsymbol{q}} \gg 1$ for $|\boldsymbol{q}| \ll \kappa$, the Thomas-Fermi screening wave-vector κ defines the boundary between the long and short wavelength regimes, and can therefore be identified with the cutoff q_{c} introduced in Chap. 2.4.3. It follows that for the Coulomb problem the bosonization approach is most accurate at high densities, where $r_{\mathrm{s}} \ll 1$ and hence $\kappa \ll k_{\mathrm{F}}$. We would like to emphasize that bosonization is *not an expansion in powers of* r_{s} [6.2]; the condition $r_{\mathrm{s}} \ll 1$ is necessary to make the higher-dimensional bosonization approach consistent.

For the Coulomb potential the dimensionless RPA interaction can be written as

---

[1] Recall that we have set $\hbar = 1$. In conventional Gaussian units we have $a_{\mathrm{B}} = \hbar^2/(me^2)$ and $\alpha = e^2/(\hbar c)$.

$$F_q^{\mathrm{RPA}} \equiv \nu f_q^{\mathrm{RPA}} = \frac{1}{(\frac{|q|}{\kappa})^{d-1} + g_d(\frac{i\omega_m}{v_F|q|})} \quad . \tag{A.56}$$

Because F_q diverges as $q \to 0$, the behavior of the collective mode for $|q| \ll \kappa$ is determined by the strong coupling limit $F_q \gg 1$, which is given in Eqs.(A.42) and (A.43). Hence, for the Coulomb interaction in d dimensions the collective plasmon mode and its weight are at long wavelengths given by

$$\omega_q = \frac{v_F \kappa}{\sqrt{d}} \left(\frac{|q|}{\kappa} \right)^{\frac{3-d}{2}} \quad , \tag{A.57}$$

$$Z_q = \frac{\nu v_F \kappa}{2\sqrt{d}} \left(\frac{|q|}{\kappa} \right)^{\frac{d+1}{2}} \quad . \tag{A.58}$$

In three dimensions this yields

$$\omega_q = \frac{v_F \kappa}{\sqrt{3}} \equiv \omega_{\mathrm{pl}} \quad , \quad d = 3 \quad , \tag{A.59}$$

$$Z_q = \frac{\nu}{2} \omega_{\mathrm{pl}} \left(\frac{q}{\kappa} \right)^2 \quad , \quad d = 3 \quad . \tag{A.60}$$

Thus, in $d = 3$ the plasmon mode approaches at long wavelengths a constant value ω_{pl}, the *plasma frequency*.

A.3.2 General singular interactions

Finally, let us consider general singular interactions of the form (6.1). Defining the screening wave-vector

$$\kappa = (g_c^2 \nu)^{1/\eta} \quad , \tag{A.61}$$

we see that the dimensionless interaction corresponding to Eq.(6.1) is

$$F_q \equiv \nu f_q = \left(\frac{\kappa}{|q|} \right)^{\eta} \mathrm{e}^{-|q|/q_c} \quad . \tag{A.62}$$

The dimensionless RPA interaction can be written as

$$F_q^{\mathrm{RPA}} = \frac{1}{(\frac{|q|}{\kappa})^{\eta} \mathrm{e}^{|q|/q_c} + g_d(\frac{i\omega_m}{v_F|q|})} \quad . \tag{A.63}$$

Assuming that $\kappa \ll q_c$, we see that $F_q \gg 1$ for $|q| \ll \kappa$. In this regime the collective mode and the associated residue are easily obtained from Eqs.(A.42) and (A.43),

$$\omega_q = \frac{v_F \kappa}{\sqrt{d}} \left(\frac{|q|}{\kappa} \right)^{1-\eta/2} \quad , \tag{A.64}$$

$$Z_q = \frac{\nu v_F \kappa}{2\sqrt{d}} \left(\frac{|q|}{\kappa} \right)^{1+\eta/2} \quad . \tag{A.65}$$

For $\eta = d - 1$ these expressions reduce to Eqs.(A.57) and (A.58).

A.4 Collective modes for finite patch number

We discuss the polarization and the dynamic structure factor for Fermi surfaces that consist of a finite number M of flat patches. The calculations in this section are valid for arbitrary Fermi surface geometries, i.e. we do not assume that for $M \to \infty$ the Fermi surface approaches a sphere.

A crucial step in higher-dimensional bosonization with linearized energy dispersion is the replacement of an arbitrarily shaped Fermi surface by a finite number of flat patches P^{α}_{Λ}. Let us assume that the number of patches is *even*, and that for each patch P^{α}_{Λ} with local Fermi velocity $\boldsymbol{v}^{\alpha}$ and density of states ν^{α} there exists an opposite patch $P^{\bar{\alpha}}_{\Lambda}$ with $\boldsymbol{v}^{\bar{\alpha}} = -\boldsymbol{v}^{\alpha}$ and $\nu^{\bar{\alpha}} = \nu^{\alpha}$. This guarantees that the inversion symmetry of the Fermi surface is not artificially broken by the patching construction (see the first footnote in Chap. 6). For simplicity let us also assume that all patch densities of states ν^{α} are identical, so that $\nu^{\alpha} = \nu/M$, where $\nu = \sum_{\alpha=1}^{M} \nu^{\alpha}$ is the global density of states (see Eqs.(4.25) and (4.28)). Then the non-interacting polarization $\Pi_0(\boldsymbol{q}, z)$ is at long wave-lengths given by (see Eqs.(4.24) and (4.36))

$$\Pi_0(\boldsymbol{q}, z) = \frac{2\nu}{M} \sum_{\alpha=1}^{M/2} \frac{(\boldsymbol{v}^{\alpha} \cdot \boldsymbol{q})^2}{(\boldsymbol{v}^{\alpha} \cdot \boldsymbol{q})^2 - z^2} = \nu \frac{P_{M-2}(\boldsymbol{q}, z)}{Q_M(\boldsymbol{q}, z)} \quad , \tag{A.66}$$

$$Q_M(\boldsymbol{q}, z) = \prod_{\alpha=1}^{M/2} (z^2 - (\boldsymbol{v}^{\alpha} \cdot \boldsymbol{q})^2) \quad , \tag{A.67}$$

$$P_{M-2}(\boldsymbol{q}, z) = \frac{2}{M} \sum_{\alpha=1}^{M/2} (\boldsymbol{v}^{\alpha} \cdot \boldsymbol{q})^2 \left[\prod_{\substack{\alpha'=1 \\ \alpha' \neq \alpha}}^{M/2} ((\boldsymbol{v}^{\alpha'} \cdot \boldsymbol{q})^2 - z^2) \right] \quad , \tag{A.68}$$

where it is understood that the sums are over all patches with $\boldsymbol{v}^{\alpha} \cdot \boldsymbol{q} \geq 0$, and in the special case $M = 2$ the product in Eq.(A.68) should be replaced by unity. The RPA polarization can then be written as

$$\Pi_{\mathrm{RPA}}(\boldsymbol{q}, z) = \nu \frac{P_{M-2}(\boldsymbol{q}, z)}{Q_M(\boldsymbol{q}, z) + F_{\boldsymbol{q}} P_{M-2}(\boldsymbol{q}, z)} \quad , \tag{A.69}$$

where as usual $F_{\boldsymbol{q}} = \nu f_{\boldsymbol{q}}$. Thus, the RPA condition for the collective density modes,

$$1 + f_{\boldsymbol{q}} \Pi_0(\boldsymbol{q}, z) = 0 \quad , \tag{A.70}$$

is equivalent with

$$Q_M(\boldsymbol{q}, z) + F_{\boldsymbol{q}} P_{M-2}(\boldsymbol{q}, z) = 0 \quad . \tag{A.71}$$

Because the left-hand side of this equation is a polynomial in z^2 with degree $M/2$, for a given $\boldsymbol{q}$ we obtain $M/2$ roots in the complex z^2-plane. The locations of the roots is easily obtained graphically by plotting the right-hand side of Eq.(A.66) as function of real z^2 and looking for the intersections with $-1/f_{\boldsymbol{q}}$. For generic $\boldsymbol{q}$ all $(\boldsymbol{v}^{\alpha} \cdot \boldsymbol{q})^2$ are different and positive, and we can order the energies such that

$$0 < (\boldsymbol{v}^{\alpha_1} \cdot \boldsymbol{q})^2 < (\boldsymbol{v}^{\alpha_2} \cdot \boldsymbol{q})^2 < \ldots < (\boldsymbol{v}^{\alpha_{M/2}} \cdot \boldsymbol{q})^2 \quad . \tag{A.72}$$

A repulsive interaction leads then $M/2$ to real roots $(\omega_{\boldsymbol{q}}^2)^{(\alpha)}$, $\alpha = 1, \ldots, M/2$, of the polynomial (A.71) (considered as function of z^2), which are located between the unperturbed poles,

$$0 < (\boldsymbol{v}^{\alpha_1} \cdot \boldsymbol{q})^2 < (\omega_{\boldsymbol{q}}^2)^{(1)} < (\boldsymbol{v}^{\alpha_2} \cdot \boldsymbol{q})^2 < (\omega_{\boldsymbol{q}}^2)^{(2)} < \\ \ldots < (\boldsymbol{v}^{\alpha_{M/2}} \cdot \boldsymbol{q})^2 < (\omega_{\boldsymbol{q}}^2)^{(M/2)} \quad . \tag{A.73}$$

Because the roots are on the positive real axis in the complex z^2-plane, they represent undamped collective modes, which give rise to δ-function peaks in the RPA dynamic structure factor. Hence, for $\omega > 0$ the dynamic structure factor has the following form[2]

$$S_{\mathrm{RPA}}(\boldsymbol{q}, \omega) = \sum_{\alpha=1}^{M/2} Z_{\boldsymbol{q}}^{\alpha} \delta(\omega - \omega_{\boldsymbol{q}}^{\alpha}) \quad , \tag{A.74}$$

with the residues given by (see also Eq.(A.33))

$$Z_{\boldsymbol{q}}^{\alpha} = \frac{1}{f_{\boldsymbol{q}}^2 \frac{\partial}{\partial z} \Pi_0(\boldsymbol{q}, z) \big|_{z=\omega_{\boldsymbol{q}}^{\alpha}}} \quad . \tag{A.75}$$

In the limit $M \to \infty$ and (at least) for sufficiently strong coupling[3] the mode $\omega_{\boldsymbol{q}}^{M/2}$ with the largest energy survives as a δ-function peak, and can be identified with the collective plasmon mode $\omega_{\boldsymbol{q}}$, see Eqs.(A.29)–(A.32). All other modes represent a quasi-continuum in the sense that they merge for $M \to \infty$ into the particle-hole continuum described by $S_{\mathrm{RPA}}^{\mathrm{sp}}(\boldsymbol{q}, \omega)$ in Eq.(A.27). For non-generic $\boldsymbol{q}$ such that $\boldsymbol{v}^{\alpha_i} \cdot \boldsymbol{q} = 0$ for some α_i or $(\boldsymbol{v}^{\alpha_i} \cdot \boldsymbol{q})^2 = (\boldsymbol{v}^{\alpha_j} \cdot \boldsymbol{q})^2$ for some $\alpha_i \neq \alpha_j$, the number of distinct modes in the quasi-continuum is reduced.

In the strong coupling limit it is easy to obtain an analytic expression for the collective plasmon mode and the associated residue. From Eq.(A.42) we expect that for $F_{\boldsymbol{q}} \equiv \nu f_{\boldsymbol{q}} \gg 1$ there exists one real solution $\omega_{\boldsymbol{q}}$ with $\omega_{\boldsymbol{q}}^2 = O(F_{\boldsymbol{q}})$. For z^2 close to $\omega_{\boldsymbol{q}}^2$ we may therefore expand $\Pi_0(\boldsymbol{q}, z)$ in powers of z^{-2}. The leading terms are

---

[2] The above simple proof that for finite M and repulsive interactions the RPA dynamic structure factor consists only of δ-function peaks can be found in the work [7.2], and is due to Kurt Schönhammer.

[3] Recall that in Sect. A.2 we have shown that for spherical Fermi surfaces in $d > 3$ the collective plasmon mode exists only for sufficiently strong interactions.

$$\Pi_0(\boldsymbol{q}, z) = -\nu \left[\frac{2}{M} \sum_{\alpha=1}^{M/2} (\boldsymbol{v}^\alpha \cdot \boldsymbol{q})^2 \right] z^{-2} - \nu \left[\frac{2}{M} \sum_{\alpha=1}^{M/2} (\boldsymbol{v}^\alpha \cdot \boldsymbol{q})^4 \right] z^{-4} + O(z^{-6}) \, . \tag{A.76}$$

Substituting this approximation into Eq.(A.70), it is easy to show that the dispersion of the plasmon mode is for large $F_{\boldsymbol{q}}$ given by

$$\omega_{\boldsymbol{q}}^2 = F_{\boldsymbol{q}} \frac{2}{M} \sum_{\alpha=1}^{M/2} (\boldsymbol{v}^\alpha \cdot \boldsymbol{q})^2 + \frac{\frac{2}{M} \sum_{\alpha=1}^{M/2} (\boldsymbol{v}^\alpha \cdot \boldsymbol{q})^4}{\frac{2}{M} \sum_{\alpha=1}^{M/2} (\boldsymbol{v}^\alpha \cdot \boldsymbol{q})^2} + O(F_{\boldsymbol{q}}^{-1}) \quad . \tag{A.77}$$

Using the fact that for a spherical Fermi surface

$$\lim_{M \to \infty} \frac{2}{M} \sum_{\alpha=1}^{M/2} (\boldsymbol{v}^\alpha \cdot \boldsymbol{q})^2 = v_{\mathrm{F}}^2 \boldsymbol{q}^2 \langle (\hat{\boldsymbol{v}}^\alpha \cdot \hat{\boldsymbol{q}})^2 \rangle_{\hat{\boldsymbol{q}}} = \frac{v_{\mathrm{F}}^2 \boldsymbol{q}^2}{d} \quad , \tag{A.78}$$

the leading term in Eq.(A.77) reduces for $M \to \infty$ to Eq.(A.42). For energies z close to $\omega_{\boldsymbol{q}}$ we may write

$$\Pi_{\mathrm{RPA}}(\boldsymbol{q}, z) \approx -\frac{Z_{\boldsymbol{q}}}{z - \omega_{\boldsymbol{q}}} \quad , \tag{A.79}$$

with

$$Z_{\boldsymbol{q}} \approx \frac{\nu}{2\sqrt{F_{\boldsymbol{q}}}} \left[\frac{2}{M} \sum_{\alpha=1}^{M/2} (\boldsymbol{v}^\alpha \cdot \boldsymbol{q})^2 \right]^{1/2} \quad , \tag{A.80}$$

where in the second line we have retained the leading term in the expansion for large large $F_{\boldsymbol{q}}$. For $M \to \infty$ and spherical Fermi surfaces we may use again Eq.(A.78) and recover our previous result (A.43) for $Z_{\boldsymbol{q}}$.

References

Chapter 1

1.1 T. Matsubara, Prog. Theor. Phys. **14**, 35 (1955).
1.2 P. C. Martin and J. Schwinger, Phys. Rev. **115**, 1342 (1959).
1.3 A. A. Abrikosov, L. P. Gorkov, and I. E. Dzyaloshinskii, *Methods of Quantum Field Theory in Statistical Physics*, (Dover, New York, 1963).
1.4 L. P. Kadanoff and G. Baym, *Quantum Statistical Mechanics*, (Benjamin, New York, 1962).
1.5 P. Nozières, *Theory of Interacting Fermi Systems*, (Benjamin, New York, 1964).
1.6 A. L. Fetter and J. D. Walecka, *Quantum Theory of Many-Particle Systems*, (McGraw-Hill, New York, 1971).
1.7 D. Pines and P. Nozières, *The Theory of Quantum Liquids*, (Addison-Wesley Advanced Book Classics, Redwood City, 1989).
1.8 L. D. Landau, Zh. Eksp. Teor. Fiz. **30**, 1058 (1956) [Sov. Phys. JETP **3**, 920 (1957)]; *ibid.* **32**, 59 (1957) [Sov. Phys. JETP **5**, 101 (1957)].
1.9 For an up-to-date review of Landau Fermi liquid theory see G. Baym and C. Pethick, *Landau Fermi liquid theory*, (Wiley, New York, 1991).
1.10 D. Vollhardt and P. Wölfle, *The superfluid phases of Helium 3*, (Taylor and Francis, London, 1990).
1.11 S. K. Ma, *Modern Theory of Critical Phenomena*, (Benjamin, Reading, 1976).
1.12 R. Shankar, Rev. Mod. Phys. **66**, 129 (1994).
1.13 J. Sólyom, Adv. Phys. **28**, 201 (1979).
1.14 V. J. Emery, in *Highly Conducting One-Dimensional Solids*, eds. J. T. Devreese, R. P. Evrard and V. E. van Doren, (Plenum, New York, 1979).
1.15 F. D. M. Haldane, J. Phys. **C 14**, 2585 (1981).
1.16 See, for example, F. H. L. Essler and V. E. Korepin, *Exactly Solvable Models of Strongly Correlated Electrons*, (World Scientific, Singapore, 1994).
1.17 S. Tomonaga, Prog. Theor. Phys. **5**, 544 (1950).
1.18 J. M. Luttinger, J. Math. Phys. **4**, 1154 (1963).
1.19 D. C. Mattis and E. H. Lieb, J. Math. Phys. **6**, 304 (1965).
1.20 V. Meden and K. Schönhammer, Phys. Rev. **B 46**, 15753 (1992); K. Schönhammer and V. Meden, Phys. Rev. **B 47**, 16205 (1993).
1.21 J. Voit, Phys. Rev. **B 47**, 6740 (1993); Rep. Progr. Phys. **58**, 977 (1995).
1.22 M. Stone, *Bosonization*, (World Scientific, Singapore, 1994).
1.23 P. W. Anderson, Phys. Rev. Lett. **64**, 1839 (1990); **65**, 2306 (1990); **66**, 3226 (1991).
1.24 For a detailed discussion of the normal-state properties of the high-temperature superconductors see, for example, the articles by P. W. Anderson and Y. Ren, by B. Batlogg, and by P. A. Lee in *High Temperature Superconductivity*, edited by K. S. Bedell, D. E. Meltzer, D. Pines, and J. R. Schrieffer (Addison-Wesley,

Reading, 1990). For more recent reviews see *Physical Properties of High Temperature Superconductors IV*, edited by D. M. Ginsberg (World Scientific, Singapore, 1994).
1.25 J. R. Engelbrecht and M. Randeria, Phys. Rev. Lett. **65**, 1032 (1990); *ibid.* **66**, 3325 (1991).
1.26 M. Fabrizio, A. Parola, and E. Tosatti, Phys. Rev. **B 44**, 1033 (1991).
1.27 D. Bohm and D. Pines, Phys. Rev. **92**, 609 (1953).
1.28 D. Pines, *The Many-Body Problem*, (Benjamin, Reading, 1961).
1.29 A. Luther, Phys. Rev. **B 19**, 320 (1979).
1.30 F. D. M. Haldane, Helv. Phys. Acta. **65**, 152 (1992); *Luttinger's Theorem and Bosonization of the Fermi surface*, in *Proceedings of the International School of Physics "Enrico Fermi"*, Course 121, 1992, edited by R. Schrieffer and R. A. Broglia (North Holland, New York, 1994).
1.31 A. Houghton and J. B. Marston, Phys. Rev. **B 48**, 7790 (1993); A. Houghton, H.-J. Kwon, and J. B. Marston, Phys. Rev. **B 50**, 1351 (1994).
1.32 A. H. Castro Neto and E. Fradkin, Phys. Rev. Lett. **72**, 1393 (1994); Phys. Rev. **B 49**, 10877 (1994).
1.33 A. Houghton, H.-J. Kwon, J. B. Marston, and R. Shankar, J. Phys. **C 6**, 4909 (1994); H.-J. Kwon, A. Houghton, and J. B. Marston, Phys. Rev. **B 52**, 8002 (1995).
1.34 A. H. Castro Neto and E. Fradkin, Phys. Rev. **B 51**, 4084 (1995).
1.35 P. Kopietz and K. Schönhammer, *Bosonization of interacting fermions in arbitrary dimensions*, Z. Phys. **B 100**, 259 (1996). This paper has originally been submitted on July 11, 1994 to Physical Review Letters.
1.36 P. Kopietz, J. Hermisson, and K. Schönhammer, *Bosonization of interacting fermions in arbitrary dimensions beyond the Gaussian approximation*, Phys. Rev. **B 52**, 10877 (1995).
1.37 P. Kopietz, *Calculation of the single-particle Green's function of interacting fermions in arbitrary dimension via functional bosonization*, in the Proceedings of the Raymond L. Orbach Symposium, pages 101–119, edited by D. Hone (World Scientific, Singapore, 1996).
1.38 P. Kopietz and G. E. Castilla, *Higher-dimensional bosonization with non-linear energy dispersion*, Phys. Rev. Lett. **76**, 4777 (1996).
1.39 J. Fröhlich, R. Götschmann, and P. A. Marchetti, J. Phys. **A 28**, 1169 (1995); J. Fröhlich and R. Götschmann, *Bosonization of Fermi Liquids*, preprint cond-mat/9606100, to appear in Phys. Rev. B.
1.40 T. Chen, J. Fröhlich and M. Seifert, *Renormalization Group Methods: Landau-Fermi liquid and BCS Superconductors*, in *Fluctuating Geometries in Statistical Mechanics and Field Theory*, edited by F. David and P. Ginsparg, (Elsevier, Amsterdam, 1996). This is part II of the lectures presented by J. Fröhlich at the 1994 Les Houches Summer School, session LXII (1994).
1.41 H. C. Fogedby, J. Phys. **C 9**, 3757 (1976).
1.42 D. K. K. Lee and Y. Chen, J. Phys. **A 21**, 4155 (1988).
1.43 J. Hermisson, Diplomarbeit, (Universität Göttingen, 1995, unpublished).
1.44 R. P. Feynman and A. R. Hibbs, *Quantum Mechanics and Path Integrals*, (McGraw-Hill, New York, 1965), chapter 9.
1.45 R. P. Feynman and F. L. Vernon, Ann. Phys. **24**, 118 (1963).
1.46 D. Schmelzer, Phys. Rev. **B 47**, 11980 (1993); D. Schmelzer and A. Bishop, *ibid.* **50**, 12733 (1994).
1.47 D. V. Khveshchenko, R. Hlubina, and T. M. Rice, Phys. Rev. **B 48**, 10766 (1994).
1.48 D. V. Khveshchenko, Phys. Rev. **B 49**, 16893 (1994); *ibid.* **52**, 4833 (1995).
1.49 Y. M. Li, Phys. Rev. **B 51**, 13046 (1995).

1.50 C. Castellani, C. Di Castro, and W. Metzner, Phys. Rev. Lett. **72**, 316 (1994).
1.51 C. Castellani and C. Di Castro, Physica **C 235-240**, 99 (1994).
1.52 W. Metzner, C. Castellani, and C. Di Castro, *Fermi Systems with Strong Forward Scattering*, to appear in Adv. Phys. (1997).
1.53 W. Kohn and J. M. Luttinger, Phys. Rev. **118**, 41 (1960); J. M. Luttinger and J. C. Ward, *ibid.* **118**, 1417 (1960).
1.54 W. Metzner and C. Castellani, Int. J. Mod. Phys. **B 9**, 1959 (1995).

Chapter 2

2.1 J. Zinn-Justin, *Quantum Field Theory and Critical Phenomena*, (Oxford University Press, Oxford, 1989).
2.2 P. W. Anderson, Phys. Rev. Lett. **71**, 1220 (1993).
2.3 J. Feldman, M. Salmhofer, and E. Trubowitz, J. Stat. Phys. **84**, 1209 (1996); M. Salmhofer, *Improved Power Counting and the Fermi Surface Renormalization*, preprint cond-mat/9607022.
2.4 V. N. Popov, *Functional Integrals in Quantum Field Theory and Statistical Physics*, (Reidel, Dordrecht, 1983).
2.5 V. N. Popov, *Functional Integrals and Collective Excitations*, (Cambridge University Press, Cambridge, 1987).
2.6 J. W. Negele and H. Orland, *Quantum Many-Particle Physics*, (Addison-Wesley, Redwood City, 1988).
2.7 J. I. Kapusta, *Finite-temperature field theory*, (Cambridge University Press, Cambridge, 1989).
2.8 J. W. Serene and D. W. Hess, Phys. Rev. **B 44**, 3391 (1991).
2.9 An elementary but clear discussion of the RPA in a homogeneous electron gas (together with many good jokes) can be found in the textbook by R. D. Mattuck, *A Guide to Feynman Diagrams in the Many-Body Problem*, chapter 10 (2nd edition, Dover, New York, 1992; 1st edition by McGraw-Hill, New York, 1967).
2.10 G. D. Mahan, *Many-Particle Physics*, (Plenum, New York, 1981).
2.11 V. D. Gorabchenko, V. N. Kohn and E. G. Maksimov, in *Modern Problems in Condensed Matter Sciences, Vol. 24: The Dielectric Function of Condensed Matter Systems,* edited by L. V. Keldysch, D. A. Kirzhnitz and A. A. Maradudin, (North Holland, Amsterdam, 1989).
2.12 K. S. Singwi, M. P. Tossi, R. H. Land, and A. Sjölander, Phys. Rev. **157**, 589 (1968).
2.13 A. K. Rajagopal and K. P. Jain, Phys. Rev. **A 5**, 1475 (1972).
2.14 M. W. C. Dharma-wardana, J. Phys. **C 9**, 1919 (1976).
2.15 A. Holas, P. K. Aravind, and K. S. Singwi, Phys. Rev. **B 20**, 4912 (1979).
2.16 J. Feldman, J. Magnen, V. Rivasseau and E. Trubowitz, Europhys. Lett. **24**, 437 and 521 (1993).
2.17 See, for example, C. Nash and S. Sen, *Topology and Geometry for Physicists*, (Academic Press, London, 1983).
2.18 G. Sterman, *An Introduction to Quantum Field Theory*, (Cambridge University Press, New York, 1993).
2.19 S. Chakravarty, B. I. Halperin, and D. R. Nelson, Phys. Rev. **B 39**, 2344 (1989).

Chapter 3

3.1 R. L. Stratonovich, Doklady Akad. Nauk S.S.S.R. **115**, 1097 (1957) [Sov. Phys. Doklady **2**, 416 (1958)]; J. Hubbard, Phys. Rev. Lett. **3**, 77 (1959).
3.2 D. J. Amit, *Field Theory, the Renormalization Group, and Critical Phenomena*, (World Scientific, Singapore, 1984), 2nd revised edition.
3.3 C. Itzykson and J. M. Drouffe, *Statistical Field Theory, Volume 1*, (Cambridge University Press, Cambridge 1989).
3.4 G. Parisi, *Statistical Field Theory*, (Addison-Wesley, Redwood City, 1988).
3.5 B. S. DeWitt, Phys. Rev. **162**, 1195 and 1239 (1967); J. Honerkamp, Nucl. Phys. **B 36**, 130 (1972); L. Alvarez-Gaumé, D. Freeman and S. Mukhi, Ann. Phys. **134**, 85 (1981).
3.6 W. E. Evenson, J. R. Schrieffer, and S. Q. Wang, J. Appl. Phys. **41**, 1199 (1970).
3.7 J. A. Hertz, Phys. Rev. **B 14**, 1165 (1976); this paper is reprinted in the book by P. W. Anderson, *Basic Notions of Condensed Matter Physics*, (Benjamin/Cummings, Menlo Park, California, 1984).
3.8 A. Gomes and P. Lederer, J. Phys. (Paris) **38**, 231 (1977).
3.9 G. Kotliar and A. E. Ruckenstein, Phys. Rev. Lett. **57**, 1362 (1986).
3.10 H. J. Schulz, Phys. Rev. Lett. **65**, 2462 (1990).

Chapter 4

4.1 I. E. Dzyaloshinskii and A. I. Larkin, Zh. Eksp. Teor. Fiz. **65**, 411 (1973) [Sov. Phys. JETP **38**, 202 (1974)].
4.2 T. Bohr, Nordita preprint 81/4, *Lectures on the Luttinger Model*, 1981 (unpublished).
4.3 J. A. Hertz and M. A. Klenin, Phys. Rev. **B 10**, 1084 (1974).
4.4 See, for example, p.100 of Ref. [2.4], or C. Itzykson and J.-B. Zuber, *Quantum Field Theory*, (McGraw-Hill, New York, 1980), p.276.
4.5 W. Metzner, unpublished notes (1995).
4.6 D. J. W. Geldart and R. Taylor, Canadian J. Phys. **48**, 150 and 167 (1970).
4.7 Y. B. Kim, A. Furusaki, X.-G. Wen, and P. A. Lee, Phys. Rev. **B 50**, 17917 (1994).
4.8 P. Kopietz and S. Chakravarty, Phys. Rev. **B 40**, 4858 (1989).
4.9 P. Kopietz, Phys. Rev. **B 40**, 4846 (1989).
4.10 A. Fleszar, A. A. Quong, and A. G. Eguiluz, Phys. Rev. Lett. **74**, 590 (1995); B. C. Larson, J. Z. Tischler, E. D. Isaacs, P. Zschack, A. Fleszar, and A. G. Eguiluz, Phys. Rev. Lett. **77**, 1346 (1996).
4.11 S. Moroni, D. M. Ceperley, and G. Senatore, Phys. Rev. Lett. **75**, 689 (1995).

Chapter 5

5.1 J. Schwinger, Phys. Rev. **128**, 2425 (1962).
5.2 R. Kubo, J. Phys. Soc. Japan **12**, 570 (1957).
5.3 L. Hedin, Phys. Rev. **139**, A 796 (1965); L. Hedin and S. Lundquist, *Effects of Electron-Electron and Electron-Phonon Interactions on the One-Electron States of Solids*, in *Solid State Physics, Vol. 23*, edited by F. Seitz, D. Turnbull, and H. Ehrenreich (Academic Press, New York, 1969).

5.4 B. Farid, *Self-consistent density-functional approach to the correlated ground states, and an unrestricted many-body perturbation theory*, preprint (1995), to appear in Phil. Mag. B (1997). I would like to thank B. Farid for giving me a copy of this work prior to publication.
5.5 P. Henrici, *Applied and Computational Complex Analysis, Volume 2*, (Wiley Classics Library Edition, New York, 1991), p. 267.
5.6 R. Caracciolo, A. Lerda, and G. R. Zemba, Phys. Lett. **B 352**, 304 (1995); M. Frau, A. Lerda, S. Sciuto, and G. R. Zemba, *Algebraic bosonization: a study of the Heisenberg and Calogero-Sutherland models*, preprint hep-th/9603112, to appear in Int. J. Mod. Phys. A.
5.7 H. Goldstein, *Klassische Mechanik*, (7th German edition, Akademische Verlagsgesellschaft, Wiesbaden 1983), p. 301.
5.8 L. D. Landau, E. M. Lifshitz, and L. P. Pitaevskii, *Electrodynamics of Continuous Media*, (Pergamon, Oxford, revised edition 1984), p. 290.
5.9 J. J. Sakurai, *Modern Quantum Mechanics*, (Addison-Wesley, Reading, revised edition 1994), p. 392.
5.10 E. S. Fradkin, Nucl. Phys. **76**, 588 (1966).
5.11 D. V. Khveshchenko and P. C. E. Stamp, Phys. Rev. Lett. **71**, 2118 (1993); Phys. Rev. **B 49**, 5227 (1994).
5.12 P. Kopietz, *Bosonization and the eikonal expansion: similarities and differences*, Int. J. Mod. Phys. **B 10**, 2111 (1996).
5.13 S. Matveenko and S. Brazovskii, Zh. Eksp. Teor. Fiz. **105**, 1653 (1994) [Sov. Phys. JETP **78**, 892 (1994)]; S. Brazovskii, S. Matveenko and P. Nozières, J. Phys. I France **4**, 571 (1994).
5.14 A. M. J. Schakel, *On Broken Symmetries in Fermi Systems*, Academisch Proefschrift, 1989 (unpublished).
5.15 A. Millis, H. Monien, and P. Pines, Phys. Rev. **B 42**, 167 (1990); P. Monthoux and D. Pines, Phys. Rev. **B 47**, 6069 (1993); *ibid.* **50**, 16015 (1994).
5.16 E. Witten, Comm. Math. Phys. **92**, 455 (1984).
5.17 I. Affleck, *Field Theory Methods and Strongly Correlated Electrons*, in: *Physics, Geometry, and Topology*, edited by H. C. Lee (Plenum, New York, 1990).
5.18 D. Schmeltzer, Phys. Rev. **B 54**, 10269 (1996).

Chapter 6

6.1 P.-A. Bares and X.-G. Wen, Phys. Rev. **B 48**, 8636 (1993).
6.2 M. Gell-Mann and K. Brueckner, Phys. Rev. **106**, 364 (1957).
6.3 I. S. Gradshteyn and I. M. Ryzhik, *Table of Integrals, Series, and Products*, (Academic Press, New York, 1980).
6.4 W. Gröbner and N. Hofreiter, *Integraltafel*, (Springer, Wien, 1961).
6.5 P. Kopietz, *Why normal Fermi systems with sufficiently singular interactions do not have a sharp Fermi surface*, J. Phys. A: Math. Gen. **28**, L571 (1995).
6.6 J. E. Hirsch, Physica **C 158** , 326 (1989).
6.7 R. Z. Bariev, A. Klümper, A. Schadschneider and J. Zittartz, J. Phys. **A 26**, 1249 and 4863 (1993).
6.8 A. Luther and I. Peschel, Phys. Rev. **B 9**, 2911 (1974).
6.9 V. A. Khodel and V. R. Shaginyan, Pis'ma Zh. Eksp. Teor. Fiz. **51**, 488 (1990) [JETP Lett. **51**, 553 (1990)]; J. Dukelsky, V. A. Khodel, P. Schuck, and V. R. Shaginyan, preprint cond-mat/9603060, to appear in Z. Phys. B.
6.10 P. Noziéres, J. Phys. I France **2**, 443 (1992).
6.11 G. E. Volovik, Pis'ma Zh. Eksp. Teor. Fiz. **53**, 208 (1991) [JETP Lett. **53**, 222 (1991)].

Chapter 7

7.1 P. Kopietz, V. Meden, and K. Schönhammer, *Anomalous scaling and spin-charge separation in coupled chains*, Phys. Rev. Lett. **74**, 2997 (1995); K. Schönhammer, V. Meden and P. Kopietz, *Spin-charge separation and anomalous scaling in coupled chains*, J. of Low Temp. Physics **99**, 587 (1995).

7.2 P. Kopietz, V. Meden, and K. Schönhammer, *Crossover between Luttinger and Fermi liquid behavior in weakly coupled chains*, preprint cond-mat/9701023 (December 1996), submitted to Phys. Rev. **B**.

7.3 B. Dardel, D. Malterre, M. Grioni, P. Weibel, Y. Baer, J. Voit, and D. Jérôme, Europhys. Lett. **24**, 687 (1993).

7.4 M. Nakamura, A. Sekiyama, H. Namatame, A. Fujimori, H. Yoshihara, T. Ohtani, A. Misu, and M. Takano, Phys. Rev. **B 49**, 16191 (1994).

7.5 D. Jérôme, Science **252**, 1509 (1991).

7.6 G.-H. Gweon, J. W. Allen, R. Claessen, J. A. Clack, D. M. Poirier, P. J. Benning, C. G. Olson, W. P. Ellis, Y.-X. Zhang, L. F. Schneemeyer, J. Markus, and C. Schlenker, J. Phys.: Cond. Mat. **8**, 9923 (1996).

7.7 H. J. Schulz, J. Phys. **C 16**, 6769 (1983); Phys. Rev. Lett. **71**, 1864 (1993).

7.8 K. Penc and J. Sólyom, Phys. Rev. **B 47**, 6273 (1993).

7.9 L. P. Gorkov and I. E. Dzyaloshinski, Zh. Eksp. Teor. Fiz. **67**, 397 (1974) [Sov. Phys. JETP **40**, 198 (1974)].

7.10 X. G. Wen, Phys. Rev. **B 42**, 6623 (1990).

7.11 C. Bourbonnais and L. G. Caron, Int. J. Mod. Phys. **B 5**, 1033 (1991); D. Boies, C. Bourbonnais, and A.-M. S. Tremblay, Phys. Rev. Lett. **74**, 968 (1995).

7.12 H. J. Schulz, Int. J. Mod. Phys. **B 5**, 57 (1991).

7.13 M. Fabrizio, A. Parola, and E. Tosatti, Phys. Rev. B **46**, 3159 (1992). M. Fabrizio and A. Parola, Phys. Rev. Lett. **70**, 226 (1993); M. Fabrizio, Phys. Rev. **B 48**, 15838 (1993).

7.14 F. V. Kusmartsev, A. Luther and A. Nersesyan, JETP Lett. **55**, 692 (1992); A. Nersesyan, A. Luther, and F. V. Kusmartsev, Phys. Lett. A **179**, 363 (1993).

7.15 C. Castellani, C. Di Castro and W. Metzner, Phys. Rev. Lett. **69**, 1703 (1992).

7.16 V. M. Yakovenko, JETP Lett. **56**, 5101 (1992).

7.17 A. Finkelstein and A. I. Larkin, Phys. Rev. B **47**, 10461 (1993).

7.18 M. Abramowitz and I. Stegun, *Handbook of Mathematical Functions*, (Dover, New York, ninth printing 1970).

7.19 V. Meden, Doktorarbeit (Universität Göttingen, 1996, unpublished).

7.20 J. M. Ziman, *Principles of the Theory of Solids* (Cambridge University Press, Cambridge, 1964), p.39.

7.21 D. C. Mattis, Phys. Rev. **B 36**, 745 (1987).

7.22 R. Hlubina, Phys. Rev. **B 50**, 8252 (1994).

7.23 A. Luther, Phys. Rev. **B 50**, 11446 (1994).

7.24 A. T. Zheleznyak, V. M. Yakovenko, and I. E. Dzyaloshinskii, preprint cond-mat/9609118.

7.25 I. T. Diatlov, V. V. Sudakov, and K. A. Ter-Martirosian, Zh. Eksp. Teor. Fiz. **32**, 767 (1957) [Sov. Phys. JETP **5**, 631 (1957)]; Y. A. Bychkov, L. P. Gorkov, and I. E. Dzyaloshinskii, Zh. Eksp. Teor. Fiz. **50**, 738 (1966) [Sov. Phys. JETP **23**, 489 (1966)]; I. E. Dzyaloshinskii and A. I. Larkin, Zh. Eksp. Teor. Fiz. **61**, 791 (1972) [Sov. Phys. JETP **34**, 422 (1972)].

7.26 D. G. Clarke and S. P. Strong, J. Phys. Codn. Mat. **8**, 10089 (1996); preprint cond-mat/9607141.

7.27 A. M. Tsvelik, preprint cond-mat/9607209.

7.28 L. Bartosch, Diplomarbeit (Universität Göttingen, 1996, unpublished).

7.29 N. Shannon, Y. Li, and N. d'Ambrumenil, preprint cond-mat/9611071.
7.30 D. G. Clarke, S. P. Strong, and P. W. Anderson, Phys. Rev. Lett. **72**, 3218 (1994).

Chapter 8

8.1 P. Kopietz, *Bosonization of coupled electron-phonon systems*, Z. Phys. **B 100**, 561 (1996).
8.2 J. H. Kim, K. Levin, R. Wentzcovitch and A. Auerbach, Phys. Rev. **B 40**, 11378 (1989); *ibid.* **44**, 5148 (1991).
8.3 M. L. Kulić and R. Zeyher, Phys. Rev. **B 49**, 4395 (1994).
8.4 M. Grilli and C. Castellani, Phys. Rev. **B 50**, 16880 (1994).
8.5 A. B. Migdal, Zh. Eksp. Teor. Fiz. **34**, 1438 (1958) [Sov. Phys. JETP **7**, 996 (1958)].
8.6 J. Bardeen and D. Pines, Phys. Rev. **99**, 1140 (1955).
8.7 S. Engelsberg and J. R. Schrieffer, Phys. Rev. **131**, 993 (1963).
8.8 V. Meden, K. Schönhammer, and O. Gunnarson, Phys. Rev. **B 50**, 11179 (1994).
8.9 D. Loss and T. Martin, Phys. Rev. **B 50**, 12160 (1994).
8.10 J. R. Schrieffer, *Theory of Superconductivity*, (Benjamin, Reading, 1964).
8.11 N. W. Ashcroft and N. D. Mermin, *Solid State Physics*, (Holt-Saunders, Philadelphia, 1976).
8.12 D. Bohm and T. Staver, Phys. Rev. **84**, 836 (1950).

Chapter 9

9.1 P. Kopietz, *Calculation of the average Green's function of electrons in a stochastic medium via higher-dimensional bosonization*, J. Phys. Cond. Mat. **8**, 10483 (1996).
9.2 S. M. Girvin and G. D. Mahan, Phys. Rev. **B 20**, 4896 (1979).
9.3 A. A. Ovchinnikov and N. S. Erikhman, Zh. Eksp. Teor. Fiz. **67**, 1474 (1974) [Sov. Phys. JETP **40**, 733 (1975)].
9.4 A. Madhukar and W. Post, Phys. Rev. Lett. **39**, 1424 (1977).
9.5 Y. Inaba, J. Phys. Soc. Jpn. **50**, 2473 (1981).
9.6 A. M. Jaynnavar and N. Kumar, Phys. Rev. Lett. **48**, 553 (1982).
9.7 J. Henrichs, Z. Phys. **B 57**, 157 (1984); *ibid.* **89**, 115 (1992).
9.8 M. Kardar and Y.-C. Zhang, Phys. Rev. Lett. **58**, 2087 (1987).
9.9 N. Lebedev, P. Maass, and S. Feng, Phys. Rev. Lett. **74**, 1895 (1995).
9.10 P. A. Lee and T. V. Ramakrishnan, Rev. Mod. Phys. **57**, 287 (1985).
9.11 B. L. Altshuler and A. G. Aronov, in *Electron-Electron Interactions in Disordered Systems*, edited by A. L. Efros and M. Pollak, (North Holland, Amsterdam, 1985).
9.12 H. Fukuyama, Prog. Theor. Phys. Suppl. **84**, 47 (1985).
9.13 P. Kleinert, J. Phys. **A**: Math. Gen. **29**, 2389 (1996).
9.14 B. Y. Hu and S. Das Sarma, Phys. Rev. **B 48**, 5469 (1993).
9.15 P. Kopietz, Int. J. Mod. Phys. **B 8**, 2593 (1994).
9.16 W. Metzner, private communication (1995).
9.17 T. Giamarchi and H. J. Schulz, Phys. Rev. **B 37**, 325 (1988).

Chapter 10

10.1 P. Kopietz and G. E. Castilla, *Quasi-particle behavior of composite fermions in the half-filled Landau level*, Phys. Rev. Lett. **78**, 314 (1997).
10.2 P. Kopietz, *Nonrelativistic fermions coupled to transverse gauge fields: The single-particle Green's function in arbitrary dimensions*, Phys. Rev. **B 53**, 12761 (1996).
10.3 T. Holstein, R. E. Norton, and P. Pincus, Phys. Rev. **B 8**, 2649 (1973).
10.4 M. Yu. Reizer, Phys. Rev. **B 39**, 1602 (1989); *ibid.* **40**, 11571 (1989); *ibid.* **44**, 5476 (1991).
10.5 A. M. Tsvelik, *Quantum Field Theory in Condensed Matter Physics*, (Cambridge University Press, Cambridge, 1995), chapter 12. This book should also be consulted by anyone who wants to see an artistic portrait of M. Yu. Reizer (and of a number of other prominent physicists).
10.6 G. Baskaran and P. W. Anderson, Phys. Rev. **B 37**, 580 (1988).
10.7 P. A. Lee, Phys. Rev. Lett. **63**, 680 (1989); N. Nagaosa and P. A. Lee, Phys. Rev. Lett. **64**, 2450 (1990); P. A. Lee and N. Nagaosa, Phys. Rev. **B 46**, 5621 (1992).
10.8 L. B. Ioffe and A. I. Larkin, Phys. Rev. **B 39**, 8988 (1989).
10.9 B. Blok and H. Monien, Phys. Rev. **B 47**, 3454 (1993).
10.10 A. Lopez and E. Fradkin, Phys. Rev. **B 44**, 5246 (1991); *ibid.* **47**, 7080 (1993).
10.11 V. Kalmeyer and S.-C. Zhang, Phys. Rev. **B 46**, 9889 (1992).
10.12 B. I. Halperin, P. A. Lee, and N. Read, Phys. Rev. **B 47**, 7312 (1993); A. Stern and B. I. Halperin, Phys. Rev. **B 52**, 5890 (1995).
10.13 J. Gan and E. Wong, Phys. Rev. Lett. **71**, 4226 (1993).
10.14 C. Nayak and F. Wilczek, Nucl. Phys. **B 417**, 359 (1994); *ibid.* **430**, 534 (1994).
10.15 M. Onoda, I. Ichinose, and T. Matsui, Nucl. Phys. **B 446**, 353 (1995).
10.16 S. Chakravarty, R. E. Norton and O. F. Syljuåsen, Phys. Rev. Lett. **74**, 1423 (1995); *ibid.* **75**, 3584 (1995).
10.17 L. B. Ioffe, D. Lidsky, and B. L. Altshuler, Phys. Rev. Lett. **73**, 472 (1994); B. L. Altshuler, L. B. Ioffe and A. J. Millis, Phys. Rev. **B 50**, 14048 (1994).
10.18 H.-J. Kwon, A. Houghton, and J. B. Marston, Phys. Rev. Lett. **73**, 284 (1994).
10.19 Y. B. Kim, P. A. Lee, and X.-G. Wen, Phys. Rev. **B 52**, 17275 (1995).
10.20 P. A. Lee, E. R. Mucciolo, and H. Smith, Phys. Rev. **B 54**, 8782 (1996).
10.21 J. Plochinski, Nucl. Phys. **B 422**, 617 (1994).
10.22 C.-O. Almbladh and L. Hedin, *Beyond the one-electron model – Many-body effects in atoms, molecules, and solids*, in *Handbook on Synchroton Radiation*, Vol.1, edited by E. E. Koch, (North-Holland, Amsterdam, 1983).
10.23 O. F. Syljuåsen, preprint cond-mat/9608051.
10.24 C. J. Pethick, G. Baym and H. Monien, Nucl. Phys. **A 498**, 313c (1989).
10.25 G. Baym, H. Monien, C. J. Pethick, and D. G. Ravenhall, Phys. Rev. Lett. **64**, 1867 (1990).
10.26 W. Kohn and J. M. Luttinger, Phys. Rev. Lett. **65**, 524 (1965).
10.27 R. B. Laughlin, Phys. Rev. Lett. **50**, 1395 (1983).
10.28 S-C. Zhang, T. H. Hansson and S. Kivelson, Phys. Rev. Lett. **62**, 82 (1989); D-H.Lee and S-C. Zhang, Phys. Rev. Lett. **66**, 1220 (1991); D-H. Lee, S. Kivelson, and S-C. Zhang, Phys. Rev. Lett. **67**, 3302 (1991).
10.29 E. Fradkin, *Field Theories in Condensed Matter Systems*, (Addison-Wesley, Redwood City, 1991).
10.30 J. Jain, Phys. Rev. Lett. **63**, 199 (1989); Phys. Rev. **B 40**, 8079 (1989); **41**, 7653 (1990).

10.31 R. L. Willett, R. R, Ruel, K. W. West, and L. N. Pfeiffer, Phys. Rev. Lett. **71**, 3846 (1993); W. Kang, H. L. Stormer, L. N. Pfeiffer, and K. W. West, *ibid.* 3850 (1993); R. R. Du, H. L. Stormer, D. C. Tsui, A. S. Yeh, L. N. Pfeiffer, and K. W. West, Phys. Rev. Lett. **73**, 3274 (1994).
10.32 D. R. Leadley, R. J. Nicholas, C. T. Foxon, and J. J. Harris, Phys. Rev. Lett. **72**, 1906 (1994).
10.33 H. C. Manoharan, M. Shayegan, and S. J. Klepper, Phys. Rev. Lett. **73**, 3270 (1994).
10.34 P. T. Coleridge, Z. W. Wasilewski, P. Zawadzki, A. S. Sachrajda, and H. A. Carmona, Phys. Rev. **B 52**, R11603 (1995).
10.35 J. H. Smet, D. Weiss, R. H. Blick, G. Lütjering, K. von Klitzing, R. Fleichmann, R. Ketzmerick, T. Geisel, and G. Weimann, Phys. Rev. Lett. **77**, 2272 (1996).
10.36 B. I. Halperin, *Fermion Chern-Simons Theory and the Unquantized Quantum Hall Effect*, in *Perspectives in Quantum Hall Effects*, edited by S. Das Sarma and A. Pinczuk (Wiley, New York, 1997).
10.37 B. L. Altshuler and L. B. Ioffe, Phys. Rev. Lett. **69**, 2979 (1992).
10.38 D. V. Khveshchenko and S. V. Meshkov, Phys. Rev. **B 47**, 12051 (1993); D. V. Khveshchenko, Phys. Rev. Lett. **77**, 1817 (1996).
10.39 A. G. Aronov, A. D. Mirlin and P. Wölfle, Phys. Rev. **B 49**, 16609 (1994); A. G. Aronov, E. Altshuler, A. D. Mirlin, and P. Wölfle, Europhys. Lett. **29**, 239 (1995).
10.40 S. C. Zhang and D. P. Arovas, Phys. Rev. Lett. **72**, 1886 (1994).
10.41 A. Ludwig, M. P. A. Fisher, R. Shankar and G. Grinstein, Phys. Rev. **B 50**, 7526 (1994).

Index